INDEXED BIBLIOGRAPHY
ON THE
FLOWERING PLANTS
OF HAWAIʻI

Bishop Museum Special Publication 82

INDEXED BIBLIOGRAPHY ON THE FLOWERING PLANTS OF HAWAI'I

SUSAN W. MILL
DONALD P. GOWING
DERRAL R. HERBST
WARREN L. WAGNER

UNIVERSITY OF HAWAII PRESS
AND
BISHOP MUSEUM PRESS

Library of Congress Cataloging-in-Publication Data
Indexed bibliography on the flowering plants of Hawai'i.

(Bishop Museum special publication ; 82)
Includes indexes.
1. Angiosperms—Hawaii—Indexes. 2. Botany—
Hawaii—Indexes. I. Mill, Susan W., 1963–
II. Series: Bernice P. Bishop Museum special publication ; 82.
Z5358.U5U34   1988   016.58213′09969   88-17257
[QK473.H4]
ISBN 0-8248-1169-0 (alk. paper)

© 1988 Bishop Museum, Honolulu
All rights reserved
Manufactured in the United States of America

∞ The paper used in this publication meets the minimum requirements of American National Standard for Information Sciences–Permanence of Paper for Printed Library Materials, ANSI z39.48–1984.

CONTENTS

INTRODUCTION 1
   Scope of Bibliography 1
   Presentation of Bibliography 2
   Brief Summary of Published Literature on Hawaiian Flowering Plants 4
   Acknowledgments 6

DEFINITIONS OF SUBJECT CATEGORIES 7

LITERATURE CITATIONS 11

SUBJECT INDEX 175
   Abiotic Environment 175
   Anatomy 175
   Autecology 175
   Bibliography 176
   Checklist/Flora 176
   Chemical Analysis 176
   Cladistics/Numerical Taxonomy 176
   Cytology/Genetics 176
   Dispersal 176
   Disturbance Event 176
   Economics 177
   Endangered Species 177
   Ethnobotany 178
   Evolutionary Biology 178
   General Work 178
   History 178
   Hybridization 178
   Management/Conservation 179
   Morphology 179

Original Citation 179
Paleobotany 180
Palynology 180
Pathology 180
Phytogeography 180
Popular Literature 180
Reproductive Biology 181
Review 181
Synecology 181
Taxonomy/Systematics 182

INDEX TO PLANT NAMES 185

INDEX TO PLACE NAMES 211

INTRODUCTION

The concept for a comprehensive bibliography of literature on native and naturalized flowering plants in Hawai'i arose from production of the *Manual of the Flowering Plants of Hawai'i* (Wagner, Herbst & Sohmer, in press). The *Manual* project was initiated in 1982 at Bishop Museum in Honolulu, under the auspices of a grant from the William G. Irwin Charity Foundation of San Francisco awarded to S. H. Sohmer, chairman of the Museum's Department of Botany. The need to bring together the literature on Hawaiian flowering plants, published in so many different places, was especially critical to the *Manual* project. A literature review was necessary when evaluating each group and when preparing the systematic treatments. In the early planning stages of the project, the *Manual* authors decided to create a computerized database to efficiently handle the literature. The database would contain the bibliographic information for each article as well as an indication of the subjects and principal taxa discussed.

Such a database, called HIBIB, was created on a WANG OIS word processor. It was designed and implemented by S. Mill, who was assisted with concepts and definitions by W. Wagner and D. Herbst. Much of the search for literature and the initial data entry were carried out by D. Gowing, without whose volunteer efforts the timely publication of this bibliography would not have been possible. The editing and finalization of the database and preparation of the manuscript was the primary responsibility of Mill, with assistance from Herbst and Wagner. The junior authors of the bibliography are listed alphabetically.

If maintained, the database conceivably could serve as an on-line resource for literature searches. It is currently used by The Nature Conservancy of Hawaii for their Hawaii Heritage Program, by the Hawaii State Department of Land and Natural Resources, and by scientists collaborating on the *Manual* project.

SCOPE OF BIBLIOGRAPHY

This work is the result of our efforts to gather and index published references on the native and naturalized flowering plants in the Hawaiian Islands, particularly those articles pertaining to taxonomy and systematics. We attempted to include all publications on Hawaiian flowering plants, from the 1784 account of Captain Cook's voyage up to and including references published through 1986. Each reference was classified according to taxa, geographical area, and content based on 29 subject categories, and then only in relation to information presented on plants in Hawai'i. Although the focus of the bibliography is on scientific, technical literature, articles of particular interest to the general public are included in the "popular literature" and "history" categories. In addition, the "economics" category contains articles on commercial aspects of plant cultivation that would be of interest to the forester, agriculturalist, and horticulturalist. However, we did not make an exhaustive search for references that are strictly agricultural in nature. Furthermore, many more references on agriculture and horticulture in the Islands are not included here because they do not deal with native or naturalized plants. Neither have we included references dealing peripherally with Hawaiian plants as hosts or food sources, nor have we included all literature dealing with insects as biological control agents for alien plants. We refer the reader to the published plant indexes of the *Proceedings of the Hawaiian Entomological Society* to locate references on these topics. Also, we have not included newspaper articles; however, indices are available for *The Honolulu Advertiser* and *Honolulu Star-Bulletin*.

Bibliographies of the serial Hawaiian floras published by Otto and Isa Degener, including *Flora Hawaiiensis*, *Degeners' Flora Hawaiiana*, and *Degeners' Flora Hawaiiensis*, were published separately in 1985 by Mill, Wagner & Herbst (*Taxon* 34: 229–259). This material was treated separately because of the great number of individual articles (1,144) included in *Flora Hawaiiensis* and the unique bibliographic problems encountered that required special consideration.

Classification of species as either native or naturalized is based upon their treatment by the authors of the *Manual*. Presently those authors estimate that there are about 860 species of alien plants now established, or naturalized, in the Hawaiian Islands and about 960 native species, in a total of 146 families. As the *Manual* does not include bryophytes, pteridophytes, or gymnosperms, the literature on these groups is not considered in the bibliography.

The compilers would appreciate having omissions and errors brought to their attention. The information may be sent to Derral Herbst, Research Associate, Department of Botany, Bishop Museum, P.O. Box 19000-A, Honolulu, Hawai'i 96817.

PRESENTATION OF BIBLIOGRAPHY

Literature Citations

The full bibliographic citation of each article is presented first, giving author, year, title, and place of publication. The entries are listed alphabetically by author(s) and, for works by the same author(s), chronologically by year, and they are numbered sequentially. Presentation of the literature citations was based primarily on 2 invaluable works, B-P-H: *Botanico-Periodicum-Huntianum* [G. H. M. Lawrence, A. F. G. Buchheim, G. S. Daniels, and H. Dolezal (eds.), 1968, Hunt Botanical Library, Pittsburgh, 1063 pp.], primarily used for journal abbreviations, and *Taxonomic Litera-*

ture: A Selective Guide to Botanical Publications and Collections with Dates, Commentaries and Types, 2nd ed. (TL–2), (F. A. Stafleu and R. S. Cowan, 1976–1986, Bohn, Scheltema & Holkema, Utrecht, Vols. 1–6), primarily useful in citing large works. Volume part numbers are given only when the parts are not sequentially numbered at least some time during the history of the series, such as for the *Occasional Papers of the Bernice P. Bishop Museum*. Typographic errors in authors' names have been corrected, whereas corrections for errors in titles are given in brackets. Diacritical marks for Hawaiian words are shown as printed in the publication. Papers presented during the 19th century at natural history society meetings, such as those of the Linnean Society of London, the Boston Society of Natural History, and the American Academy of Arts and Sciences, and later published in their proceedings were often untitled. The compilers have abstracted titles from the moderator's introductory remarks and they are presented here in brackets. In addition to providing one general entry for large works, some are broken down into the individual parts that were published separately. The plant name index refers to the particular part in which the plant is mentioned.

SUBJECT INDEXES

The subject indexes are presented following the literature citations. The entry numbers of all references pertinent to a given subject are listed under that category heading. Articles are classified by subject according to the Definitions of Subject Categories section. Because of the lengthiness of some of the subject lists, they are most useful when used adjunctly with the plant name or place name index. All subject lists can be cross-checked to the references cited for a certain plant or locality. For example, to locate references on cytology of a particular species, one would find the references shared by the cytology index and the index entry for the specific plant.

INDEX TO PLANT NAMES

The index to scientific names for families, genera, and species discussed in the literature is presented after the subject indexes. Names of genera and species are presented as they appear in the literature; however, misspellings and orthographic variants have been corrected according to the guidelines of the International Code of Botanical Nomenclature, except when they appear in the title of the publication. Family names follow A. Cronquist's *An Integrated System of Classification of Flowering Plants* (1981, Columbia Univ. Press, New York, 1261 pp.). Generic names are cross-referenced. In searching for literature on a particular species, however, one should refer to all names considered synonymous.

Not all references to a plant are indexed. Each article was subjectively analyzed to determine the principal species treated; incidental mentions of species were not indexed. Moreover, most articles providing lists of taxa were not indexed. Most general taxonomic works, however, are indexed to species with original citations. A large number of papers on the Hawaiian flora providing notes on and original citations of taxa in many different groups, for example, Fosberg's series of "Miscellaneous notes on Hawaiian plants" or Sherff's "Notes upon certain new or otherwise interesting plants of the Hawaiian Islands and Colombia," give no indication in their titles of the plants discussed. We have emphasized indexing these references because the information on these plants would otherwise be difficult to retrieve.

To review all of the literature on a particular species, 3 areas of the bibliography should be searched: (1) under the generic name in the index for monographic works or articles that are essentially comprehensive in discussing the Hawaiian species of the genus; (2) under the specific name for articles dealing specifically with the plant in question; and (3) under general works dealing with many species in Hawai'i, such as Hillebrand's *Flora of the Hawaiian Islands* or Carlquist's *Hawaii, A Natural History*. Many of the general works are either strictly taxonomic, describing or enumerating Hawaiian species, or deal with a broad subject and give little specific information about a particular species.

If only vernacular names are given in an article, the Latin name as listed in the *Manual* is indexed; for example, 'ōhi'a is indexed as *Metrosideros*, koa as *Acacia koa*.

INDEX TO PLACE NAMES

The place names were derived from 2 geographical fields in the database, one for island and one for specific location. Specific locations were indexed for articles dealing with a particular area, whether a particular valley or ridge or a larger area such as a district or national park. Articles that deal with the plants of a particular island are indexed by island name. Articles that discuss more than one main island are not geographically indexed. Articles referring to more than one of the leeward islands, but not any of the main islands, are indexed under the category "Northwestern Hawaiian Islands." Spellings in the geographical index are uniformly presented according to *Place Names of Hawaii*, revised and expanded edition (M. K. Pukui, S. H. Elbert, and E. T. Mookini, 1974, Univ. Press Hawaii, Honolulu, 289 pp.) or *Atlas of Hawaii* [R. W. Armstrong (ed.), 1973, Univ. Press Hawaii, Honolulu, 222 pp.].

BRIEF SUMMARY OF PUBLISHED LITERATURE ON HAWAIIAN FLOWERING PLANTS

In general, the history of botany in Hawai'i, as elsewhere, can be divided into 3 overlapping phases: (1) exploration; (2) description; and (3) organismal biology and synthesis. The most important contributions of the exploratory period were the collections of specimens made primarily by individuals associated with the great exploring expeditions of the 18th and 19th centuries. In addition, several naturalists, such as Jules Rémy and David Douglas, made independent visits to Hawai'i. The collections were the basis for publications by many of the botanists active during this period.

Some of the most important expeditions that visited Hawai'i were: (1) Captain James Cook's third world voyage on the *Discovery* (1778–1779 in Hawai'i), which resulted in publications such as those by J. E. Smith, G.

Bentham, W. T. Aiton, and H. St. John (entry nos. 2429 and 2444); (2) the voyage of the *Rurik* (1816), whose plant collections were made by L. A. Chamisso and J. Eschscholtz and later published on by Chamisso, sometimes co-authored with D. Schlechtendahl; (3) the voyages of the *Uranie* (1819) and *Bonite* (1841), whose collections were made by C. Gaudichaud-Beaupré, who later published on them himself; (4) the voyage of the *Blonde* (1825) with J. Macrae as botanist, resulting in publications by G. Bentham and others; (5) Captain Beechey's voyage of the *Blossom* (1826–1827) with G. Lay and A. Collie as collectors and the botany published by W. J. Hooker and G. A. W. Arnott; and (6) the United States Exploring Expedition (1840–1841) with botanical collections by W. Rich, C. Pickering, and W. Brackenridge, and publications on the flowering plants later made primarily by A. Gray.

The second phase, which actually began with the early explorations, was primarily concerned with the description of new taxa and comprises the bulk of the technical publications in this bibliography. The papers range from simple or brief descriptions of new taxa to detailed revisions of genera. Most of these papers were merely based on descriptions of herbarium specimens and seldom included information on the biology of the organisms involved. Rarely did the works utilize experimental techniques. The best and most significant publications of this period were those by Wilhelm Hillebrand and Joseph F. Rock. Hillebrand was a practicing physician in Hawai'i from 1851 to 1871 and wrote the only complete flora of the Hawaiian Islands. Rock was a self-taught botanist, employed as a forester and teacher in Hawai'i from 1907 to 1929. He returned to Hawai'i in 1953 after being expelled from mainland China, and he lived here until his death in 1962. He published numerous articles and books concerning the Hawaiian flora. Much of his herbarium work and many of his field studies have been difficult to improve upon, especially his classic monograph of the Hawaiian lobelioids.

The most productive botanists of this period surely have been O. and I. Degener, F. R. Fosberg, H. St. John, and E. E. Sherff. Together they have published over 400 articles and books on Hawaiian plants. C. N. Forbes, A. A. Heller, H. Mann, C. Skottsberg, and H. Wawra also made significant contributions to descriptive botany in Hawai'i.

Some of the first work of the organismal and synthesis stage consisted of Skottsberg's efforts in the areas of phytogeography, reproductive biology, and cytology. Other early work of this period includes O. Selling's palynological studies. This period really began in fuller scope in the 1950s and 1960s with work by S. Carlquist and, later, G. Gillett. Carlquist, especially, has published numerous articles and books on insular evolution, dispersal, reproductive biology, and anatomy. Gillett's work included early, and at least partially successful, efforts to apply biosystematic techniques to Hawaiian plants. This type of work has been continued largely by mainland workers who visit Hawai'i, including important studies by F. Ganders, R. Gardner, T. Lowrey, R. Pearcy, and R. Robichaux. The most notable work done by a resident researcher is the elegant work of G. Carr on the Hawaiian silversword alliance (*Argyroxiphium*, *Dubautia*, and *Wilkesia*). The current

Manual of the Flowering Plants of Hawai'i (Wagner, Herbst & Sohmer, in press) will provide a new base-line synthesis of the flowering plants as well as help to advance the study of biosystematics and evolution of the fascinating and unique Hawaiian flora.

Some chemical studies have also been performed on Hawaiian plants, most of which have focused on screening the plants primarily for alkaloids. Detailed analytical work on compounds in several native species has been done by P. Scheuer and collaborators.

Much of the literature included in this bibliography that is not basically taxonomic or biosystematic falls into one of 3 areas: (1) popular or economic works; (2) conservation and/or management; and (3) ecological. The study of community-level interactions as well as the ecology of single species has been of great importance in understanding island biology and conservation of the biota. For example, quantitative analysis of the impact of feral animals (e.g., Spatz & Mueller-Dombois, 1973; Mueller-Dombois, 1981; Scowcroft & Giffin, 1983) is critical in understanding the full effects of such impact on island ecosystems. The International Biological Program (IBP) Island Ecosystems Integrated Research Program Technical Reports series and the Cooperative National Park Resources Studies Unit Technical Reports have been important outlets for ecological studies, especially the work of D. Mueller-Dombois and collaborators.

ACKNOWLEDGMENTS

The contributions of SWM and WLW were made possible by a grant from the Irwin Charity Foundation of San Francisco awarded to S. H. Sohmer, chairman, Department of Botany, Bishop Museum, and the contribution of DRH was made possible by the U.S. Fish and Wildlife Service, Pacific Islands Office, Honolulu, Hawai'i. We want to express our appreciation to S. H. Sohmer for his support and encouragement. The resources of the libraries of Bishop Museum and the University of Hawaii, Mānoa, and supplementarily the Missouri Botanical Garden Library, Hawaiian Sugar Planters' Association Library, and Hawaii State Archives were used. We gratefully acknowledge the help of staff members of these institutions and particularly would like to thank the following individuals for assistance in obtaining references and/or for guidance in citation procedures: Marguerite Ashford, Neal Evenhuis, Peter Hoch, Kathy Nystrom, Margaret Purk, C. Marie Riley, Janet Short, Janice Wilson, and Constance Wolf. We thank Charles Lamoureux, who made available his extensive literature card file, and Harold St. John, who made available an unpublished bibliography compiled with the late Edwin H. Bryan, Jr., and provided other bibliographic information. We would also like to thank Porter Lowry II and Thomas Lammers for their help in tracking down specific references. We are especially grateful to Clyde Imada for his assistance in indexing references and for helping with each step of the manuscript preparation. Finally, we appreciate the monetary gift from Dr. and Mrs. Daniel Palmer in support of the publication of this work.

DEFINITIONS OF SUBJECT CATEGORIES

Three references were especially helpful in defining the subject headings that we used: (1) S. Blackmore and E. Tootill (eds.), 1984, *The Facts on File Dictionary of Botany*, Market House Books Ltd., Aylesbury, 391 pp.; (2) P. H. Davis and V. H. Heywood, 1973, *Principles of Angiosperm Taxonomy*, Robert E. Krieger Publ. Co., Huntington, N.Y., 558 pp.; and (3) R. J. Little and C. E. Jones, 1980, *A Dictionary of Botany*, Van Nostrand Reinhold Co., New York, 400 pp.

A number of these categories, although separately defined, necessarily overlap in certain areas, such as reproductive biology and autecology, genetics and evolutionary biology, or pathology and disturbance event. When a reference falls into such an area of overlap it is usually listed under both categories.

Abiotic Environment. The abiotic, physical factors that affect the growth and development of a plant; includes geologic, geographic, and climatic factors such as substrate, temperature, light, and precipitation.

Anatomy. The study of the internal structure and form of plants; includes histology.

Autecology. The study of individual species in relation to their environment, i.e., influences of abiotic and biotic factors; involves investigation of the life cycle (e.g., phenology), demography (population dynamics), insect host specificity and utilization by other organisms, and aspects of physiology such as germination, growth, and development, especially as influenced by the environment (physiological ecology).

Bibliography. Books or articles that include an extensive listing of references on a particular subject.

Checklist/Flora. A checklist is a list of the species that occur in a particular geographical area, sometimes with additional notes on distribution, abundance, or nomenclature. A flora is an enumeration of all the species that grow in a particular region, providing keys to and descriptions and nomenclature of the plants as well as information on geographical distribution and ecology.

Chemical Analysis. The analysis of chemical components of plants, such as studies involving chromatography; includes chemotaxonomy.

Cladistics/Numerical Taxonomy. Cladistics, or phylogenetic systematics, is an area of biology concerned with construction of branching diagrams, or cladograms, that summarize the sequence of changes in characters during the evolutionary history of a group of organisms. Cladograms estimate recency of common ancestry through hypotheses of genealogical history and are often used in development of classification as well as analysis of geographical patterns. Numerical taxonomy is a method of classification based on the statistical analysis of the variation of a large number of characters, weighted or unweighted, in a group of plants. Numerical techniques estimate the overall similarity of individuals in populations or of species and may also assess character correlations.

Cytology/Genetics. Cytology is the study of the internal structure and function of the cell, here primarily including chromosome number and structure. Genetics is the study of inheritance, such as the study of DNA structure and replication, genetic coding, and regulation of gene expression; includes electrophoretic studies, cytotaxonomy, cytogenetics, population genetics, and biochemical genetics.

Dispersal. The process of plant migration, primarily involving the mechanisms of propagule distribution.

Disturbance Event. Documentation of the effects of alien organisms on native plants or ecosystems, including discussion of the introduction and spread of alien plants, the activities of man and other introduced animals, and other factors of habitat modification; includes herbivory by introduced insects.

Economics. This category includes references that discuss the value of plants in forestry, agriculture, horticulture, and other commercial ventures, and information on propagation techniques. Note, however, the limitation of references included in the bibliography to those treating native or naturalized species.

Endangered Species. This category comprises references on the endangered Hawaiian flora, giving the status and endangerment of, past and present threats to, and management proposals for threatened and endangered taxa. We are using the terms "threatened" and "endangered" in a broad sense, as used by the author(s) of a reference, and not according to formal definitions, such as those used by the U.S. Fish and Wildlife Service.

Ethnobotany. The study of the relationship between plants and humans, including their influence on and uses of plants, especially noncommercial uses, in ancient or modern times.

Evolutionary Biology. A vast field embracing many facets of biological study as applied to processes of organic diversification; includes the study of the phylogeny (evolutionary history) of an organism or group of organisms and the relationships between taxa, natural selection, speciation and adaptive radiation, reproductive isolation, and eletrophoretic measurement of genetic differentiation.

General Work. A taxonomic work that deals with many Hawaiian taxa or a reference that covers a broad subject in relation to the flora as a whole, such as natural history or biogeography.

History. The history of botany in Hawai'i, including collectors and their itineraries.

Hybridization. The natural or artificial process of crossing individuals with different genetic makeup; includes observations of progeny obtained in breeding experiments (e.g., seed viability and pollen fertility of hybrids) and hypotheses of putative natural hybridization.

Management/Conservation. Discussions and proposals of conservation measures and management techniques for native plants and their habitats; includes regeneration/restoration studies and methods of biological control of alien plants (e.g., use of chemicals, pathogens, and herbivorous insects).

Morphology. The study of the external form and structure of plants, including comparative and developmental studies.

Original Citation. The publication of new nomenclatural combinations and/or description of new taxa.

Paleobotany. The study of the plant life of the geologic past.

Palynology. The study of living or fossil pollen grains or spores.

Pathology. The study of diseases of native and naturalized plants, their effects and treatment.

Phytogeography. The study of the geographical and ecological distribution of plants and how it relates to evolutionary history and the influences of climate and geology; includes discussion of the affinities and derivation of the Hawaiian flora.

Popular Literature. Nontechnical literature written for the general public, or articles of particular interest to nonbotanists; includes casual observations of the native or introduced flora by naturalists.

Reproductive Biology. The study of the reproductive strategies (breeding systems) of plants; includes the study of pollination mechanisms, fertilization, and self-compatibility studies. [See also autecology (phenology, flowering periods) and dispersal.]

Review. An article that provides a summary or synopsis of a reference dealing with the Hawaiian flora.

Synecology. The study of the interactions between organisms within a community and the effects upon them of nonliving/abiotic factors in the environment; includes phytosociology (physiognomy and floristic composition) and succession.

Taxonomy/Systematics. Taxonomy, the science of the classification of organisms, includes the study of the principles, procedures, and rules for identifying, naming, and describing plant taxa. Systematics is the scientific study and description of the variation in organisms and the relationships between them, as well as their subsequent classification; it is a broader term than "taxonomy" and includes identification, taxonomy, classification, and nomenclature.

LITERATURE CITATIONS

1. Anonymous. n.d. The banana poka caper starring the "banana poka gang." Hawaii Dept. Land Nat. Resources, Div. Forestry, Honolulu, 16 pp.
2. ———. n.d. Kipuka Puaulu trail guide, Hawai'i Volcanoes National Park. U.S. Dept. Interior, Natl. Park Serv., 8 pp.
3. ———. n.d. Self-guiding nature trail, Kipuka Puaulu (Bird Park). U.S. Dept. Interior, Natl. Park Serv., 7 pp.
4. ———. 1895. Indigo plant. Hawaiian Planters' Monthly 14: 486.
5. ———. 1899. Recent additions to systematic agrostology. U.S.D.A. Div. Agrostology Circ. 15: 1–10.
6. ———. 1908. Our supplementary illustration. Gard. Chron. III. 44: 412.
7. ——— (W. T.). 1915. *Hibiscus waimeae*. Gard. Chron. III. 57: 8.
8. ———. 1920. The vegetation of the Sandwich Islands. Gard. Chron. III. 68: 235.
9. ———. 1923. Rules and regulations Hawaii National Park, Hawaii, U.S.A. U.S. Dept. Interior, Natl. Park Serv., 15 pp.
10. ———. 1930. Exploring Hawaii's natural botanical gardens. Mid-Pacific Mag. 40: 317–322.
11. ———. 1930. Circular of general information regarding Hawaii National Park. U.S. Dept. Interior, Natl. Park Serv., 20 pp.
12. ———. 1934. Hawaii National Park. U.S. Dept. Interior, Natl. Park Serv., 28 pp.
13. ———. 1935. Sandalwood, Hawaii's most valuable tree. Science 82: 7–8.
14. ———. 1937. Hawaii National Park. Rev. ed. U.S. Dept. Interior, Natl. Park Serv., 34 pp.
15. ———. 1938. Hawaii National Park, Hawaii. U.S. Dept. Interior, Natl. Park Serv., 34 pp.
16. ———. 1940. Bibliography. Hawaii National Park. Hawaii Natl. Park Nat. Hist. Bull. 5: 1–56.
17. ———. 1942. Midway plants. Two rare varieties found on desolate islands. Sci. Amer. 167: 170.
18. ———. 1942. Vegetation on Midway. Science 96: 11.
19. ———. 1956. Plants commonly grown in Honolulu gardens. Garden Club of Honolulu, 18 pp.
20. ———. 1970. The taro collection. Harold L. Lyon Arbor., Univ. Hawaii, Honolulu, 8 pp.

21. ———. 1970. A checklist of palms. Harold L. Lyon Arbor., Univ. Hawaii, Honolulu, 24 pp.
22. ———. 1972. Endangered Hawaiian flora and associated fauna. Elepaio 33: 7–8.
23. ———. 1973. Species trials on Kahoolawe. Newslett. Hawaiian Bot. Soc. 12: 12–13.
24. ———. 1974. Champion trees of Hawaii. Amer. Forests 80(5): 26–31, 34–35.
25. ———. 1975. Threatened or endangered fauna or flora: review of status of vascular plants and determination of "critical habitat." Fed. Reg. 40: 27823–27924.
26. ———. 1975. Honolulu Botanic Gardens inventory, 1975. Friends of Foster Garden Press, Honolulu, 195 pp.
27. ———. 1975. Endangered species at Waimea Arboretum. Notes Waimea Arbor. 2(1): 2.
28. ———. 1975. Propagation. *Sesbania tomentosa* Hook [Hook.] & Arn. (74s920). Notes Waimea Arbor. 2(2): 7–8.
29. ———. 1975. Report on endangered and threatened plant species of the United States. U.S. Gov. Printing Off., Washington, Serial No. 94-A: 1–200.
30. ———. 1976. Endangered and threatened species: plants. Fed. Reg. 41: 24523–24572.
31. ———. 1976. Supplement to Honolulu Botanic Gardens inventory 1975. Friends of Foster Garden Press, Honolulu, 35 pp.
32. ———. 1977. *Wedelia trilobata* (L.) Hitchc. Notes Waimea Arbor. & Bot. Gard. 4(1): 3.
33. ———. 1977. *Alectryon mahoe*: germination results. Notes Waimea Arbor. & Bot. Gard. 4(1): 8.
34. ———. 1977. Timber and watershed management research in Hawaii. U.S.D.A. Forest Serv. PSW-1251: 1–2.
35. ———. 1977. Maintenance of native Hawaiian forest ecosystems. U.S.D.A. Forest Serv. PSW-1752: 1–2.
36. ———. 1977. Hawaii forest insect and disease research. U.S.D.A. Forest Serv. PSW-2207: 1–2.
37. ———. 1979. Service lists 32 plants. Endangered Species Techn. Bull. 4(11): 1, 5–8.
38. ———. 1979. Native coastal plants brought by wind, sea, birds. Hawai'i Coastal Zone News 4(2): 4–5.
39. ———. 1979. *Kokia cookei*—extinction or survival? Notes Waimea Arbor. & Bot. Gard. 6(1): 2–5.
40. ———. 1980. Endangered and threatened wildlife and plants: review of plant taxa for listing as endangered or threatened species. Fed. Reg. 45: 82479–82569.
41. ———. 1980. Endangered and threatened plants. Field Mus. Nat. Hist. Bull. 51(10): 11.

42. ———. 1981. Emblems of our islands. Ampersand (Honolulu) 14(1): 9–19.
43. ———. 1981. Native forest nature trail. Hamakua Distr. Developm. Council, Honokaa, Hawaii, 12 pp.
44. ———. 1981. Native Hawaiian plants at Makapu'u. Malamalama 5(1): 1.
45. ———. 1981. Makapu'u treasures: plants of old Hawaii. Malamalama 5(1): 2.
46. ———. 1981. *Kokia cookei*—progress report. Notes Waimea Arbor. & Bot. Gard. 8(1): 8.
47. ———. 1981. Conservation of Hawaii's coastal plants. Notes Waimea Arbor. & Bot. Gard. 8(1): 14.
48. ———. 1981. A biological evaluation of *ohia* decline on the island of Hawaii. U.S.D.A. Forest Serv., Pacific Southw. Region, 22 pp.
49. ———. 1982. Radiating silverswords. Nat. Hist. 91(12): 36–39.
50. ———. 1982. Majesty. The exceptional trees of Hawaii. Outdoor Circle, Honolulu, 72 pp.
51. ———. 1982. National list of scientific plant names. Vol. 1. List of plant names. U.S.D.A. Soil Conservation Serv., SCS-TP-159: 1–416.
52. ———. 1982. National list of scientific plant names. Vol. 2. Synonymy. U.S.D.A. Soil Conservation Serv., SCS-TP-159: 1–438.
53. ———. 1982. A selection of Hawaiian coastal endemic, indigenous, & Polynesian introduced plants suitable for cultivation. Waimea Arbor. Found. Educational Ser. 1: 1–6.
54. ———. 1983. Wahiawa Botanic Garden self-guided tour. Dept. Parks & Recreation, City & County Honolulu, pp. 1–5.
55. ———. 1985. Endangered classification proposed for four plants. Endangered Species Techn. Bull. 10(8): 1, 4–5.
56. ———. 1985. Endangered and threatened wildlife and plants; review of plant taxa for listing as endangered or threatened species; notice of review. Fed. Reg. 50: 39525–39584.
57. ———. 1986. Species-specific fungus approved for use against Coster's curse. 'Elepaio 46: 194–195.
58. ———. 1986. Endangered and threatened wildlife and plants. Dept. Interior, U.S. Fish & Wildlife Serv., Washington, D.C., 30 pp.
59. ———. 1986. Haleakala silversword. Newslett. Hawaiian Bot. Soc. 25: 89.
60. ———. 1986. Mullein discovered in Haleakala National Park. Newslett. Hawaiian Bot. Soc. 25: 89.
61. ———. 1986. Amaranthaceae—*Achyranthes*. Notes Waimea Arbor. & Bot. Gard. 13(1): 7–8.
62. Abbott, I. A. 1977. The influence of the major food crops on the social system of old Hawai'i. Newslett. Hawaiian Bot. Soc. 16: 78–79.
63. ———. 1982. The ethnobotany of Hawaiian taro. Native Planters: Ho'okupu Kalo 1: 17–22.
64. ———. 1983. Hawaiian plant resources for tapa: the record is faulty. Newslett. Hawaiian Bot. Soc. 22: 23–24.

65. Adams, J. D. 1983. Using LANDSAT images to study plant succession: a case in Hawaii (Abstr.). Amer. J. Bot. 70(5), pt. 2: 43.
66. Adee, K. 1980. Canopy structure in the 'ōhi'a decline zone of Mauna Kea and Mauna Loa, Hawai'i (Abstr.). Proc. 3rd Conf. Nat. Sci., Hawaii Volcanoes Natl. Park, p. 1.
67. Aellen, P. 1929. Beitrag zur Systematik der *Chenopodium*-Arten Amerikas, vorwiegend auf Grund der Sammlung des United States National Museum in Washington, D.C. II. Repert. Spec. Nov. Regni Veg. 26: 119–160.
68. ———. 1933. Nomenklatorische Bemerkungen zu einigen Chenopodien. Ostenia (Montevideo) 1933: 98–101.
69. Affonso, G. T. 1895. The forest of Hamakua. Paradise Pacific 8(1): 3.
70. Agee, H. P. 1920. The kudzu—an interesting legume. Hawaiian Pl. Rec. 22: 215–217.
71. Aiton, W. T. 1811. Hortus kewensis; or, a catalogue of the plants cultivated in the Royal botanic garden at Kew. 2nd ed. Vol. II. Longman, Hurst, Rees, Orme, and Brown, London, 432 pp.
72. Akamine, E. K. 1942. Methods of increasing the germination of *koa haole* seed. Hawaii Agric. Exp. Sta. Circ. 21: 1–14.
73. ———. 1944. Germination of Hawaiian range grass seeds. Hawaii Agric. Exp. Sta. Univ. Hawaii Techn. Bull. 2: 1–60.
74. ———. 1951. Viability of Hawaiian forest tree seeds in storage at various temperatures and relative humidities. Pacific Sci. 5: 36–46.
75. ———. 1952. Germination of seed of *koa haole* [*Leucaena glauca* (L.) Benth.]. Pacific Sci. 6: 51–52.
76. ———, M. Aragaki, J. H. Beaumont, F. A. I. Bowers, R. A. Hamilton, T. Nishida, G. D. Sherman, K. Shoji, W. B. Storey, A. P. Martinez, W. Y. J. Yee, T. Onsdorff, and T. N. Shaw. 1972. Passion fruit culture in Hawaii. Univ. Hawaii Coop. Extens. Serv. Circ. 345: 1–35.
77. Albert, H. 1986. Structure of a disturbed forest community replanted with *Eucalyptus robusta* on Wai'alae Nui Ridge, O'ahu, Hawai'i. Newslett. Hawaiian Bot. Soc. 25: 60–69.
78. Alexander, A. B. 1912. How to use Hawaiian fruit and food products. Paradise-Pacific Print., Honolulu, 73 pp.
79. Alexander, J. M. 1883. Mountain climbing on West Maui. Hawaiian Almanac and Annual for 1884: 32–34.
80. Alexander, W. D. 1892. Scientific expedition to Mauna Kea.—A week on the summit. Friend 50: 74–75.
81. Allen, M. S. 1981. An analysis of the Mauna Kea adze quarry archaeobotanical assemblage. Master's thesis, Univ. Hawaii, Honolulu, 162 pp.
82. Allen, O. N., and E. K. Allen. 1933. The manufacture of poi from taro in Hawaii: with special emphasis upon its fermentation. Hawaii Agric. Exp. Sta. Univ. Hawaii Techn. Bull. 70: 1–32.
83. ——— and ———. 1936. Root nodule bacteria of some tropical legumi-

nous plants: I. Cross-inoculation studies with *Vigna sinensis*. Soil Sci. 42: 61–77.
84. ——— and ———. 1936. Plants in the sub-family Caesalpinioideae observed to be lacking nodules. Soil Sci. 42: 87–91.
85. ——— and ———. 1939. Root nodule bacteria of some tropical leguminous plants: II. Cross-inoculation tests within the cowpea group. Soil Sci. 47: 63–76.
86. Allen, W. F. 1938. Kahoolawe in 1858. Paradise Pacific 50(5): 22, 27.
87. Amerson, A. B., Jr. 1971. The natural history of French Frigate Shoals, Northwestern Hawaiian Islands. Atoll Res. Bull. 150: 1–383.
88. ———. 1975. Species richness on the nondisturbed Northwestern Hawaiian Islands. Ecology 56: 435–444.
89. ———, R. B. Clapp, and W. O. Wirtz, II. 1974. The natural history of Pearl and Hermes Reef, Northwestern Hawaiian Islands. Atoll Res. Bull. 174: 1–306.
90. André, E. 1879. *Pritchardia macrocarpa*, Linden. Ill. Hort. 26: 105.
91. Andrews, E. C. 1941. Origin of the Pacific insular floras. Proc. 6th Pacific Sci. Congr., California 4: 613–620.
92. Andrews, H. C. 1810. *Schinus dentata*. Bot. Repos. 10: *pl. 620.*
93. Anslijn, N. 1834. Beschrijving der Washingtons- en Sandwich-eilanden, en van het Pittcairn-eiland. Mortier en Zoon, Leyden, 176 pp.
94. Apfelbaum, S. I., J. P. Ludwig, and C. E. Ludwig. 1983. Ecological problems associated with disruption of dune vegetation dynamics by *Casuarina equisetifolia* L. at Sand Island, Midway Atoll. Atoll Res. Bull. 261: 1–19.
95. Apple, R. A. 1976. *Koa* and *lehua* timber harvesting and product utilization: religio-ecological relationships in Hawaii, A.D. 1778 (Abstr.). Proc. 1st Conf. Nat. Sci., Hawaii Volcanoes Natl. Park, p. 9.
96. Aragaki, M., F. F. Laemmlen, and W. T. Nishijima. 1972. Collar rot of koa caused by *Calonectria crotalariae*. Pl. Dis. Reporter 56: 73–74.
97. Arakawa, J. M., and W. P. Char. 1976. Report on plant collection trip to Puu Kukui, West Maui. Bull. Pacific Trop. Bot. Gard. 6: 25–35.
98. Archer, K. M., and J. D. Parrish. 1984. Leaf litter decomposition in Hawaiian streams. Proc. 5th Conf. Nat. Sci., Hawaii Volcanoes Natl. Park, pp. 1–39.
99. Arnold, H. L., Jr. 1931. Poisonous plants found in Hawaii. Queen's Hospital Bull. 7(9): 2–5.
100. ———. 1941. *Kahili* flower (*Grevillea banksii*) dermatitis. A preliminary report. Hawaii Med. J. 1: 15–18.
101. ———. 1942. Dermatitis due to the blossom of *Grevillea banksii*. A newly recognized and common entity in Hawaii. Arch. Dermatol. Syphilology 45: 1037–1051.
102. ———. 1944. Poisonous plants of Hawaii. Tongg Publ. Co., Honolulu, 71 pp.
103. ———. 1968. Poisonous plants of Hawaii. Charles E. Tuttle Co., Rutland, Vt., 71 pp.

104. Arrigoni, E. 1977. A nature walk to Ka'ena, O'ahu. Univ. Hawaii UNIHI-SEAGRANT-AR-77-02: 1–112.
105. ———. 1978. A nature walk to Ka'ena Point. Topgallant Publ. Co., Ltd., Honolulu, 159 pp.
106. Arthur, J. C. 1922. Changes in phanerogamic names. Torreya 22: 30.
107. Artschwager, E. 1948. Vegetative characteristics of some wild forms of *Saccharum* and related grasses. Techn. Bull. U.S.D.A. 951: 1–69.
108. ———. 1951. Structure and taxonomic value of the dewlap in sugarcane. Techn. Bull. U.S.D.A. 1038: 1–12.
109. Atkinson, I. A. E. 1970. Successional trends in the coastal and lowland forest of Mauna Loa and Kilauea volcanoes, Hawaii. Pacific Sci. 24: 387–400.
110. Auld, W., and A. Jaeger. 1889. Hawaiian varieties of bananas. Hawaiian Almanac and Annual for 1890: 79–81.
111. Ausen, T. T. de. 1966. Coastline ecosystems in Oahu, Hawaii. Master's thesis, Univ. Hawaii, Honolulu, 202 pp.
112. Ayensu, E. S., and R. A. DeFilipps. 1978. Endangered and threatened plants of the United States. Smithsonian Inst., Washington, 403 pp.
113. ——— and W. L. Stern. 1964. Systematic anatomy and ontogeny of the stem in Passifloraceae. Contr. U.S. Natl. Herb. 34: 45–72.
114. Ayers, P., and R. Kariel. 1984. Honolulu Botanic Gardens inventory, 1984. Friends of Foster Garden Press. Honolulu, 200 pp.
115. Baas, P. 1973. The wood anatomical range in *Ilex* (Aquifoliaceae) and its ecological and phylogenetic significance. Blumea 21: 193–258.
116. ———. 1975. Vegetative anatomy and the affinities of Aquifoliaceae, *Sphenostemon*, *Phelline*, and *Oncotheca*. Blumea 22: 311–407.
117. Babbitt, S. C. 1940. Some of Honolulu's imported trees. Bull. Gard. Club Amer. 7(7): 37–42.
118. Babcock, E. B., and G. L. Stebbins, Jr. 1937. The genus *Youngia*. Publ. Carnegie Inst. Wash. 484: 1–108.
119. Bacigalupi, R. 1931. Taxonomic studies in *Cuphea*. Contr. Gray Herb. 95: 3–26.
120. Bacon, J. D. 1984. Chromosome numbers and taxonomic notes in the genus *Nama* (Hydrophyllaceae). II. Sida 10: 269–275.
121. Baehni, C. 1938. Mémoires sur les Sapotacées I. Système de classification. Candollea 7: 394–508.
122. ———. 1942. Mémoires sur les Sapotacées II. Le genre *Pouteria*. Candollea 9: 147–476.
123. Bailey, E. 1887. The flora and fauna of the Hawaiian Islands. Hawaiian Almanac and Annual for 1888: 49–54.
124. Bailey, H. E., and V. L. Bailey. 1941. Forests and trees of the western national parks. U.S. Conservation Bull. 6: 1–129.
125. Bailey, L. H., and E. Z. Bailey. 1976. Hortus third. A concise dictionary of plants cultivated in the United States and Canada. Rev. ed. Macmillan Publ. Co., Inc., New York, 1290 pp.

126. Baillon, H. E. 1862. Deuxieme memoire sur les Loranthacées. Adansonia 3: 50–128.
127. ———. 1866–1895. Histoire des plantes. Vols. 1–13. Librairie Hachette et Cie, Paris.
128. ———. 1876–1892. Dictionnaire de botanique. Vols. 1–4. Librairie Hachette et Cie, Paris.
129. ———. 1880. Sur la tribu des Labordiées. Bull. Mens. Soc. Linn. Paris 1: 238–240.
130. ———. 1886 [1885]. Campanulacées. Hist. pl. 8: 317–374.
131. ———. 1891. Observations sur les Sapotacées de la Nouvelle-Calédonie. Bull. Mens. Soc. Linn. Paris 2: 945–949, 963–965.
132. ———. 1892 [1891]. Sapotacées. Hist. pl. 11: 255–304.
133. Baker, E. G. 1890–1893. Synopsis of genera and species of Malveae. J. Bot. 28: 15–18, 140–145, 207–213, 239–243, 339–343, 367–371 (1890); 29: 49–53, 164–172, 362–366 (1891); 30: 71–78, 136–142, 235–240, 290–296, 324–332 (1892); 31: 68–76, 212–217, 267–273, 334–338, 361–368 (1893).
134. Baker, G. E., and P. H. Dunn. 1972. Phylloplane mycota as a means of measuring coevolution of fungi and endemic plants. U.S. IBP Island Ecosystems IRP Techn. Rep. 2: 98–110.
135. ———, ———, and J. Meeker. 1973. Progress report on the roles of fungi in island ecosystems. U.S. IBP Island Ecosystems IRP Techn. Rep. 21: 6.13–6.16.
136. ———, ———, and W. A. Sakai. 1974. The roles of fungi in Hawaiian island ecosystems I. Fungal communities associated with leaf surfaces of three endemic vascular plants in Kilauea Forest Reserve and Hawaii Volcanoes National Park, Hawaii. U.S. IBP Island Ecosystems IRP Techn. Rep. 42: 1–46.
137. Baker, H. G. 1967. Support for Baker's law—as a rule. Evolution 21: 853–856.
138. ——— and I. Baker. 1982. Chemical constituents of nectar in relation to pollination mechanisms and phylogeny: *in* Nitecki, M. H. (ed.), Biochemical aspects of evolutionary biology. Univ. Chicago Press, Chicago, pp. 131–171.
139. ——— and ———. 1983. Floral nectar sugar constituents in relation to pollinator type: *in* Jones, E. C., and R. J. Little (eds.), Handbook of experimental pollination biology. Sci. & Acad. Ed., New York, pp. 117–141.
140. ——— and P. A. Cox. 1984. Further thoughts on dioecism and islands. Ann. Missouri Bot. Gard. 71: 244–253.
141. Baker, I., and H. G. Baker. 1982. Some chemical constituents of floral nectars of *Erythrina* in relation to pollinators and systematics. Allertonia 3: 25–37.
142. Baker, J. K. 1980. The plant genus *Hibiscadelphus* in Hawai'i. A history, analysis of problems, and a management plan for trees in Hawaii

Volcanoes National Park. Coop. Natl. Park Resources Stud. Unit, Hawaii, Techn. Rep. 34: 19–31.
143. ———. 1981. Research on endangered Hawaiian species at the Hawaii Field Research Center, Hawaii. Newslett. Hawaiian Bot. Soc. 20: 5–8.
144. ——— and M. S. Allen. 1976. Hybrid *Hibiscadelphus* (Malvaceae) from Hawaii. Phytologia 33: 276.
145. ——— and ———. 1976. Studies on the endemic Hawaiian genus: *Hibiscadelphus* (*hau-kuahiwi*). Proc. 1st Conf. Nat. Sci., Hawaii Volcanoes Natl. Park, pp. 19–28.
146. ——— and ———. 1977 [1978]. Hybrid *Hibiscadelphus* (Malvaceae) in the Hawaiian Islands. Pacific Sci. 31: 285–291.
147. ——— and ———. 1978. Roof rat depredations on *Hibiscadelphus* (Malvaceae) trees. Proc. 2nd Conf. Nat. Sci, Hawaii Volcanoes Natl. Park, pp. 2–5.
148. ——— and D. W. Reeser. 1972. Goat management problems in Hawaii Volcanoes National Park. U.S. Dept. Interior, Natl. Park Serv., Nat. Resources Rep. 2: 1–22.
149. Baker, R. F. 1980. An inventory of palms in Hawaii. Principes 24: 55–64.
150. Baker, R. J. 1936. Familiar Hawaiian flowers. Publ. privately, Honolulu, 133 pp.
151. Bakhuizen van den Brink, R. C., Jr. 1936–1955. Revisio Ebenacearum Malayensium. Bull. Jard. Bot. Buitenzorg, sér. 3, 15: 1–515.
152. Balakrishnan, N. 1982. Nutrient studies in the '*ohi'a* rain forest, Hawaii. Proc. 4th Conf. Nat. Sci., Hawaii Volcanoes Natl. Park, p. 11.
153. ———. 1985. Vegetation patterns and nutrient regimes in a tropical montane rain forest ecosystem, Hawaii. Ph.D. dissertation, Univ. Hawaii, Honolulu, 537 pp.
154. ——— and D. Mueller-Dombois. 1983 [1984]. Nutrient studies in relation to habitat types and canopy dieback in the montane rain forest ecosystem, island of Hawai'i. Pacific Sci. 37: 339–359.
155. Baldwin, C. W. 1908. Geography of the Hawaiian Islands. Amer. Book Co., New York, 128 pp.
156. Baldwin, D. D., and W. Auld. 1890. List of indigenous Hawaiian woods, trees and large shrubs. Hawaiian Almanac and Annual for 1891: 87–91.
157. Baldwin, E. D. 1889. A trip to the summit of Mauna Kea. Hawaiian Almanac and Annual for 1890: 54–58.
158. Baldwin, P. H. 1940. Environmental relationships of birds in the Kilauea section, Hawaii National Park. Hawaii Natl. Park Nat. Hist. Bull. 6: 1–26.
159. ——— and G. O. Fagerlund. 1943. The effect of cattle grazing on *koa* reproduction in Hawaii National Park. Ecology 24: 118–122.
160. Baldwin, R. E. 1979. Hawaii's poisonous plants. Petroglyph Press, Hilo, Hawaii, 112 pp.
161. Baldwin, W. D. 1930. The avocado in Hawaii. Mid-Pacific Mag. 39: 417–422.

162. Balfour, I. B. 1880. Observations on the genus *Pandanus* (screw-pines); with an enumeration of all species described or named in books, herbaria, and nurserymen's catalogues; together with their synonyms and native countries as far as these have been ascertained. J. Linn. Soc., Bot. 17: 33–68.
163. Balgooy, M. M. J. van. 1960. Preliminary plant-geographical analysis of the Pacific as based on the distribution of Phanerogam genera. Blumea 10: 385–430.
164. ———. 1969. A study on the diversity of island floras. Blumea 17: 139–178.
165. ———. 1971. Plant-geography of the Pacific as based on a census of Phanerogam genera. Blumea Suppl. 6: 1–222.
166. ——— (ed.). 1975. Pacific plant areas. Vol. 3. Rijksherbarium, Leiden, 386 pp.
167. ——— (ed.). 1984. Pacific plant areas. Vol. 4. Rijksherbarium, Leiden, 270 pp.
168. Ball, A. A. 1916. The oil of *Santalum freycinetianum* Gaud.: *in* Rock, J. F., The sandalwoods of Hawaii. A revision of the Hawaiian species of the genus *Santalum*. Hawaii, Board Agric. Forest. Bot. Bull. 3: 15.
169. Ball, P. W., and V. H. Heywood. 1964. A revision of the genus *Petrorhagia*. Bull. Brit. Mus. (Nat. Hist.), Bot. 3: 119–172.
170. Baretta-Kuipers, T. 1982. Wood structure of the genus *Erythrina*. Allertonia 3: 53–69.
171. Barneby, R. C., and B. A. Krukoff. 1982. Notes on the species of *Erythrina*. XVI. Allertonia 3: 7–9.
172. Barnhart, P. T. 1931. The Royal Hawaiian Hotel gardens. J. Pan-Pacific Res. Inst. 6(3): 10–11.
173. Barr, M. E. 1983. On *Diaporthopsis metrosideri*. Mycologia 75: 930–931.
174. Barrau, J. 1953. Taro. A short annotated list of relevant publications in the library of the South Pacific Commission. S. Pacific Commiss. Quart. Bull. 3(4): 31–32.
175. Barry, D., Jr. 1951. Bromeliads in the Hawaiian Islands. Bull. Bromeliad Soc. 1: 13.
176. Bartlett, H. H. 1913. The purpling chromogen of a Hawaiian *Dioscorea*. U.S.D.A. Bur. Pl. Industr. Bull. 264: 1–19.
177. Bates, D. M. 1965. Notes on the cultivated Malvaceae 1. *Hibiscus*. Baileya 13: 56–130.
178. Bawa, K. S. 1980. Evolution of dioecy in flowering plants. Annual Rev. Ecol. Syst. 11: 15–39.
179. ———. 1982. Outcrossing and the incidence of dioecism in island floras. Amer. Naturalist 119: 866–871.
180. Beardsley, J. W., Jr. 1978. Biological control of wildland weed pests in Hawai'i—is it a feasible solution? Proc. 2nd Conf. Nat. Sci., Hawaii Volcanoes Natl. Park, pp. 26–29.
181. ——— and J. R. Leeper. 1972. Progress report on effects of sap-sucking

Homoptera on Hawaiian ecosystems. U.S. IBP Island Ecosystems IRP Techn. Rep. 2: 138–140.
182. ―――― and ――――. 1973. Progress report on effects of sap-sucking Homoptera on Hawaiian ecosystems. U.S. IBP Island Ecosystems IRP Techn. Rep. 21: 6.29–6.30.
183. Beaumont, J. H. 1939. Fruit and nut growing in Hawaii. Paradise Pacific 51(12): 100–103.
184. Beccari, O. 1889. Le palme del genere *Pritchardia*. Malesia 3: 281–317.
185. ――――. 1907. Le palme Americane della tribu della *Corypheae*. Webbia 2: 1–343.
186. ――――. 1910. Palmae Australasiche nuove opoco note. Webbia 3: 131–165.
187. ――――. 1913. Contributi alla conoscenza delle Palme. Webbia 4: 143–240.
188. ―――― (U. Martelli, ed.). 1931. Asiatic palms—*Corypheae*. Ann. Roy. Bot. Gard. (Calcutta) 13: 1–354.
189. ―――― and J. F. Rock. 1921. A monographic study of the genus *Pritchardia*. Mem. Bernice P. Bishop Mus. 8: 1–77.
190. Bechtinger, J. 1869. Ein Jahr auf den Sandwich-Inseln. (Hawaiische Inseln.). Publ. privately, Vienna, 204 pp.
191. Beck, G. 1888. Itinera principum S. Coburgi. Vol. II. Carl Gerold's Sohn, Vienna, 205 pp.
192. Becker, W. 1916. Violae Asiaticae et Australenses. I. Beih. Bot. Centralbl. 34: 208–266.
193. Beckley, E. J. 1885. Medicinal plants (these plants are indigenous): *in* The Hawaiian exhibit at the World's Exposition, New Orleans. Hyman Smith, New Orleans, pp. 16–18.
194. Beetle, A. A. 1941. Studies in the genus *Scirpus* L. III. The American species of the section *Lacustres* Clarke. Amer. J. Bot. 28: 691–700.
195. ――――. 1942. Studies in the genus *Scirpus* L. IV. The section *Bolboschoenus* Palla. Amer. J. Bot. 29: 82–88.
196. ――――. 1942. Studies in the genus *Scirpus* L. V. Notes on the section *Actaeogeton* Reich. Amer. J. Bot. 29: 653–656.
197. ――――. 1943. A key to the North American species of the genus *Scirpus* based on achene characters. Amer. Midl. Naturalist 29: 533–538.
198. ――――. 1947. Cyperaceae. N. Amer. Fl. 18(8): 479–504.
199. ――――. 1949. Annotated list of original descriptions in *Scirpus*. Amer. Midl. Naturalist 41: 453–493.
200. Bega, R. V. 1974. *Phytophthora cinnamomi*: its distribution and possible role in *ohia* decline on the island of Hawaii. Pl. Dis. Reporter 58: 1069–1073.
201. ――――. 1979. Heart and root rot fungi associated with deterioration of *Acacia koa* on the island of Hawaii. Pl. Dis. Reporter 63: 682–684.
202. Belcher, R. O. 1956. A revision of the genus *Erechtites* (Compositae), with inquiries into *Senecio* and *Arrhenechthites*. Ann. Missouri Bot. Gard. 43: 1–85.

203. Belt, S. C. 1979. A craftsman's garden. Bull. Pacific Trop. Bot. Gard. 9: 59–62.
204. Benl, G. 1937. Eigenartige Verbreitungseinrichtungen bei der Cyperaceengattung *Gahnia* Forst. Flora 131: 369–386.
205. ———. 1940. Die Systematik der Gattung *Gahnia* Forst. Bot. Arch. 40: 151–257.
206. ———. 1940. Nomina nova vel emendata generis Gahniae Forst. Repert. Spec. Nov. Regni Veg. 39: 30–34.
207. ———. 1952. Zur Systematik der Cyperaceengattung *Gahnia* Forst. Bot. Jahrb. Syst. 75: 82–89.
208. Bennett, F. D. 1840. Narrative of a whaling voyage round the globe, from the year 1833 to 1836, comprising sketches of Polynesia, California, the Indian Archipelago, etc. with an account of southern whales, the sperm whale fishery, and the natural history of the climates visited. Vol. 2. Richard Bentley, London, 395 pp.
209. Bennett, G. 1832. An account of the sandal wood tree (*Santalum*) with observations on some of the botanical productions of the Sandwich Islands. Ann. Mag. Nat. Hist. 5: 255–261.
210. Bennett, T. M. 1985. Palynology of selected horizons from the Ewa Coastal Plain, Oahu, Hawaii. Master's thesis, Univ. Hawaii, Honolulu, 107 pp.
211. Benson, L. 1982. The cacti of the United States and Canada. Stanford Univ. Press, Stanford, 1044 pp.
212. Bentham, G. 1830. *Phyllostegia*. Edwards's Bot. Reg. 15: 1292.
213. ———. 1831. Labiatae: *in* Chamisso, L. C. A. von, and D. F. L. Schlechtendal, De plantis in expeditione speculatoria Romanzoffiana observatis. Linnaea 6: 76–82.
214. ———. 1832–1836. Labiatarum genera et species: or, a description of the genera and species of plants of the order Labiatae; with their general history, characters, affinities, and geographical distribution. James Ridgway and Sons, London, 783 pp.
214a. ———. 1842. Notes on Mimoseae. London J. Bot. 1: 318–392; not seen.
215. ———. 1848. Labiatae. Prodr. 12: 27–603, 697–701.
216. ———. 1871. Revision of the genus *Cassia*. Trans. Linn. Soc. London 27: 503–591.
217. ———. 1877. *Stenogyne rotundifolia*, A. Gray. Hooker's Icon. Pl. 13: 37–38.
218. ——— and J. D. Hooker. 1862–1883. Genera plantarum ad exemplaria imprimis in herbariis kewensibus servata definita. Vols. 1–3. Lovell Reeve & Co., London.
219. Berger, A. J. 1975. Hawaii's dubious distinction. Defenders 50: 491–496.
220. ———, J. Beardsley, R. Burkhart, P. Higashino, W. J. Hoe, C. W. Smith, and H. E. Smith. 1975. Haleakala National Park resources basic inventory 1975. Narrative report. Coop. Natl. Park Resources Stud. Unit, Hawaii, Techn. Rep. 9: 1–151.

221. Bergman, H. F. 1932. Intracarpellary fruits and other central proliferations of the floral axis in *Hibiscus*. Amer. J. Bot. 19: 600–603.
221a. Besser, W. S. J. G. von. 1834. *Artemisia* Linnaei. Nouv. Mèm. Soc. Imp. Nat. Moscou 3: 21–89; not seen.
222. Bird, B. K. 1982. Bibliography of taro and other edible aroids (supplement: 1977–1982). Hawaii Inst. Trop. Agric. Human Resources Res. Ser. 18: 1–123.
223. Bird, I. L. 1876. The Hawaiian Archipelago. Six months among the palm groves, coral reefs, and volcanoes of the Sandwich Islands. J. Murray, London, 318 pp.
224. ———. 1964. Six months in the Sandwich Islands. Univ. Hawaii Press, Honolulu, 278 pp.
225. Birgham, F. 1876. Die Flora des hawaiischen Archipels. Natur (Halle) 25: 211–214, 225–227.
226. Birindelli, D. 1968. Notes from Koke'e and the *'I'iwi* feeding on the banana passion flower. Elepaio 29: 32–33.
227. Bishop, L. E. 1971. Botanist's report: *in* Donaghho, W. R., Kohala Mountains scientific expedition. Elepaio 31: 102.
228. ———. 1973. Honolulu Botanic Gardens inventory 1972. Friends of Foster Garden Press, Honolulu, 293 pp.
229. ——— and D. R. Herbst. 1973. A new *Hibiscadelphus* (Malvaceae) from Kauai. Brittonia 25: 290–293.
230. Bishop, M. B. 1940. Hawaiian life of the pre-European period. The Southworth-Anthoensen Press, Portland, Maine, 105 pp.
231. Bitter, F. A. G. 1900. Die phanerogamische Pflanzenwelt der Insel Laysan. Abh. Naturwiss. Vereine Bremen 16: 430–439.
232. ———. 1910–1911. Die Gattung *Acaena*. Vorstudien zu einer Monographie. Biblioth. Bot. 74: 1–336.
233. ———. 1921. *Solana africana*. III. Bot. Jahrb. Syst. 57: 248–286.
234. Blackman, L. G. 1903. The fibres of the Hawaiian Islands. Occas. Pap. Bernice P. Bishop Mus. 2(1): 37–64.
235. ———. 1904. The fibre industry in Hawaii. No. 3.—sisal. Hawaiian Forester Agric. 1: 226–242.
236. Blake, S. F. 1921. Revision of the genus *Acanthospermum*. Contr. U.S. Natl. Herb. 20: 383–392.
237. ——— and A. C. Atwood. 1942. Geographical guide to floras of the world. U.S.D.A. Misc. Publ. 401: 1–336.
238. Blake, S. T. 1971. A revision of *Plectranthus* (Labiatae) in Australasia. Contr. Queensland Herb. 9: 1–120.
239. Blitman, H. E. 1972. *Na Ho'olu'u O Hawai'i*. Dyes of Hawai'i. 'Ohana Books, Honolulu, 33 pp.
240. Bloxam, A. 1925. Diary of Andrew Bloxam, naturalist of the *"Blonde"* on her trip from England to the Hawaiian Islands 1824–25. Special Publ. Bernice P. Bishop Mus. 10: 1–96.
241. Böckeler, O. 1874. Die Cyperaceen des Königlichen Herbariums zu Berlin. Linnaea 38: 223–355.

242. ———. 1875. Diagnosen neuer Cyperaceen. Flora 58: 257–266.
243. ———. 1878. Diagnosen theils neuer, theils ungenügend beschriebener bekannter Cyperaceen. Flora 61: 33–41.
244. ———. 1880. Diagnosen neuer Cyperaceen. Flora 63: 435–440.
245. Boeke, J. E. 1942. On quantitative statistical methods in taxonomy; subdivision of a polymorphous species: *Planchonella sandwicensis* (Gray) Pierre. Blumea 5: 47–65.
246. Boewig, H. 1915. The histology and development of *Cassytha filiformis*, L. Bot. Contr. Univ. Penn. 2: 399–416.
247. Boissier, P. E. 1862. Euphorbiaceae, subordo I. Euphorbieae. Prodr. 15(2): 3–188.
248. ———. 1866. Icones Euphorbiarum ou figures de cent vingt-deux espèces du genre *Euphorbia*, dessinées et gravées par Heyland avec des considérations sur la classification et la distribution géographique des plantes de ce genre. Victor Masson et fils, Paris, 24 pp.
249. Boone, R. S. 1966. Dry-wood termite attacks in a 55-year-old display of Hawaii-grown wood. Pacific Sci. 20: 524–527.
250. Boynton, D. 1980. Interpreting for the general public. Proc. 3rd Conf. Nat. Sci., Hawaii Volcanoes Natl. Park, pp. 19–20.
251. Brackenridge, W. D. 1886. Notes on the flora of the Sandwich Islands. Gard. Monthly & Hort. 28: 83–85.
252. Braid, K. W. 1925. Revision of the genus *Alphitonia*. Bull. Misc. Inform. 1925: 168–186, 320.
253. Brand, A. 1913. Hydrophyllaceae. Pflanzenr. IV. 251 (Heft 59): 1–210.
254. Brattsten, I. 1945. Stammens och bladets anatomiska byggnad hos *Plantago fernandezia* Bert. och *P. pachyphylla* Gray. Acta Horti Gothob. 16: 165–184.
255. Braun, A. 1864. Revision of the genus *Najas* of Linnaeus. J. Bot. 2: 274–279.
256. Breedlove, D. E., P. E. Berry, and P. H. Raven. 1982. The Mexican and Central American species of *Fuchsia* (Onagraceae) except for sect. *Encliandra*. Ann. Missouri Bot. Gard. 69: 209–234.
257. Brennan, J. P. M. 1955. Notes on Mimosoideae: I. Kew Bull. 1955: 161–192.
258. Brewbaker, J. L. 1966. Enzyme fingerprints for the plant detective. Newslett. Hawaiian Bot. Soc. 5: 1–3.
259. ———. 1976. 'Hawaiian Giant' *koa haole*. Hawaii Agric. Exp. Sta. Misc. Publ. 125: 1–4.
260. ——— and J. W. Hylin. 1965. Variations in mimosine content among *Leucaena* species and related Mimosaceae. Crop Sci. 5: 348–349.
261. ———, D. L. Plucknett, and V. Gonzalez. 1972. Varietal variation and yield trials of *Leucaena leucocephala* (*koa haole*) in Hawaii. Hawaii Agric. Exp. Sta. Res. Bull. 166: 1–29.
262. ——— and B. K. Styles. 1984. Economically important nitrogen fixing tree species planted in Hawaii. Newslett. Hawaiian Bot. Soc. 23: 30–35.

263. Bridges, K., and D. Mueller-Dombois. 1973. Temporal relations in island ecosystems. U.S. IBP Island Ecosystems IRP Techn. Rep. 21: 3.1–3.7.
264. Brigham, W. T. 1868. The Hawaiian flora. Hawaiian Club Pap. (Boston), Oct. 1868: 45–48.
265. ———. 1868. Notes on *Hesperomannia*, a new genus of Hawaiian Compositae. Mem. Boston Soc. Nat. Hist. 1: 527–528.
266. ———. 1868. [The results of Mr. Mann's study of the Hawaiian flora]. Proc. Boston Soc. Nat. Hist. 12: 158–161.
267. ———. 1906. Fruits for the Hawaiian Islands. Hawaiian Forester Agric. 3: 289–311.
268. ———. 1906. Mat and basket weaving of the old Hawaiians, with illustrations of similar work from other parts of the Pacific. Mem. Bernice P. Bishop Mus. 2(1): 1–105.
269. ———. 1911. *Ka hana kapa*. The making of bark-cloth in Hawaii. Mem. Bernice P. Bishop Mus. 3: 1–273.
270. ———. 1914. Hualalai. Mid-Pacific Mag. 8: 224–229.
271. Briquet, J. 1897. Labiatae. Nat. Pflanzenfam. IV. 3a: 183–375.
272. Britten, E. J. 1954. Cytology of Hawaiian plants (Abstr.). Proc. Hawaiian Acad. Sci. 29: 8.
273. ———. 1955. White clover in Hawaii (Abstr.). Proc. Hawaiian Acad. Sci. 30: 8.
274. ———. 1959. Genetic-environmental control of flowering in *Trifolium repens* (Abstr.). Proc. Hawaiian Acad. Sci. 34: 29–30.
275. ———. 1960. Genetic and environmental control of flowering in *Trifolium repens* in the tropics. Science 131: 100–101.
276. ———. 1961. An interesting growth relationship between two specimens of *Erythrina sandwicensis*. Amer. Midl. Naturalist 66: 504.
277. ———. 1962. Hawaii as a natural laboratory for research on climate and plant response. Pacific Sci. 16: 160–169.
278. Bromdep, A. 1966. Studies of nodulation of selected species of legumes in Hawaiian soils and photoperiod effects on flowering in *Desmodium* species. Master's thesis, Univ. Hawaii, Honolulu, 38 pp.
279. Brongniart, A. 1829. Phanérogamie: *in* Duperrey, L. I., Voyage autour du monde, exécuté par ordre du roi, sur la Corvette de Sa Majesté, La Coquille, pendant les années 1822, 1823, 1824 et 1825, sous le ministère et conformément aux instructions de S. E. M. le Marquis de Clermont-Tonnerre, ministre de la marine. Arthus Bertrand, Paris, 232 pp.
280. ——— and A. Gris. 1861. Note sur le genre *Joinvillea* de Gaudichaud et sur la famille des Flagellariées. Bull. Soc. Bot. France 8: 264–269.
281. Brown, B. 1986. Saving Hawaii's endangered plants. Maui Inc. September/October 1986: 42–44.
282. Brown, F. B. H. 1921. Origin of the Hawaiian flora. Proc. 1st Pan-Pacific Sci. Congr., Honolulu, pp. 131–142.

283. ———. 1922. The secondary xylem of Hawaiian trees. Occas. Pap. Bernice P. Bishop Mus. 8(6): 217–371.
284. ———. 1930. New Polynesian plants. Occas. Pap. Bernice P. Bishop Mus. 9(4): 1–23.
285. ———. 1931. Flora of southeastern Polynesia I. Monocotyledons. Bernice P. Bishop Mus. Bull. 84: 1–194.
286. ———. 1935. Flora of southeastern Polynesia III. Dicotyledons. Bernice P. Bishop Mus. Bull. 130: 1–386.
287. ——— and E. D. W. Brown. 1933. A discussion of representative Pacific genera with evidence bearing on their origin and migration (Abstr.). Proc. Hawaiian Acad. Sci. 8: 17–18.
288. Brown, N. E. 1914. Notes on the genera *Cordyline*, *Dracaena*, *Pleomele*, *Sansevieria* and *Taetsia*. Bull. Misc. Inform. 1914: 273–279.
289. Brown, W. V. 1977. The Kranz syndrome and its subtypes in grass systematics. Mem. Torrey Bot. Club 23(3): 1–97.
290. ——— and B. N. Smith. 1975. The genus *Dichanthelium* (Gramineae). Bull. Torrey Bot. Club 102: 10–13.
291. Bryan, E. H., Jr. 1926. Introduction: *in* Bryan, E., H., Jr., and collaborators, Insects of Hawaii, Johnston Island and Wake Island. Bernice P. Bishop Mus. Bull. 31: 3–16.
292. ———. 1928. Guide to the plant groups in Hawaii. J. Pan-Pacific Res. Inst. 3(2): 3–11.
293. ———. 1929. The background of Hawaiian botany. Mid-Pacific Mag. 37: 33–40.
294. ———. 1931. Kahoolawe, the island of dust (Abstr.). Proc. Hawaiian Acad. Sci. 6: 13–14.
295. ———. 1933. A nature trail at Hauula. Mid-Pacific Mag. 46: 133–138.
296. ———. 1933. Hawaiian nature notes. Publ. privately, Honolulu, 285 pp.
297. ———. 1933. Fiberwork: *in* Handy, E. S. C., K. P. Emory, E. H. Bryan, Jr., P. H. Buck, J. H. Wise, and others, Ancient Hawaiian civilization. A series of lectures delivered at The Kamehameha Schools. The Kamehameha Schools, Honolulu, pp. 125–129. (Also publ. in Handy et al., rev. ed., 1965, Charles E. Tuttle Co., Rutland, Vt., pp. 129–133).
298. ———. 1933. Nature's balance in Hawaii: *in* Handy, E. S. C., K. P. Emory, E. H. Bryan, Jr., P. H. Buck, J. H. Wise, and others, Ancient Hawaiian civilization. A series of lectures delivered at The Kamehameha Schools. The Kamehameha Schools, Honolulu, pp. 265–271. (Also publ. in Handy et al., rev. ed., 1965, Charles E. Tuttle Co., Rutland, Vt., pp. 269–275).
299. ———. 1934. The contributions of Bishop Museum to Polynesian biogeography. Mém. Soc. Biogéogr. 4: 279–288.
300. ———. 1938. Ancient Hawaiian life. Advertiser Publ. Co., Honolulu, 113 pp.
301. ———. 1938. Lisianski, an island of Hawaii. Paradise Pacific 50(2): 31, 33–34.

302. ———. 1938. Kaula—an island of Hawaii. Paradise Pacific 50(4): 27, 38–39.
303. ———. 1938. Laysan, an island of Hawaii. Paradise Pacific 50(5): 21, 28–30.
304. ———. 1938. Midway Island, U.S.A. Paradise Pacific 50(6): 7, 29–30.
305. ———. 1939. We see only the peaks of Hawaii. Paradise Pacific 51(12): 13–16.
306. ———. 1940. Hawaiian fiber plants. Paradise Pacific 52(1): 15–16.
307. ———. 1941. American Polynesia. Coral islands of the Central Pacific. Tongg Publ. Co., Honolulu, 208 pp.
308. ———. 1942. American Polynesia and the Hawaiian chain. Tongg Publ. Co., Honolulu, 253 pp.
309. ———. 1949. Plants on Popoia Island. Elepaio 10: 28–29.
310. ———. 1954. The Hawaiian chain. Bishop Mus. Press, Honolulu, 71 pp.
311. ———. 1956. Plants recorded from Laysan and Midway islands: *in* Bailey, A. M., Birds of Midway and Laysan islands. Denver Mus. Nat. Hist., Mus. Pictorial 12: 18–21.
312. ———. 1963. Carl Skottsberg, 1880–1963. Newslett. Hawaiian Bot. Soc. 2: 99–100.
313. ———. 1965. Marie Catherine Neal, 1889–1965. Newslett. Hawaiian Bot. Soc. 4: 19–21.
314. ———. 1978. The northwestern Hawaiian Islands. An annotated bibliography. Dept. Interior, U.S. Fish & Wildlife Service, Honolulu, 190 pp.
315. ——— (ed.). 1979. Mauna Kea here we come. The inside story of a scientific expedition. Publ. privately, Honolulu, 79 pp.
316. Bryan, L. W. 1924. New trees on the Hilo coast. Hawaiian Forester Agric. 21: 4–5.
317. ———. 1927. The Mauna Kea Forest Reserve. Paradise Pacific 40(12): 105–106.
318. ———. 1929. Reforesting with *koa* by the seed-spot method. Hawaiian Forester Agric. 26: 136–137.
319. ———. 1932. The Hilo Forest Reserve. Hawaiian Pl. Rec. 36: 279–321.
320. ———. 1939. Forestry in Hawaii. Gilmore's Hawaii Sugar Manual—1938–39: 24–27.
321. ———. 1947. Twenty-five years of forestry work on the island of Hawaii. Hawaiian Pl. Rec. 51: 1–80.
322. ———. 1948. *Ahinahina*, the unique silversword. Paradise Pacific 60(12): 40, 125.
323. ———. 1949. Indigenous orchids of Hawaii. Hawaii Orchid Soc. Bull. 1(1): 7.
324. ———. 1957. The genus *Pritchardia—loulu* palms. Principes 1: 159–161.
325. ———. 1965. Recollections of an Hawaiian botanist. Newslett. Hawaiian Bot. Soc. 4: 21–24.
326. ———. 1966. Forestry in the Hawaiian Islands (Abstr.): *in* Forest and forest products. Proc. 11th Pacific Sci. Cong., Tokyo 6: 1.

327. ———. 1971. Native Hawaiian plants. Newslett. Hawaiian Bot. Soc. 10: 38–42.
328. ———. 1973. *Ahinahina*. Newslett. Hawaiian Bot. Soc. 12: 1–2.
329. ———. 1984. [Letter to the editor]. Newslett. Hawaiian Bot. Soc. 23: 21.
330. ——— and C. M. Walker. 1962. A provisional check list of some common native and introduced forest plants in Hawaii. Pacific Southw. Forest Range Exp. Sta. Misc. Pap. 69: 1–36. (Rev. ed., 1966, 34 pp.)
331. Bryan, W. A. 1906. Report of a visit to Midway Island. Occas. Pap. Bernice P. Bishop Mus. 2(4): 291–299.
332. ———. 1911. Laysan Island. A visit to Hawaii's bird reservation. Mid-Pacific Mag. 2: 302–315.
333. ———. 1915. Natural history of Hawaii. Publ. privately, Honolulu, 596 pp.
334. ———. 1921. Honolulu, a flower garden. Mid-Pacific Mag. 22: 75–79.
335. ———. 1921. A ramble in Honolulu palm gardens. Mid-Pacific Mag. 22: 560–565.
336. ———. 1921. Hawaiian fauna and flora. Proc. 1st Pan-Pacific Sci. Conf., Honolulu, pp. 153–158.
337. ———. 1922. The trees of Honolulu. Mid-Pacific Mag. 23: 164–169.
338. Buck, M. G. 1982. Hawaiian treefern harvesting affects forest regeneration and plant succession. U.S.D.A. Forest Serv. Res. Note PSW-355: 1–8.
339. ——— and T. E. Paysen. 1984. A system of vegetation classification applied to Hawaii. U.S.D.A. Forest Serv. Gen. Techn. Rep. PSW-76: 1–11.
340. Bullock, A. A. 1958. Nomenclatural notes: VI. Kew Bull. 13: 97–100.
341. Burgan, R. E. 1976. Correlation of plant moisture in Hawaii with the Keetch-Byram drought index. U.S.D.A. Forest Serv. Res. Note PSW-307: 1–6.
342. ——— and R. E. Nelson. 1972. Decline of *ohia lehua* forests in Hawaii. U.S.D.A. Forest Serv. Gen. Techn. Rep. PSW-3: 1–4.
343. ——— and W. H. C. Wong, Jr. 1971. Species trials at the Waiakea Arboretum . . . tree measurements in 1970. U.S.D.A. Forest Serv. Res. Note PSW-240: 1–6.
344. ———, ———, R. G. Skolmen, and H. L. Wick. 1971. Guide to log defect indicators in *koa*, *ohia* . . . preliminary rules for volume deductions. U.S.D.A. Forest Serv., GPO 980-775: 1–4.
345. Burkart, A. 1976. A monograph of the genus *Prosopis* (Leguminosae subfam. Mimosoideae). J. Arnold Arbor. 57: 219–249, 450–525.
346. Burkill, I. H. 1926. [An account of the vegetation found on lava surfaces of various ages in the crater of Kilauea, Hawaii]. Proc. Linn. Soc. London 138: 53–54.
347. Burton, P. J. 1980. Light regimes and *Metrosideros* regeneration in a Hawaiian montane rain forest. Master's thesis, Univ. Hawaii, Honolulu, 378 pp.

348. ———. 1980. Plant invasion into an *'ōhi'a*-tree fern rain forest following experimental canopy opening. Proc. 3rd Conf. Nat. Sci., Hawaii Volcanoes Natl. Park, pp. 21–39.
349. ———. 1982. The effect of temperature and light on *Metrosideros polymorpha* seed germination. Pacific Sci. 36: 229–240.
350. ——— and D. Mueller-Dombois. 1984. Response of *Metrosideros polymorpha* seedlings to experimental canopy opening. Ecology 65: 779–791.
351. Burtt, B. L. 1966. Cyrtandras on Oahu, Hawaiian Islands. Science 152: 936.
352. Bushnell, O. A., M. Fukuda, and T. Makinodan. 1950. The antibacterial properties of some plants found in Hawaii. Pacific Sci. 4: 167–183.
353. Byron, G. A. 1826. Voyage of H.M.S. *Blonde* to the Sandwich Islands, in the years 1824–1825. John Murray, London, 260 pp.
354. Cable, W. J. 1973. The potassium requirement of taro [*Colocasia esculenta* (L.) Schott] in solution culture. Master's thesis, Univ. Hawaii, Honolulu, 131 pp.
355. Cabrera, A. L. 1966. The genus *Lagenophora* (Compositae). Blumea 14: 285–308.
356. Calder, C. C. 1919. The species of *Oxalis* now wild in India. Rec. Bot. Serv. India 6: 325–341; not seen.
357. Cambage, R. H. 1924. The evolution of the genus *Acacia*. Proc. Pan-Pacific Sci. Congr., Australia 1: 297–307.
358. Campbell, D. H. 1892–1893. A vacation in the Hawaiian Islands. Bot. Gaz. (Crawfordsville) 17: 411–416, 18: 19–25.
359. ———. 1893. A botanist's vacation in the Hawaiian Islands. Paradise Pacific 6(2): 19–20.
360. ———. 1911. The embryo-sac of *Pandanus*. Ann. Bot. (London) 25: 773–789.
361. ———. 1916. Some problems of the Pacific floras. Proc. Natl. Acad. U.S.A. 2: 434–437.
362. ———. 1918. The origin of the Hawaiian flora. Mem. Torrey Bot. Club 17: 90–96.
363. ———. 1919. The derivation of the flora of Hawaii. Stanford Univ. Press, Stanford, 34 pp.
364. ———. 1920. Some botanical and environmental aspects of Hawaii. Ecology 1: 257–269.
365. ———. 1926. An outline of plant geography. The Macmillan Co., New York, 392 pp.
366. ———. 1928. The Australasian element in the Hawaiian flora. Amer. J. Bot. 15: 215–221.
367. ———. 1928. The Australasian element in the Hawaiian flora. Proc. 3rd Pan-Pacific Sci. Congr., Tokyo 1: 938–946.
368. ———. 1932. Some problems of the Hawaiian flora (Abstr.). Science 76: 544.

369. ———. 1933. The flora of the Hawaiian Islands. Quart. Rev. Biol. 8: 164–184.
370. ———. 1934. Exotic vegetation of the Pacific regions. Proc. 5th Pacific Sci. Congr., Canada 1: 785–790.
371. ———. 1943. Continental drift and plant distribution. Publ. privately, 43 pp.
372. Candolle, A. C. P. de. 1866. Piperaceae novae. J. Bot. 4: 132–147, 161–167, 210–219.
373. ———. 1869. Piperaceae. Prodr. 16(1): 235^{65}–471.
374. ———. 1898. Piperaceae novae. Annuaire Conserv. Jard. Bot. Genève 2: 252–288.
375. ———. 1912. Piperaceae: *in* Hochreutiner, B. P. G., Plantae Hochreutineraneae, étude systématique et biologique des collections faites par l'auteur au cours de son voyage aux Indes Néerlandaises et autour du monde pendant les années 1903 à 1905. Fascicule I. Annuaire Conserv. Jard. Bot. Genève 15: 231–235.
376. ———. 1912. Meliaceae: *in* Hochreutiner, B. P. G., Plantae Hochreutineraneae, étude systématique et biologique des collections faites par l'auteur au cours de son voyage aux Indes Néerlandaises et autour du monde pendant les années 1903 à 1905. Fascicule I. Annuaire Conserv. Jard. Bot. Genève 15: 245–247.
377. ———. 1913. The Hawaiian peperomias. Coll. Hawaii Publ. Bull. 2: 1–38.
378. ———. 1923. Piperacearum clavis analytica. Candollea 1: 65–415.
379. Candolle, A. L. P. P. de. 1841. Second mémoire sur la famille des Myrsinéacées. Ann. Sci. Nat. II, Bot. 16: 65–97.
380. ———. 1844. Myrsineaceae. Prodr. 8: 75–140, 668–670.
381. ———. 1844. Ebenaceae. Prodr. 8: 209–243, 672–673.
382. ———. 1844. Apocynaceae. Prodr. 8: 317–489, 675–677.
383. ———. 1845. Loganiaceae. Prodr. 9: 1–37, 560–561.
383a. ———. 1847. Myoporaceae. Prodr. 11: 701–716.
384. ———. 1857. Santalaceae. Prodr. 14: 619–692.
385. ———. 1868. Gunnereae. Prodr. 16(2): 596–600.
386. ———. 1878. Smilacées. Monogr. phan. 1: 1–217.
387. ——— and A. C. P. de Candolle. 1878–1896. Monographiae phanerogamarum Prodromi nunc continuatio, nunc revisio auctoribus Alphonso et Casimir de Candolle allisque botanicis ultra memoratis. Vols. 1–9. G. Masson, Paris.
388. Candolle, A. P. de. 1824. Capparideae. Prodr. 1: 237–254.
389. ———. 1824. Sapindaceae. Prodr. 1: 601–618.
390. ———. 1825. Leguminosae. Prodr. 2: 93–524.
391. ———. 1825. Rosaceae. Prodr. 2: 525–639.
392. ———. 1830. Rubiaceae. Prodr. 4: 341–622.
393. ———. 1836. Compositae. Prodr. 5: 4–706.
394. ———. 1838. Continuatio Compositarum Senecionidearum. Prodr. 6: 1–687.

395. ———. 1838. Mantissa Compositarum. Prodr. 7(1): 263–308.
396. ———. 1839. Lobeliaceae. Prodr. 7(2): 339–413, 784–786.
397. ———. 1839. Goodenovieae. Prodr. 7(2): 502–520.
398. ———. 1839. Vaccinieae. Prodr. 7(2): 552–579.
399. ———. 1839. Epacrideae. Prodr. 7(2): 734–771.
400. ———. 1845. Cyrtandraceae. Prodr. 9: 258–286, 564.
401. ———. 1845. Borragineae. Prodr. 9: 466–559, 565–566.
402. ——— and A. L. P. P. de Candolle. 1823–1874. Prodromus systematis naturalis regni vegetabilis, sive enumeratio contracta ordinum, generum, specierumque plantarum hucusque cognitarum, juxta methodi naturalis normas digesta. Vols. 1–17. Treuttel and Würtz, Paris.
403. Canfield, J. E. 1982. Bog vegetation of Alakai Swamp, Kauai: communities and substrates (Abstr.). Proc. 4th Conf. Nat. Sci., Hawaii Volcanoes Natl. Park, p. 33.
404. ———. 1984. Water, water everywhere, but . . . : tissue water relations of bog and rainforest plants of Alakai Swamp, Kauai (Abstr.). Proc. 5th Conf. Nat. Sci., Hawaii Volcanoes Natl. Park, p. 40.
405. ———. 1986. The role of edaphic factors and plant water relations in plant distribution in the bog/wet forest complex of Alakaʻi Swamp, Kauaʻi, Hawaiʻi. Ph.D. dissertation, Univ. Hawaii, Honolulu, 280 pp.
406. ———. 1986. Edaphic characteristics and plant distribution in the bog/wet forest complex of Alakai Swamp, Kauai (Abstr.). Proc. 6th Conf. Nat. Sci., Hawaii Volcanoes Natl. Park, p. 2.
407. ——— and R. L. Stemmermann. 1980. Vascular plants of Kīpahulu Valley below 2000 feet: *in* Smith, C. W. (ed.), Resources base inventory of Kīpahulu Valley below 2000 feet. Coop. Natl. Park Resources Stud. Unit, Univ. Hawaii, Honolulu, pp. 11–44.
408. Carlquist, S. 1955. Maui, Kauai, and five silverswords. Pacific Disc. 8(3): 4–9.
409. ———. 1956. Documented chromosome numbers of plants. Madroño 13: 205–206.
410. ———. 1957. Leaf anatomy and ontogeny in *Argyroxiphium* and *Wilkesia* (Compositae). Amer. J. Bot. 44: 696–705.
411. ———. 1957. Systematic anatomy of *Hesperomannia*. Pacific Sci. 11: 207–215.
412. ———. 1957. Wood anatomy of *Mutisieae* (Compositae). Trop. Woods 106: 29–45.
413. ———. 1957. The genus *Fitchia* (Compositae). Univ. Calif. Publ. Bot. 29: 1–144.
414. ———. 1958. Structure and ontogeny of glandular trichomes of *Madinae* (Compositae). Amer. J. Bot. 45: 675–682.
415. ———. 1958. Wood anatomy of *Heliantheae* (Compositae). Trop. Woods 108: 1–30.
416. ———. 1959. Studies on *Madinae*: anatomy, cytology, and evolutionary relationships. Aliso 4: 171–236.

417. ———. 1959. Vegetative anatomy of *Dubautia*, *Argyroxiphium*, and *Wilkesia* (Compositae). Pacific Sci. 13: 195–210.
418. ———. 1961. Wood anatomy of *Inuleae* (Compositae). Aliso 5: 21–37.
419. ———. 1962. Ontogeny and comparative anatomy of thorns of Hawaiian Lobeliaceae. Amer. J. Bot. 49: 413–419.
420. ———. 1962. *Trematolobelia*: seed dispersal; anatomy of fruit and seeds. Pacific Sci. 16: 126–134.
421. ———. 1962. A theory of paedomorphosis in dicotyledonous woods. Phytomorphology 12: 30–45.
422. ———. 1964. Plant morphology in the Pacific Basin. Newslett. Hawaiian Bot. Soc. 3: 15–17.
423. ———. 1965. Island life. A natural history of the islands of the world. The Natural History Press, Garden City, N.Y., 451 pp.
424. ———. 1966. Wood anatomy of *Anthemideae*, *Ambrosieae*, *Calenduleae*, and *Arctotideae* (Compositae). Aliso 6(2): 1–23.
425. ———. 1966. The biota of long-distance dispersal. I. Principles of dispersal and evolution. Quart. Rev. Biol. 41: 247–270.
426. ———. 1966. The biota of long-distance dispersal. II. Loss of dispersibility in Pacific Compositae. Evolution 20: 30–48.
427. ———. 1966. The biota of long-distance dispersal. III. Loss of dispersibility in the Hawaiian flora. Brittonia 18: 310–335.
428. ———. 1966. The biota of long-distance dispersal. IV. Genetic systems in the floras of oceanic islands. Evolution 20: 433–455.
429. ———. 1967. The biota of long-distance dispersal. V. Plant dispersal to Pacific Islands. Bull. Torrey Bot. Club 94: 129–162.
430. ———. 1969. Wood anatomy of Goodeniaceae and the problem of insular woodiness. Ann. Missouri Bot. Gard. 56: 358–390.
431. ———. 1969. Wood anatomy of Lobelioideae (Campanulaceae). Biotropica 1: 47–72.
432. ———. 1970. Wood anatomy of Hawaiian, Macaronesian, and other species of *Euphorbia*: *in* Robson, N. K. B., D. F. Cutler, and M. Gregory (eds.), New research in plant anatomy. J. Linn. Soc., Bot. 63, Suppl. 1: 181–193.
433. ———. 1970. Hawaii, a natural history. Geology, climate, native flora and fauna above the shoreline. The Natural History Press, Garden City, N.Y., 463 pp.
434. ———. 1972. Island biology: we've only just begun. Bioscience 22: 221–225.
435. ———. 1974. Introduction to the new edition: *in* Rock, J. F., The indigenous trees of the Hawaiian Islands. Charles E. Tuttle Co., Rutland, Vt., pp. xi–xvi.
436. ———. 1974. Island biology. Columbia Univ. Press, New York, 660 pp.
437. ———. 1978. Wood anatomy and relationships of Bataceae, Gyrostemonaceae, and Stylobasiaceae. Allertonia 1: 297–330.
438. ———. 1980. Further concepts in ecological wood anatomy, with

comments on recent work in wood anatomy and evolution. Aliso 9: 499–553.
439. ———. 1980. Hawaii, a natural history. Geology, climate, native flora and fauna above the shoreline. 2nd ed. Pacific Tropical Botanical Garden, Lawai, Hawaii, 468 pp.
440. ———. 1981. Wood anatomy of Pittosporaceae. Allertonia 2: 355–392.
441. ———. 1981. Chance dispersal. Amer. Sci. 69: 509–516.
442. ———. 1982. The first arrivals. Nat. Hist. 91(12): 20–30.
443. ———. 1982. Hawaii: a museum of evolution. Nature Conservancy News 32(3): 4–11.
444. ———. 1983. Hawaii: a museum of evolution. Bull. Pacific Trop. Bot. Gard. 13: 33–39.
445. ——— and D. R. Bissing. 1976. Leaf anatomy of Hawaiian geraniums in relation to ecology and taxonomy. Biotropica 8: 248–259.
446. Carlson, N. K. 1952. Three grasses' struggle for supremacy on the island of Molokai. J. Range Managem. 5: 8–12.
447. ———. 1952. Grazing land problems, Molokai Island, Territory of Hawaii. J. Range Managem. 5: 230–242.
448. ———. 1973. The Kamehameha Schools/Bernice Pauahi Bishop Estate and the forests of the Big Island. Newslett. Hawaiian Bot. Soc. 12: 16–19.
449. ——— and L. W. Bryan. 1959. Hawaiian timber for the coming generations. Bernice P. Bishop Estate, Honolulu, 112 pp.
450. Carolin, R. C. 1964. The genus *Geranium* L. in the south western Pacific area. Proc. Linn. Soc. New South Wales 89: 326–361.
451. ———. 1966. Seeds and fruit of the Goodeniaceae. Proc. Linn. Soc. New South Wales 91: 68–83.
452. ———. 1970. The trichomes of the Goodeniaceae. Proc. Linn. Soc. New South Wales 96: 8–22.
453. Carpenter, F. L. 1976. Plant-pollinator interactions in Hawaii: pollination energetics of *Metrosideros collina* (Myrtaceae). Ecology 57: 1125–1144.
454. ———. 1976. Plant-pollinator interactions in Hawaii: pollination energetics of *Metrosideros collina* (Myrtaceae). U.S. IBP Island Ecosystems IRP Techn. Rep. 76: 1–62.
455. ——— and R. E. MacMillen. 1973. Interactions between Hawaiian honeycreepers and *Metrosideros collina* on the island of Hawaii. U.S. IBP Island Ecosystems IRP Techn. Rep. 33: 1–23.
456. ——— and ———. 1975. Pollination energetics and foraging strategies in a *Metrosideros*-honeycreeper association. U.S. IBP Island Ecosystems IRP Techn. Rep. 63: 1–9.
457. ——— and ———. 1976. Threshold model of feeding territoriality and test with a Hawaiian honeycreeper. Science 194: 639–642.
458. Carr, G. D. 1978. Chromosome numbers of Hawaiian flowering plants and the significance of cytology in selected taxa. Amer. J. Bot. 65: 236–242.

459. ———. 1978. Adaptive radiation in the silversword alliance—an overview. Newslett. Hawaiian Bot. Soc. 17: 64–68.
460. ———. 1978. Hybridization in the Hawaiian silversword complex. Proc. 2nd Conf. Nat. Sci., Hawaii Volcanoes Natl. Park, pp. 37–40.
461. ———. 1979. Uniform culture and propagation of Hawaiian tarweeds. Newslett. Hawaiian Bot. Soc. 18: 3–5.
462. ———. 1979 [1980]. Two new species in the Hawaiian endemic genus *Dubautia* (Compositae). Pacific Sci. 33: 233–237.
463. ———. 1985. Prospectus of a monograph of the Hawaiian *Madiinae* (Asteraceae): *Argyroxiphium*, *Dubautia*, and *Wilkesia*. Newslett. Hawaiian Bot. Soc. 24: 39–40.
464. ———. 1985. Monograph of the Hawaiian *Madiinae* (Asteraceae): *Argyroxiphium*, *Dubautia*, and *Wilkesia*. Allertonia 4: 1–123.
465. ———. 1985. Additional chromosome numbers of Hawaiian flowering plants. Pacific Sci. 39: 302–306.
466. ———. 1985. Habital variation in the Hawaiian *Madiinae* (*Heliantheae*) and its relevance to generic concepts in the Compositae. Taxon 34: 22–25.
467. ——— and J. K. Baker. 1977. Cytogenetics of *Hibiscadelphus* (Malvaceae): a meiotic analysis of hybrids in Hawaii Volcanoes National Park. Pacific Sci. 31: 191–194.
468. ——— and D. W. Kyhos. 1980. Adaptive radiation in the Hawaiian silversword alliance (Compositae) (Abstr.). Abstr. 2nd Int. Congr. Syst. Evol. Bio., Univ. British Columbia, Vancouver, p. 157.
469. ——— and ———. 1981. Adaptive radiation in the Hawaiian silversword alliance (Compositae–*Madiinae*). I. Cytogenetics of spontaneous hybrids. Evolution 35: 543–556.
470. ——— and ———. 1986. Chromosome evolution in the Hawaiian silversword alliance (Compositae–*Madiinae*) (Abstr.). Amer. J. Bot. 73: 755–756.
470a. ——— and ———. 1986. Adaptive radiation in the Hawaiian silversword alliance (Compositae–Madiinae). II. Cytogenetics of artificial and natural hybrids. Evolution 40: 959–976.
471. ——— and A. Meyrat. 1982. The status of the Mauna Kea silversword. Proc. 4th Conf. Nat. Sci., Hawaii Volcanoes Natl. Park, pp. 34–39.
472. ———, E. A. Powell, and D. W. Kyhos. 1986. Self-incompatibility in the Hawaiian *Madiinae* (Compositae): an exception to Baker's rule. Evolution 40: 430–434.
473. Carson, H. L. 1973. Genetic variation within island species in relation to environment. U.S. IBP Island Ecosystems IRP Techn. Rep. 21: 5.1–5.5.
474. ———. 1984. Speciation and the founder effect on a new oceanic island: *in* Radovsky, F. J., P. H. Raven, and S. H. Sohmer (eds.), Biogeography of the Pacific. Special Publ. Bernice P. Bishop Mus. 72: 45–54.
475. ———, D. E. Hardy, H. T. Spieth, and W. S. Stone. 1970. The evolutionary biology of the Hawaiian Drosophilidae: *in* Hecht, M. K.,

and W. C. Steere, Essays in evolution and genetics in honor of Theodosius Dobzhansky. Appleton-Century-Crofts, New York, pp. 437–543.

476. ——— and A. T. Ohta. 1981. Origin of the genetic basis of colonizing ability: *in* Scudder, G. G. E., and J. L. Reveal (eds.), Evolution today. Proc. 2nd Int. Cong. Syst. Evol. Biol., Vancouver, pp. 365–370.

477. Caum, E. L. 1918. A new weed. Hawaiian Pl. Rec. 19: 347–349.

478. ———. 1930. Notes on the flora of Molokini: *in* Palmer, H. S., Geology of Molokini. Occas. Pap. Bernice P. Bishop Mus. 9(1): 15–18.

479. ———. 1930. New Hawaiian plants. Occas. Pap. Bernice P. Bishop Mus. 9(5): 1–30.

480. ———. 1933. The bindweed. Hawaiian Pl. Rec. 37: 19–25.

481. ———. 1933. Notes on *Pteralyxia*. Occas. Pap. Bernice P. Bishop Mus. 10(8): 1–24.

482. ———. 1936. Notes on the flora and fauna of Lehua and Kaula Islands. Occas. Pap. Bernice P. Bishop Mus. 11(21): 1–17.

483. ——— and E. Y. Hosaka. 1936. A new species of *Schiedea*. Occas. Pap. Bernice P. Bishop Mus. 11(23): 1–5.

484. Chamberlain, D. F. 1977. Catalogue of the names published by Hector Léveillé: X. Notes Roy. Bot. Gard. Edinburgh 35: 247–264.

485. Chamisso, L. A. von. 1821. Remarks and opinions, of the naturalist of the expedition, continued: *in* Kotzebue, O. von, A voyage of discovery, into the South Sea and Beering's Straits, for the purpose of exploring a north-east passage, undertaken in the years 1815–1818, at the expense of his highness the Chancellor of the empire, Count Romanzoff, in the ship *Rurick*, under the command of the lieutenant in the Russian Imperial Navy, Otto von Kotzebue. Vol. 3. Longman, Hurst, Rees, Orme, and Brown, London, 442 pp.

486. ———. 1829. Boragineae Juss.: *in* Chamisso, L. A. von, and D. F. L. von Schlechtendal, De plantis in expeditione speculatoria Romanzoffiana observatis. Linnaea 4: 435–496.

487. ———. 1829. Aquaticae quaedam diversae affinitatis: *in* Chamisso, L. A. von, and D. F. L. von Schlechtendal, De plantis in expeditione speculatoria Romanzoffiana observatis. Linnaea 4: 497–508.

488. ———. 1830. Notices respecting the botany of certain countries visited by the Russian voyage of discovery under the command of Capt. Kotzebue. Bot. Misc. 1: 305–323.

489. ———. 1830. Rutaceae: *in* Chamisso, L. A. von, and D. F. L. von Schlechtendal, De plantis in expeditione speculatoria Romanzoffiana observatis. Linnaea 5: 43–59.

490. ———. 1831. Arcticae: *in* Chamisso, L. A. von, and D. F. L. von Schlechtendal, De plantis in expeditione speculatoria Romanzoffiana observatis. Linnaea 6: 528–592.

491. ———. 1833. Lobeliaceae: *in* Chamisso, L. A. von, and D. F. L. von Schlechtendal, De plantis in expeditione speculatoria Romanzoffiana observatis. Linnaea 8: 201–223.

492. ———. 1833. Goodenoviae R. B.: *in* Chamisso, L. A. von, and D. F. L. von Schlechtendal, De plantis in expeditione speculatoria Romanzoffiana observatis. Linnaea 8: 224–228.

493. ———. 1836. Reise um die Welt mit der Romanzoffischen Entdeckungs-Expedition in den Jahren 1815–18 auf der Brigg *Rurik*, Capitain Otto v. Kotzebue. Vols. 1–2. Tagebuch, Bemerkungen und Ansichten, Leipzig.

494. ———. 1862. Remarks and opinions respecting the Sandwich Islands. Friend 19: 9–11, 14–16.

495. ——— and D. F. L. von Schlechtendal. 1826–1836. De plantis in expeditione speculatoria Romanzoffiana observatis rationem dicunt. Linnaea Vols. 1–10.

496. ——— and ———. 1826. De plantis in expeditione speculatoria Romanzoffiana observatis. Linnaea 1: 1–73.

497. ——— and ———. 1826. De plantis in expeditione speculatoria Romanzoffiana observatis. Linnaea 1: 165–226.

498. ——— and ———. 1826. De plantis in expeditione speculatoria Romanzoffiana observatis. Linnaea 1: 511–570.

499. ——— and ———. 1827. De plantis in expeditione speculatoria Romanzoffiana observatis. Linnaea 2: 1–37.

500. ——— and ———. 1827. De plantis in expeditione speculatoria Romanzoffiana observatis. Linnaea 2: 145–233.

501. ——— and ———. 1827. De plantis in expeditione speculatoria Romanzoffiana observatis. Linnaea 2: 345–379.

502. ——— and ———. 1827. De plantis in expeditione speculatoria Romanzoffiana observatis. Linnaea 2: 541–611.

503. ——— and ———. 1828. De plantis in expeditione speculatoria Romanzoffiana observatis. Linnaea 3: 1–63.

504. ——— and ———. 1828. De plantis in expeditione speculatoria Romanzoffiana observatis. Linnaea 3: 199–233.

505. ——— and ———. 1829. De plantis in expeditione speculatoria Romanzoffiana observatis. Linnaea 4: 1–36.

506. ——— and ———. 1829. De plantis in expeditione speculatoria Romanzoffiana observatis. Linnaea 4: 129–202.

507. Chance, G. D., and J. D. Bacon. 1984. Systematic implications of seed coat morphology in *Nama* (Hydrophyllaceae). Amer. J. Bot. 71: 829–842.

508. Chang, A. T., and G. D. Sherman. 1953. The nickel content of some Hawaiian soils and plants and the relation of nickel to plant growth. Hawaii Agric. Exp. Sta. Univ. Hawaii Techn. Bull. 19: 1–25.

509. Char, W. P. 1976. Field studies of the *Sesbania* complex on the island of Hawaii. Bull. Pacific Trop. Bot. Gard. 6: 41.

510. ———. 1977. Strand vegetation of Hawaiʻi. Hawaii Coastal Zone News 1(12): 4–6.

511. ———. 1981. Strand ecosystems in Hawaiʻi (Abstr.): *in* Conserving

Hawaii's coastal ecosystems. Program and presentation summaries. Univ. Hawaii Sea Grant Marine Advisory Program, pp. 1–3.
512. ———. 1983. A revision of the Hawaiian species of *Sesbania* (Leguminosae). Master's thesis, Univ. Hawaii, Honolulu, 183 pp.
513. Chase, A. 1938. The carpet grasses. J. Wash. Acad. Sci. 28: 178–182.
514. Chock, A. T. K. 1953. A taxonomic revision of the Hawaiian species of the genus *Sophora* Linnaeus (family Leguminosae). Master's thesis, Univ. Hawaii, Honolulu, 102 pp.
515. ———. 1956. A taxonomic revision of the Hawaiian species of the genus *Sophora* Linnaeus (family Leguminosae). Pacific Sci. 10: 136–158.
516. ———. 1963. J. F. Rock, 1884–1962. Newslett. Hawaiian Bot. Soc. 2: 1–9.
517. ———. 1963. Kokee. Newslett. Hawaiian Bot. Soc. 2: 37–39.
518. ———. 1963. Botany at Bernice P. Bishop Museum. Newslett. Hawaiian Bot. Soc. 2: 57–65.
519. ———. 1963. J. F. Rock, 1884–1962. Taxon 12: 89–102.
520. ———. 1968. Hawaiian ethnobotanical studies I. Native food and beverage plants. Econ. Bot. 22: 221–238.
521. ———, E. H. Bryan, Jr., and Mrs. L. Marks. 1963. Bibliography of J. F. Rock. Newslett. Hawaiian Bot. Soc. 2: 10–13.
522. Choisy, J. D. 1841. De convolvulaceis dissertatio tertia, complectens Cuscutarum hucusque cognitarum methodicam enumerationem et descriptionem, necnon et brevum gallicam de Cuscutis praefactionem. Mém. Soc. Phys. Genève 9: 261–288; not seen.
523. Chow, K. H. 1968. Interspecific hybridization in the genus *Desmodium*. Master's thesis, Univ. Hawaii, Honolulu, 117 pp.
524. ———. 1971. Morphological variation and breeding behavior of some tropical and subtropical legumes. Ph.D. dissertation, Cornell Univ., Ithaca, 344 pp.
525. ———. 1982. Inheritance of rugose leaf in *Desmodium*. Pacific Sci. 36: 221–228.
526. ——— and L. V. Crowder. 1973. Hybridization of *Desmodium* species. Euphytica 22: 399–404.
527. ——— and ———. 1974. Flowering behavior and seed development in four *Desmodium* species. Agron. J. 66: 236–238.
528. ——— and ———. 1974. Putative parents of a *Desmodium* selection examined morphologically and by isozyme patterns. Bot. Gaz. (Crawfordsville) 135: 180–184.
529. ——— and ———. 1975. Esterase isozyme patterns of some tropical and subtropical herbaceous legumes. Pacific Sci. 29: 365–369.
530. Christensen, C. 1920. Dansk botanisk Forening. Bot. Tidsskr. 37: 148–159.
531. Christensen, C. 1979. Propagating Kauai's *Brighamia*. Bull. Pacific Trop. Bot. Gard. 9: 2–4.
532. Christophersen, E. 1931. Notes on *Joinvillea*. Occas. Pap. Bernice P. Bishop Mus. 9(12): 1–7.

533. ——. 1934. A new Hawaiian *Abutilon*. Occas. Pap. Bernice P. Bishop Mus. 10(15): 1–7.
534. —— and E. L. Caum. 1931. Vascular plants of the Leeward Islands, Hawaii. Bernice P. Bishop Mus. Bull. 81: 1–41.
535. Chudnoff, M. 1984. Tropical timbers of the world. U.S.D.A. Forest Serv. Agric. Handbook 607: 1–464.
536. Chung, H. L. 1923. The sweet potato in Hawaii. Hawaii Agric. Exp. Sta. Bull. 50: 1–20.
537. —— and J. C. Ripperton. 1929. Utilization and composition of oriental vegetables in Hawaii. Hawaii Agric. Exp. Sta. Bull. 60: 1–64.
538. Chung, N. 1938. Tress from everywhere on a cosmopolitan campus. Paradise Pacific 50(10): 15–16.
539. Clapp, R. B. 1972. The natural history of Gardner Pinnacles, Northwestern Hawaiian Islands. Atoll Res. Bull. 163: 1–29.
540. —— and W. O. Wirtz, II. 1975. The natural history of Lisianski Island, Northwestern Hawaiian Islands. Atoll Res. Bull. 186: 1–196.
541. Clark, C. A., and F. W. Gould. 1978. *Dichanthelium* subgenus *Turfosa* (Poaceae). Brittonia 30: 54–59.
542. Clarke, C. B. 1883. *Cyrtandreae* (Gesneracearum tribus). Monogr. phan. 5: 1–303.
543. Clarke, F. L. 1874. Decadence of Hawaiian forests. Hawaiian Almanac and Annual for 1875: 19–20.
544. Clarke, G. 1978. The distribution of *Myrica faya* and other selected problem exotics within Hawaii Volcanoes National Park (Abstr.). Proc. 2nd Conf. Nat. Sci., Hawaii Volcanoes Natl. Park, p. 51.
545. Clausen, R. T. 1945. A botanical study of the yam beans (*Pachyrrhizus*). Cornell Univ. Agric. Exp. Sta. Mem. 264: 1–38.
546. Clay, H. F. 1956. Orchids in Hawaii. Islands to host international conference in 1957. Paradise Pacific 68(7): 14–15.
547. ——. 1961. Narrative report of botanical field work on Kure Island. Atoll Res. Bull. 78: 1–4.
548. —— and J. C. Hubbard. 1962. Trees for Hawaiian gardens. Univ. Hawaii, Coop. Extens. Serv. Bull. 67: 1–103.
549. —— and ——. 1977. The Hawai'i garden. Tropical exotics. Univ. Press Hawaii, Honolulu, 267 pp.
550. —— and ——. 1977. The Hawai'i garden. Tropical shrubs. Univ. Press Hawaii, Honolulu, 295 pp.
551. Cogniaux, A. 1881. Cucurbitaceae. Monogr. phan. 3: 325–954.
552. Colozza, A. 1908. Studio anatomico sulle "Goodeniaceae." Nuovo Giorn. Bot. Ital. 15: 6–92.
553. Conant, S. 1985. Ecosystem monitoring, restoration, and management in Hawai'i: a summary: *in* Stone, C. P., and J. M. Scott (eds.), Hawai'i's terrestrial ecosystems: preservation and management. Coop. Natl. Park Resources Stud. Unit, Univ. Hawaii, Honolulu, pp. 475–480.
554. ——. 1985. Recent observations on the plants of Nihoa Island, Northwestern Hawaiian Islands. Pacific Sci. 39: 135–149.

555. ———, C. C. Christensen, P. Conant, W. C. Gagné, and M. L. Goff. 1984. The unique terrestrial biota of the Northwestern Hawaiian Islands: *in* Grigg, R. W., and K. Y. Tanoue (eds.), Resource investigations in the Northwestern Hawaiian Islands. Vol. 2. Univ. Hawaii Sea Grant Coll. Program (UNIHI-SEAGRANT-MR-84-01), Honolulu, pp. 77-94.

556. ——— and D. R. Herbst. 1983. A record of *Nephrolepis multiflora* from Nihoa Island, Northwestern Hawaiian Islands. Newslett. Hawaiian Bot. Soc. 22: 17-19.

557. Conn, B. J. 1980. A taxonomic revision of *Geniostoma* subg. *Geniostoma* (Loganiaceae). Blumea 26: 245-364.

558. Conrad, C. E., and P. G. Scowcroft. 1984. Structure, productivity, and nutrient cycling—important but little known parameters of Hawaii's *koa-ohia* forests (Abstr.). Proc. 5th Conf. Nat. Sci., Hawaii Volcanoes Natl. Park, p. 52.

559. Constance, L. 1963. Chromosome number and classification in Hydrophyllaceae. Brittonia 15: 273-285.

560. Conter, F. E. 1903. The cultivation of sisal in Hawaii. Hawaii Agric. Exp. Sta. Press Bull. 4: 1-31.

561. Cook, J. 1784. A voyage to the Pacific Ocean, undertaken, by the command of his majesty, for making discoveries in the Northern Hemisphere. Vol. II. G. Nicol & T. Cadell, London, 549 pp.

562. Cook, O. F. 1915. *Glaucothea*, a new genus of palms from Lower California. J. Wash. Acad. Sci. 5: 236-241.

563. Cooke, C. M., Jr., and M. C. Neal. 1928. Flowering time of common plants in Honolulu. Hawaii, Xmas number: 6-7.

564. ——— and ———. 1955. Approximate flowering time of Honolulu plants. Bishop Mus. Press, Honolulu, 4 pp.

565. Cooperrider, T. S., and M. M. Galang. 1965. A *Pluchea* hybrid from the Pacific. Amer. J. Bot. 52: 1020-1026.

566. Cooray, R. G. 1974. Stand structure of a montane rain forest on Mauna Loa, Hawaii. Master's thesis, Univ. Hawaii, Honolulu, 165 pp.

567. ———. 1974. Stand structure of a montane rain forest on Mauna Loa, Hawaii. U.S. IBP Island Ecosystems IRP Techn. Rep. 44: 1-98.

568. ———. 1975. Stand structure and map of a montane rain forest on Mauna Loa, Hawaii (Abstr.). Proc. 13th Pacific Sci. Congr., Canada 1: 96.

569. Copeland, E. B. 1948. The origin of the native flora of Polynesia. Pacific Sci. 2: 293-296.

570. Copeland, H. F. 1954. Observations on certain Epacridaceae. Amer. J. Bot. 41: 215-222.

571. Cordero, J. 1980. Propagation of *Hibiscus brackenridgei* var. *mokuleiana* at Waimea. Notes Waimea Arbor. & Bot. Gard. 7(1): 12-13.

572. Corn, C. A. 1972. Seed dispersal methods in Hawaiian "*Metrosideros*": *in* Behnke, J. A. (ed.), Challenging biological problems: directions toward their solution. Oxford Univ. Press, New York, pp. 422-435.

573. ———. 1972. Genecological studies of *Metrosideros*. U.S. IBP Island Ecosystems IRP Techn. Rep. 2: 88–95.
574. ———. 1972. Seed dispersal methods in Hawaiian *Metrosideros*. U.S. IBP Island Ecosystems IRP Techn. Rep. 6: 1–19.
575. ———. 1976. Variation of Hawaiian *Metrosideros* along the south and east flanks of Mauna Loa, Hawaii. Proc. 1st Conf. Nat. Sci., Hawaii Volcanoes Natl. Park, pp. 75–88.
576. ———. 1978. Experimental hybridizations in Hawaiian *Metrosideros*. Proc. 2nd Conf. Nat. Sci., Hawaii Volcanoes Natl. Park, pp. 77–85.
577. ———. 1979. Variation in Hawaiian *Metrosideros*. Ph.D. dissertation, Univ. Hawaii, Honolulu, 294 pp.
578. ———. 1982. The Melbourne Hillebrand collection of Hawaiian plants. Proc. 4th Conf. Nat. Sci., Hawaii Volcanoes Natl. Park, pp. 49–50.
579. ———. 1983. The Melbourne Hillebrand collection of Hawaiian plants. Newslett. Hawaiian Bot. Soc. 22: 20–21.
580. ——— and G. C. Ashton. 1970. Genecological studies on *Metrosideros*. U.S. IBP Island Ecosystems IRP Techn. Rep. 1: 80–85.
581. ——— and W. M. Hiesey. 1973. Altitudinal variation in Hawaiian *Metrosideros*. Amer. J. Bot. 60: 991–1002.
582. ——— and ———. 1973. Altitudinal ecotypes in Hawaiian *Metrosideros*. U.S. IBP Island Ecosystems IRP Techn. Rep. 18: 1–19.
583. ———, C. H. Lamoureux, and D. R. Herbst. 1981. Floristic changes on Green Island, Kure Atoll. Abstr. 13th Int. Bot. Congr., Sydney, p. 110.
584. Cornell, R. D. 1931. The tree lover's Hawaii. Asia 31: 776–784, 808–810.
585. Corner, E. J. H. 1962. The classification of Moraceae. Gard. Bull. Straits Settlem. 19: 187–252.
586. ———. 1966. The natural history of palms. Univ. Calif. Press, Berkeley, 393 pp.
587. ———. 1975. The evolution of *Streblus* Lour. (Moraceae): with a new species of sect. *Bleekrodea*. Phytomorphology 25: 1–12.
588. Corum, A. K. 1985. Folk remedies from Hawai'i. Bess Press, Honolulu, 137 pp.
589. Cory, C. S. 1984. Pollination biology of two species of Hawaiian Lobeliaceae (*Clermontia kakeana* and *Cyanea angustifolia*) and their presumed co-evolved relationship with native honeycreepers (Drepanididae). Master's thesis, Calif. State Univ., Fullerton, 83 pp.
590. ———. 1986. Saving endangered plants. Notes Waimea Arbor. & Bot. Gard. 13(1): 5–6.
591. Coryell, J. 1986. *Kukui*. Mānoa Winter 1986: 12–15.
592. Coulter, J. W. 1941. The relation of soil erosion to land utilization in the Territory of Hawaii. Proc. 6th Pacific Sci. Congr., California 4: 897–907.
593. Cowan, R. S. 1949. A taxonomic revision of the genus *Neraudia* (Urticaceae). Pacific Sci. 3: 231–270.
594. Cox, C. B., I. N. Healey, and P. D. Moore. 1976. Biogeography. An

ecological and evolutionary approach. 2nd ed. John Wiley & Sons, New York, 194 pp.
595. Cox, P. A. 1981. Vertebrate pollination and the maintenance of unisexuality in *Freycinetia*. Ph.D. dissertation, Harvard Univ., Cambridge; not seen.
596. ———. 1982. Vertebrate pollination and the maintenance of dioecism in *Freycinetia*. Amer. Naturalist 120: 65–80.
597. ———. 1983. Extinction of the Hawaiian avifauna resulted in a change of pollinators for the *ieie*, *Freycinetia arborea*. Oikos 4l: 195–199.
598. ———, B. Wallace, and I. Baker. 1984. Monoecism in the genus *Freycinetia* (Pandanaceae). Biotropica 16: 313–314.
599. Cracknell, R. 1969. '*Awa*: stone age to space age. Newslett. Hawaiian Bot. Soc. 8(2): 1–4.
600. Cranwell, L. M. 1984. Lehua Maka Noe—an endangered Hawaiian bog. Newslett. Hawaiian Bot. Soc. 23: 3–6.
601. Crawford, D. J. 1985. Electrophoretic data and plant speciation. Syst. Bot. 10: 405–416.
602. Crawford, D. L. 1937. Hawaii's crop parade. A review of useful products derived from the soil in the Hawaiian Islands, past and present. Advertiser Publ. Co., Honolulu, 305 pp.
603. Cribb, P. J., and C. Z. Tang. 1982. *Spathoglottis* (Orchidaceae) in Australia and the Pacific Islands. Kew Bull. 36: 721–729.
604. Croft, L., D. E. Hemmes, and J. D. Macneil. 1976. Puukohola Heiau National Historic Site plant survey. Newslett. Hawaiian Bot. Soc. 15: 81–94.
605. Croizat, L. C. M. 1941. A discussion of new and critical synonymy. J. Arnold Arbor. 22: 133–142.
606. ———. 1952. Manual of phytogeography or an account of plant-dispersal throughout the world. W. Junk, The Hague, The Netherlands, 587 pp.
607. Crosby, W., and E. Y. Hosaka. 1955. Vegetation: *in* Cline, M. G. (ed.), Soil survey of the Territory of Hawaii. U.S.D.A. Soil Survey Ser. 1939, 25: 28–34.
608. Cuddihy, L. W. 1978. Effects of cattle grazing on the mountain parkland ecosystem, Mauna Loa, Hawaii. Master's thesis, Univ. Hawaii, Honolulu, 198 pp.
609. ———. 1984. Effects of cattle grazing on the mountain parkland ecosystem, Mauna Loa, Hawaii. Coop. Natl. Park Resources Stud. Unit, Hawaii, Techn. Rep. 51: 1–135.
610. Cumbie, B. G., and D. Mertz. 1962. Xylem anatomy of *Sophora* (Leguminosae) in relation to habit. Amer. J. Bot. 49: 33–40.
611. Dadswell, H. E., and H. D. Ingle. 1947. The wood anatomy of the Myrtaceae, I. A note on the genera *Eugenia*, *Syzygium*, *Acmena*, and *Cleistocalyx*. Trop. Woods 90: 1–7.
612. Dahlberg, K. 1914. The story of the papaia. Mid-Pacific Mag. 8: 261–265.
613. Dalton, H. C. 1983. Ethnobotany of the sweet potato in Hawaii. Bull. Pacific Trop. Bot. Gard. 13: 81–84.

614. Danser, B. H. 1936. Loranthaceae: *in* Hochreutiner, B. P. G., Plantae Hochreutineranae, étude systématique et biologique des collections faites par l'auteur au cours de son voyage aux Indes Néerlandaises et autour du monde pendant les années 1903 à 1905. Candollea 6: 457–459.
615. ———. 1937. A revision of the genus *Korthalsella*. Bull. Jard. Bot. Buitenzorg III, 14: 115–159.
616. Darwin, S. P. 1979. A synopsis of the indigenous genera of Pacific Rubiaceae. Allertonia 2: 1–44.
617. Datton, E. K. 1984. Visiting the native forest of Moloka'i. Bull. Pacific Trop. Bot. Gard. 14: 82–86.
618. Davis, B. D. 1978. Human settlement and environmental change at Barbers Point, O'ahu. Proc. 2nd Conf. Nat. Sci., Hawaii Volcanoes Natl. Park, pp. 87–97.
619. ———. 1982. Horticultural adaptation and ecological change in southwestern O'ahu: preliminary evidence from Barbers Point. Proc. 4th Conf. Nat. Sci., Hawaii Volcanoes Natl. Park, pp. 51–59.
620. Davis, C. J. 1966. Pest control in Hawaii as practiced by the State Department of Agriculture. Elepaio 27: 45–47.
621. ———. 1970. Black twig borer threatens native trees. Newslett. Hawaiian Bot. Soc. 9: 38–39.
622. ———. 1970. Communication letter to Park Superintendent on *Acacia* psyllid infestation, Mauna Loa transect. U.S. IBP Island Ecosystems IRP Techn. Rep. 1: 95.
623. ———. 1972. Progress report: Cerambycid studies on *Sapindus saponaria* (Sapindaceae). U.S. IBP Island Ecosystems IRP Techn. Rep. 2: 128–129.
624. Davis, T. A., and J. C. Selvaraj. 1964. Floral asymmetry in Malvaceae. J. Bombay Nat. Hist. Soc. 61: 402–409.
625. Davison, C. 1927. Hawaiian medicine. Queen's Hospital Bull. 4(3): 1–4.
626. Dawson, J. W. 1970. Pacific capsular Myrtaceae 2. The *Metrosideros* complex: *M. collina* group. Blumea 18: 441–445.
627. Decaisne, J. 1871–1872. Le jardin fruitier du Muséum ou iconographie de toutes les espèces et variétés d'arbres fruitiers cultivés dans cet établissement, avec leur description, leur histoire, leur synonymie. Vol. 2. Firmin Didot, Paris; not seen.
628. Degen, J. L., R. A. Criley, and D. A. McLain. 1973. Ground covers for Hawaii landscapes. Univ. Hawaii Coop. Extens. Serv. Circ. 457: 1–40.
629. Degener, O. 1925. Plant collecting in Hawaii. Torreya 25: 18–19.
630. ———. 1929. The genus *Bidens* (*Campylotheca*) in Hawaii (Abstr.). Proc. Hawaiian Acad. Sci. 4: 6–7.
631. ———. 1930. Illustrated guide to the more common or noteworthy ferns and flowering plants of Hawaii National Park. Publ. privately, Honolulu, 312 pp.

632. ———. 1932. *Kokoolau*, the Hawaiian tea. J. Pan-Pacific Res. Inst. 7(2): 1–16.
633. ———. 1932. The Flora Hawaiiensis or New illustrated flora of the Hawaiian Islands. J. Pan-Pacific Res. Inst. 7(4): 2–16.
634. ———. 1932–1980. Flora Hawaiiensis or New illustrated flora of the Hawaiian Islands. Publ. privately, Honolulu. [See Mill et al., Taxon 34: 229–259 (1985) for citation of each of the 1,144 articles included in this work].
635. ———. 1937. Miscelanea. Videant consules [Review of "Hawaii's crop parade" by D. L. Crawford]. Revista Sudamer. Bot. 6: 38–41.
636. ———. 1943. *Stenogyne sherffii* Degener, a new mint from Hawaii. Brittonia 5: 58–59.
637. ———. 1945. Plants of Hawaii National Park illustrative of plants and customs of the South Seas. [Rev. ed. of Illustrated guide to the more common or noteworthy ferns and flowering plants of Hawaii National Park, 1930]. Edwards Brothers, Inc., Ann Arbor, Mich., 314 pp.
638. ———. 1945. Tropical plants the world around I. J. New York Bot. Gard. 46: 74–91.
639. ———. 1945. Tropical plants the world around II. J. New York Bot. Gard. 46: 110–125.
640. ———. 1945. Tropical plants the world around. III. J. New York Bot. Gard. 46: 132–143.
641. ———. 1945. Tropical plants the world around. IV. J. New York Bot. Gard. 46: 158–167.
642. ———. 1947. Hawaii's native flowers. Paradise Pacific 59(6): 15.
643. ———. 1948. Silversword. Hawaiian Digest 3(15): 23–27.
644. ———. 1950. *Hillebrandia*. Begonian 17: 123–124.
645. ———. 1950. Botanists' adventure in Kalalau Valley. Paradise Pacific 62(3): 18–20.
646. ———. 1951. Kalalau Valley. Island of Kauai, Hawaii. Gard. J. New York Bot. Gard. 1: 3–5.
647. ———. 1961. *Scaevola* misconceptions. Taxon 10: 227–228.
648. ———. 1963. Botanists' expedition on Lanai. Newslett. Hawaiian Bot. Soc. 2: 107–108.
649. ———. 1966. Book review. Phytologia 13: 369–370.
650. ———. 1968. Earl Edward Sherff (1886–1966). Taxon 17: 189–198.
651. ———. 1970. Caveat emptor. Newslett. Hawaiian Bot. Gard. Found. 4(7): 1–4.
652. ———. 1972. Axis deer damages. Elepaio 32: 105–106.
653. ———. 1977. Help save the dwindling endemic flora of the Hawaiian Islands at least as herbarium specimens for museums of the world. Phytologia 37: 281–284.
654. ——— and I. Degener. 1958. The Hawaiian beach *Scaevola* (Goodeniaceae). Phytologia 6: 321.
655. ——— and ———. 1961. Past, present, and future of the Hawaiian flora. Abstr. Symp. Pap., 10th Pacific Sci. Congr., Honolulu, pp. 130–131.

656. ―― and ――. 1961. A new Hawaiian variety of *Capparis*. Phytologia 7: 369.
657. ―― and ――. 1961. A new *Dodonaea* from Molokai, Hawaii. Phytologia 7: 465.
658. ―― and ――. 1961. *Gouldia* in Hawaii. Phytologia 7: 465–467.
659. ―― and ――. 1961. Green Hawaii. Past, present, and future of an island flora. Pacific Disc. 14(5): 15–17.
660. ―― and ――. 1963. Kaena Point, Oahu. Newslett. Hawaiian Bot. Soc. 2: 77–78.
661. ―― and ――. 1964. Hawaii's vanishing native plants. Hist. Nat. Pro Nat. 2(8): 32.
662. ―― and ――. 1965. Orchids of Hawaii Nei. Bull. Pacific Orchid Soc. Hawaii 23: 12–15.
663. ―― and ――. 1965. The Hawaiian genus *Neowimmeria* (Lobeliaceae). Phytologia 12: 73.
664. ―― and ――. 1966. E. E. Sherff, 1866–1966. Newslett. Hawaiian Bot. Soc. 5: 32.
665. ―― and ――. 1966. Yes, thank you; we love ferns. Phytologia 13: 449–452.
666. ―― and ――. 1967. Partial review of Doty & Mueller-Dombois' "Atlas," and new taxa in Hawaiian Rubiaceae. Phytologia 14: 213–215.
667. ―― and ――. 1967. Partial review of Doty & Mueller-Dombois' "Atlas," and new taxa in Hawaiian Rubiaceae, II. Phytologia 15: 42–52.
668. ―― and ――. 1967. *Scaevola gaudichaudiana* & *S. mollis*. Phytologia 15: 160–162.
669. ―― and ――. 1968. Review. Phytologia 17: 113.
670. ―― and ――. 1968. The eruption in Hiiaka Crater, island of Hawaii. Phytologia 17: 343.
671. ―― and ――. 1968. Review of F. E. Wimmer, Campanulaceae–*Lobelioideae* supplementum et Campanulaceae–*Cyphioideae*. Das Pflanzenreich, IV. 276c (108. Heft). I–X, 816–1024; with description of *Trematolobelia wimmeri* Deg. & Deg., sp. nov. Phytologia 17: 369–371.
672. ―― and ――. 1969. Review [of "On the flora and vegetation of the Hawaiian Islands" by Y. Mäkinen]. Phytologia 19: 47–49.
673. ―― and ――. 1970. Book review. The genus *Pelea*, with pertinent and impertinent remarks. Phytologia 19: 313–319.
674. ―― and ――. 1971. *Schiedea* and *Pleomele*—comments by Otto and Isa Degener. Newslett. Hawaiian Bot. Soc. 10: 9.
675. ―― and ――. 1971. Postscripts and notes about *Acacia* on Lanai. Newslett. Hawaiian Bot. Soc. 10: 27–28.
676. ―― and ――. 1971. *Scaevola kilaueae* var. *powersii* Deg. & Deg. Phytologia 21: 72.
677. ―― and ――. 1971. *Rumex* of Hawaii. Phytologia 21: 139–146.

678. —— and ——. 1971. Some *Aleurites* taxa in Hawaii and a note regarding *Argemone*. Phytologia 21: 315–319.
679. —— and ——. 1971. *Pritchardia* and *Cocos* in the Hawaiian Islands. Phytologia 21: 320–326.
680. —— and ——. 1971. Review and comments about a thing. Phytologia 21: 369–374.
681. —— and ——. 1971. *Sophora* in Hawaii. Phytologia 21: 411–416.
682. —— and ——. 1971. Numata & Asano, "Biological flora of Japan" and remarks about *Paederia, Phryma, Rabdosia, Rapanea, Sigesbeckia* & *Vitex*. Phytologia 22: 210–214.
683. —— and ——. 1972. 'Ohi'a infection. Elepaio 32: 65.
684. —— and ——. 1972. *Wikstroemia pulcherrima* var. *petersonii* Deg. & Deg., from Hawaii. Phytologia 24: 151–154.
685. —— and ——. 1973. *Santalum paniculatum* var. *chartaceum* Deg. & Deg. Phytologia 27: 145–147.
686. —— and ——. 1974. An explanation and an appeal. Newslett. Hawaiian Bot. Soc. 13: 6–7.
687. —— and ——. 1974. To save a rare *naupaka*. Newslett. Hawaiian Bot. Soc. 13: 16.
688. —— and ——. 1974. Notes from Drs. Otto & Isa Degener. Newslett. Hawaiian Bot. Soc. 13: 17–18.
689. —— and ——. 1974. *Coix lacryma-jobi*. Newslett. Hawaiian Bot. Soc. 13: 23.
690. —— and ——. 1974. *Spathodea* in Hawaii. Phytologia 28: 419–420.
691. —— and ——. 1974. Appraisal of Hawaiian taxonomy. Phytologia 29: 240–246.
692. —— and ——. 1974. [Degeners' Flora Hawaiiensis, Leaflet No. 1,] Prodromus of *Galeatella* and *Neowimmeria*. Publ. privately, Honolulu, 13 pp.
693. —— and ——. 1974–1985. Degeners' Flora Hawaiiensis. Publ. privately, Honolulu. [See Mill et al., Taxon 34: 229–259 (1985) for citation of articles included in this work. See also entry no. 697].
694. —— and ——. 1975. Silverswords & the Blue Data Book. Notes Waimea Arbor. 2(1): 3–6.
695. —— and ——. 1975. *Rapanea*, Myrsinaceae, in the Pacific. Phytologia 31: 21.
696. —— and ——. 1975. Degeners' Flora Hawaiiensis, Leaflet No. 2, *Myrsine, Rapanea* and *Suttonia*. Publ. privately, Honolulu, 2 pp.
697. —— and ——. 1975. Degeners' Leaflet No. 3. Concerning a magazine article. Publ. privately, Honolulu, 6 pp. [Presumed to be Leaflet No. 3 of Degeners' Flora Hawaiiensis. Inadvertently not included in Mill et al., Bibliography of Otto and Isa Degeners' Hawaiian floras (1985).]
698. —— and ——. 1975. [Review of] List of flowering plants in Hawaii. Sida 6: 120–122.

699. —— and ——. 1976. Some Hawaiian specimens in Leningrad and in Geneva. Newslett. Hawaiian Bot. Soc. 15: 34–35.
700. —— and ——. 1976. McBride's "Practical folk medicine of Hawaii," and opinions about *Tacca hawaiiensis* versus *Tacca leontopetaloides* and other taxa. Phytologia 34: 1–4.
701. —— and ——. 1976. *Wikstroemia perdita* Deg. & Deg., an extinct (?) endemic of a paradise lost by exotic primates. Phytologia 34: 28–32.
702. —— and ——. 1977. *Hibiscadelphus* number KK-HX-1. An international treasure in Hawaii. Phytologia 35: 385–396.
703. —— and ——. 1977. Some taxa of red-flowered *Hibiscus* endemic to the Hawaiian Islands. Phytologia 35: 459–470.
704. —— and ——. 1978. Kaena Point, Hawaiian Islands and a prodromus regarding some taxa in *Sesbania* (Leguminosae). Phytologia 39: 147–160.
705. —— and ——. 1981. No *hala* hallucination! Notes Waimea Arbor. & Bot. Gard. 8(1): 2–5.
706. —— and ——. 1982. *Pandanus tectorius* again and *Delissea* anew. Notes Waimea Arbor. & Bot. Gard. 9(1): 8–12.
707. —— and ——. 1983. Plants of Hawaii National Parks illustrative of plants and customs of the South Seas. Page proof of ed. 3. Publ. privately, Volcano, Hawaii, 314 pp.
708. —— and ——. 1984. To whom it may concern regarding Kahauale'a geothermal project. Notes Waimea Arbor. & Bot. Gard. 11(2): 6–12.
709. ——, ——, and H. Hörmann. 1969. *Cyanea carlsonii* Rock and the unnatural distribution of *Sphagnum palustre* L. Phytologia 19: 1–3.
710. ——, ——, and J. H. Kern. 1964. A new Hawaiian *Gahnia* (Cyperaceae). Blumea 12: 349–351.
711. ——, ——, K. Sunada, and Mrs. K. Sunada. 1976. *Argyroxiphium kauense*, the Kau silversword. Phytologia 33: 173–177.
712. ——, ——, and H. Ziegenspeck. 1956. Vorkommen von *Drosera anglica* Huds in Hawaii und die Samenverbreitung mancher *Drosera* Arten in Hawaii und Europa. Publ. privately, 6 pp.
713. —— and A. B. H. Greenwell. 1951. A new variety of *Perrottetia* (Celastraceae) from the Hawaiian Islands. Revista Sud-Amer. Bot. 10(1): 25.
714. —— and E. Y. Hosaka. 1940. *Straussia sessilis*, a new species from Hawaii. Bull. Torrey Bot. Club 67: 301.
715. —— and C. Skottsberg. 1937. A new Hawaiian species of Rutaceae. Brittonia 2: 362.
716. DeLisle, D. G. 1963. Taxonomy and distribution of the genus *Cenchrus*. Iowa State J. Sci. 37: 259–351.
717. Department of Agriculture, Div. of Plant Industry (T. K. Tagawa, Consultant). 1979. Foreign noxious weed survey. Hawaii, Dept. Agric., Div. Plant Industry, and U.S.D.A. Animal, Plant Health Inspection Serv., Plant Protection, Quarantine, Honolulu, 130 pp.

718. DeWreede, R. E. 1968. A preliminary report on the ecological conditions of Kipahulu Valley, Maui: *in* Warner, R. M. (ed.), Scientific report of the Kipahulu Valley Expedition. Nature Conservancy, San Francisco, pp. 9–22.
719. Dibble, S. 1909. A history of the Sandwich Islands. Thos. G. Thrum, Honolulu, 428 pp.
720. Diels, L. 1910. Menispermaceae. Pflanzenr. IV. 94 (Heft 46): 1–345.
721. Dietrich, W. 1977. The South American species of *Oenothera* sect. *Oenothera* (*Raimannia, Renneria*; Onagraceae). Ann. Missouri Bot. Gard. 64: 425–626.
722. Dillingham, E. L. 1913. Indigenous trees. Friend 71: 39.
723. Doerr, J. E., Jr. 1931. Blackberries. Nature Notes, Hawaii Natl. Park 1(1): 8.
724. ———. 1931. Nature study. Nature Notes, Hawaii Natl. Park 1(2): 13–15.
725. ———. 1931. The Halemaumau Trail. "The world's weirdest walk." Nature Notes, Hawaii Natl. Park 1(4): 25–36.
726. Dole, G. H. 1875. The Koloa Swamp. Islander 1: 58–59.
727. Domke, W. 1934. Untersuchungen über die systematische und geographische Gliederung der Thymelaeaceen nebst einer neubeschreibung ihrer Gattungen. Biblioth. Bot. 111: 1–151.
728. Don, G. 1831–1838. A general history of the dichlamydeous plants, comprising complete description of the different orders . . . Vols. 1–4. J. G. & F. Rivington et al., London.
729. ———. 1834. Rubiaceae. Gen. hist. 3: 453–665.
730. ———. 1834. Lobeliaceae. Gen. hist. 3: 697–719.
731. ———. 1834. Goodenoviae. Gen. hist. 3: 722–731.
732. Donaghho, W. R. 1951–1952. Journal of ornithological work during the summer of 1937. Elepaio 11: 50–52, 56–58, 62–65, 72–73; 12: 6–7, 13–15, 21–23, 26–28, 42–44 (1951), 46–48 (1952).
733. ———. 1970. Destruction of virgin *'ohi'a* and *koa* forest on Hawaii by the Division of Forestry. Elepaio 30: 67.
734. Doria, J. J. 1979. Haleakala's silversword has a chance. Natl. Parks & Conservation Mag., Environmental J. 53(12): 14–16.
735. Doty, M. S. 1967. Contrast between the pioneer populating process on land and shore. Bull. S. Calif. Acad. Sci. 66: 175–194.
736. ——— and D. Mueller-Dombois. 1966. Atlas for bioecology studies in Hawaii Volcanoes National Park. Univ. Hawaii, Hawaii Bot. Sci. Pap. 2: 1–507. [Also published as Hawaii Agric. Exp. Sta. Misc. Publ. 89].
737. ——— and B. C. Stone. 1966. Two new species of *Halophila* (Hydrocharitaceae). Brittonia 18: 303–306.
738. ——— and ———. 1967. Typification for the generic name *Halophila* Thouars. Taxon 16: 414–418.
739. Doty, R. D. 1980. Groundwater conditions in the *'ōhi'a* rain forest near Hilo. Proc. 3rd Conf. Nat. Sci., Hawaii Volcanoes Natl. Park, pp. 101–111.

740. ———. 1982 [1983]. Annual precipitation on the island of Hawaii between 1890 and 1977. Pacific Sci. 36: 421–425.
741. ———. 1983. Stream flow in relation to *ohia* forest decline on the island of Hawaii. Water Resources Bull. 19: 217–221.
742. Douglas, D. 1836. Mr. Douglas' voyage from the Columbia to the Sandwich Islands, and the ascent of Mouna Roa: *in* Hooker, W., A brief memoir of Mr. David Douglas, with extracts from his letters. Companion Bot. Mag. 2: 161–177.
743. ———. 1914. Journal kept by David Douglas during his travels in North America 1823–1827. Together with a particular description of thirty-three species of American oaks and eighteen species of *Pinus* . . . William Wesley & Son, London, 364 pp.
744. Douglas, G. 1969. Draft check list of Pacific oceanic islands. Micronesica 5: 327–463.
745. Drake del Castillo, E. 1886–1892. Illustrationes florae insularum maris pacifici. Parts 1–7. G. Masson, Paris, 458 pp. (Facsimile ed., 1977, J. Cramer, Vaduz).
746. ———. 1886. Illustrationes florae insularum maris pacifici. Part 2. G. Masson, Paris, pp. 33–48.
747. ———. 1887. Note sur deux genres intéressants de la famille des Composeés: *Fitchia* Hook. f. et *Remya* Hillebr. Centenaire Soc. Philomathique, Paris, pp. 229–234.
748. ———. 1887. Illustrationes florae insularum maris pacifici. Part 3. G. Masson, Paris, pp. 49–64.
749. ———. 1888. Illustrationes florae insularum maris pacifici. Part 4. G. Masson, Paris, pp. 65–80.
750. ———. 1890. Remarques sur la flore de la Polynésie et sur ses rapports avec celle des terres voisines. G. Masson, Paris, 52 pp.
751. ———. 1890. Illustrationes florae insularum maris pacifici. Part 6. G. Masson, Paris, pp. 105–216.
752. ———. 1892. Illustrationes florae insularum maris pacifici. Part 7. G. Masson, Paris, pp. 217–458.
753. Drenth, E. 1972. A revision of the family Taccaceae. Blumea 20: 367–406.
754. Druce, G. C. 1917. Nomenclatorial notes: chiefly African and Australian. Bot. Exch. Club Brit. Isles Rep. 4: 601–653.
755. Drude, O. 1889. Palmae. Nat. Pflanzenfam. II. 3: 1–93.
756. ———. 1890. Handbuch der pflanzengeographie. J. Engelhorn, Stuttgart, 582 pp.
757. Dubard, M. 1912. Les Sapotacées du groupe des Syderoxylinées. Ann. Inst. Bot.-Géol. Colon. Marseille II, 10: 1–90.
758. Dunal, F. 1852. Solanaceae. Prodr. 13(1): 1–690.
759. Edmondson, C. H. 1941. Viability of coconut seeds after floating in sea. Occas. Pap. Bernice P. Bishop Mus. 16(12): 293–304.
760. Eggler, W. A. 1971. Quantitative studies of vegetation on sixteen young lava flows on the island of Hawaii. Trop. Ecol. 12: 66–100.

761. Egler, F. E. 1937. A new species of Hawaiian *Portulaca*. Occas. Pap. Bernice P. Bishop Mus. 13(15): 167–170.
762. ———. 1938. Reduction of *Portulaca caumii* F. Brown to *P. villosa* Chamisso. Repert. Spec. Nov. Regni Veg. 44: 264–265.
763. ———. 1939. Vegetation zones of Oahu, Hawaii. Empire Forest. J. 18: 44–57.
764. ———. 1939. *Santalum ellipticum*, a restatement of Gaudichaud's species. Occas. Pap. Bernice P. Bishop Mus. 14(21): 349–357.
765. ———. 1940. A key to the common Leguminosae of the Hawaiian Islands based upon characters of fruit and leaf. New York State Coll. Forest., Syracuse, N.Y., 42 pp.
766. ———. 1942. Indigene versus alien in the development of arid Hawaiian vegetation. Ecology 23: 14–23.
767. ———. 1944. Hawaiian vegetation zones [Review of "Vegetation zones of Hawaii" by J. C. Ripperton]. Geogr. Rev. 34: 333–335.
768. ———. 1947. Arid southeast Oahu vegetation, Hawaii. Ecol. Monogr. 17: 383–435.
769. Egli, R. E. 1971. Effects of fire on the coastal grasslands of southeastern Oahu, Hawaii. Master's thesis, Univ. Hawaii, Honolulu, 41 pp.
770. Ehrendorfer, F. 1979. Reproductive biology in island plants: *in* Bramwell, D. (ed.), Plants and islands. Academic Press, New York, pp. 293–306.
771. Elliott, M. E. 1981. Wetlands and wetland vegetation of the Hawaiian Islands. Master's thesis, Univ. Hawaii, Honolulu, 228 pp.
772. Ellshoff, Z. E. 1986. Symposium on control of introduced plants in native ecosystems of Hawai'i: summary of presentations. Newslett. Hawaiian Bot. Soc. 25: 79–88.
773. Elschner, C. 1915. The Leeward Islands of the Hawaiian Group. Contributions to the knowledge of the islands of Oceania. Reprinted from Sunday Advertiser, Honolulu, 68 pp.
774. Ely, C. A., and R. B. Clapp. 1973. The natural history of Laysan Island, Northwestern Hawaiian Islands. Atoll Res. Bull. 171: 1–361.
775. Emory, K. P. 1924. The island of Lanai. A survey of native culture. Bernice P. Bishop Mus. Bull. 12: 1–129.
776. ———. 1928. Archaeology of Nihoa and Necker islands. Bernice P. Bishop Mus. Bull. 53: 1–124.
777. ———. 1943. South sea lore. Special Publ. Bernice P. Bishop Mus. 36: 1–75. (Rev. ed., 1944, 79 pp.).
778. Endlicher, S. L. 1833. Atakta botanika. Nova genera et species plantarum descripta et iconibus illustrata. F. Beck, Vienna, 26 pp.
779. ———. 1836. Bemerkungen über die flora der Südseeinseln. Ann. Wiener Mus. Naturgesch. 1: 129–190.
780. ———. 1836–1841. Genera plantarum secundum ordines naturales disposita. Parts 1–18. Fr. Beck, Vienna.
781. Engard, C. J. 1944. Organogenesis in *Rubus*. Univ. Hawaii Res. Publ. 21: 1–234.

782. ———. 1945. Habit of growth of *Rubus rosaefolius* Smith in Hawaii. Amer. J. Bot. 32: 536–538.
783. Engelmann, C. 1859. Systematic arrangement of the species of the genus *Cuscuta*, with critical remarks on old species and descriptions of new ones. Trans. Acad. Sci. St. Louis 1: 453–532.
784. Engler, A. 1882. Burseraceae et Anacardiaceae. Monogr. phan. 4: 1–500, 536–540.
785. ———. 1888. Urticaceae. Nat. Pflanzenfam. III. 1: 98–118.
786. ———. 1890. Beiträge zur Kenntnis der Sapotaceae. Bot. Jahrb. Syst. 12: 496–525.
787. ———. 1896. Rutaceae. Nat. Pflanzenfam. III. 4: 95–201.
788. ———. 1897. Loranthaceae. Nat. Pflanzenfam. Nachtr. III. 1: 124–140.
789. ———. 1900–1953. Das Pflanzenreich. Regni vegetabilis conspectus. Nos. 1–107. Wilhelm Engelmann, Leipzig.
790. ———. 1908. Myrsinaceae. Nat. Pflanzenfam. Nachtr. IV. 1: 269–278.
791. ———. 1931. Rutaceae. Nat. Pflanzenfam. ed. 2, 19a: 187–359.
792. ——— and K. A. E. Prantl. 1887–1915. Die natürlichen Pflanzenfamilien nebst ihren Gattungen und wichtigeren Arten insbesondere den Nutzpflanzen, bearbeitet unter Mitwirkung zahlreicher hervorragender Fachgelehrten. Wilhelm Engelmann, Leipzig.
793. ——— and ———. 1924–1959. Die natürlichen Pflanzenfamilien nebst ihren Gattungen und wichtigsten Arten, insbesondere den Nutzpflanzen. Unter Mitwirkung zahlreicher hervorragender Fachgelehrten begründet von A. Engler et K. Prantl. 2nd ed. Vols. 1–26. Wilhelm Engelmann, Leipzig.
794. Epling, C. 1941. The distribution of American Labiatae. Proc. 6th Pacific Sci. Congr., California 4: 571–575.
795. ———. 1948. A synopsis of the tribe *Lepechinieae* (Labiatae). Brittonia 6: 352–364.
796. Erdtman, G. 1952. Pollen morphology and plant taxonomy. Chronica Botanica Co., Waltham, Mass., 539 pp.
797. Eriksen, F. I., and A. S. Whitney. 1982. Growth and N fixation of some tropical forage legumes as influenced by solar radiation regimes. Agron. J. 74: 703–709.
798. Eshleman, A. M. 1973. A preliminary survey of lead and mercury in the Hawaiian environment. Ph.D. dissertation, Univ. Hawaii, Honolulu, 154 pp.
799. Evenson, W. E. 1978. A mathematical model of 'ohi'a dieback as a natural phenomenon. Proc. 2nd Conf. Nat. Sci., Hawaii Volcanoes Natl. Park, pp. 105–113.
800. ———. 1982. Climate analysis in 'ōhi'a dieback area on the island of Hawaii (Abstr.). Proc. 4th Conf. Nat. Sci., Hawaii Volcanoes Natl. Park, p. 61.
801. ———. 1983 [1984]. Climate analysis in 'ōhi'a dieback area on the island of Hawai'i. Pacific Sci. 37: 375–384.

802. Eyde, R. H., and C. C. Tseng. 1969. Flower of *Tetraplasandra gymnocarpa*: hypogyny with epigynous ancestry. Science 166: 506–508.
803. ―――― and ――――. 1971. What is the primitive floral structure of Araliaceae? J. Arnold Arbor. 52: 205–239.
804. Fagerlind, F. 1940. Zytologie und gametophytenbildung in der gattung *Wikstroemia*. Hereditas 24: 23–50.
805. ――――. 1949. Some reflections on the history of the climate and vegetation of the Hawaiian Islands. Svensk Bot. Tidskr. 43: 73–81.
806. ――――. 1959. Development and structure of the flower and gametophytes in the genus *Exocarpos*. Svensk Bot. Tidskr. 53: 257–282.
807. Fagerlund, G. O. 1944. *Hau kuahiwi*, sole survivor. Unique tree is one of the many wonders of Hawaii National Park. Paradise Pacific 56(1): 21, 32.
808. ――――. 1947. The exotic plants of Hawaii National Park. Hawaii Natl. Park Nat. Hist. Bull. 10: 1–62.
809. ―――― and A. L. Mitchell. 1944. A checklist of the plants, Hawaii National Park, Kilauea–Mauna Loa section, with a discussion of the vegetation. Hawaii Natl. Park Nat. Hist. Bull. 9: 1–76.
810. Farquhar, J. K. M. L. 1900. Gardens, fields, and wilds of the Hawaiian Islands. Trans. Mass. Hort. Soc. 1900: 51–57.
811. Fay, J. J. 1979. Endangered and threatened wildlife and plants; determination that *Kokia cookei* is an endangered species. Fed. Reg. 44: 62470–62471.
812. ――――. 1979. Hawaii: extinction unmerciful. Garden 2(4): 22–27.
813. ――――. 1980. Endangered and threatened wildlife and plants; proposed endangered status for the 'Ewa plains *'akoko* (*Euphorbia skottsbergii* var. *kalaeloana*). Fed. Reg. 45: 58166–58168.
814. ――――. 1982. Endangered and threatened wildlife and plants; determination that *Euphorbia skottsbergii* var. *kalaeloana* ('Ewa Plains *'akoko*) is an endangered species. Fed. Reg. 47: 36846–36849.
815. ――――. 1985. Endangered and threatened wildlife and plants; proposed rule to determine endangered status for *Scaevola coriacea* (dwarf *naupaka*). Fed. Reg. 50: 28878–28881.
816. ――――. 1985. Endangered and threatened wildlife and plants; proposed rule to determine endangered status for *Mezoneuron kavaiense* (*uhiuhi*). Fed. Reg. 50: 31632–31635.
817. Fedde, F. 1911. Vermischte neue Diagnosen. Repert. Spec. Nov. Regni Veg. 9: 571–576.
818. Fenzl, E. 1833. *Schiedea ligustrina*: in Endlicher, S. L., Atakta botanika. Nova genera et species plantarum descripta et iconibus illustrata. F. Beck, Vienna, pp. 13–16.
819. Fernald, M. L., and A. E. Brackett. 1929. The representatives of *Eleocharis palustris* in North America. Rhodora 31: 57–77. [Rep. in Contr. Gray Herb. 83 without change in pagination.]
820. Fisher, H. I. 1949. Populations of birds on Midway and the man-made factors affecting them. Pacific Sci. 3: 103–110.

821. ——— and P. H. Baldwin. 1946. War and the birds of Midway Atoll. Condor 48: 3–15.
822. Fisher, J. B. 1980. The vegetative and reproductive structure of papaya (*Carica papaya*). Lyonia 1: 191–208.
823. Fisher, W. K. 1906. Birds of Laysan and the Leeward Islands, Hawaiian group. Bull. U.S. Fish Commiss. 23: 769–807.
824. Fleming, D. 1935. Wildflowers grow in profusion on Maui. Friend 105: 468–469.
825. Fletcher, W. B. 1968. New hope for the sandalwood. Natl. Parks Mag. 42(255): 16–18.
826. Floyd, M. E. 1977. Autecological study and phenolic analysis of *Pelea anisata* Mann (*mokihana*). Master's thesis, Univ. Hawaii, Honolulu, 67 pp.
827. ———. 1979 [1980]. Phenolic chemotaxonomy of the genus *Pelea* A. Gray (Rutaceae). Pacific Sci. 33: 153–160.
828. ———. 1979 [1980]. Some aspects of reproduction of *mokihana* (*Pelea anisata* Mann) in the forests of Kōkeʻe. Pacific Sci. 33: 161–164.
829. Flynn, T. W. 1984. One of Hawaii's native orchids. Bull. Pacific Trop. Bot. Gard. 14: 8.
830. Folkers, K., and F. Koniuszy. 1939. *Erythrina* alkaloids. III. Isolation and characterization of a new alkaloid, erythramine. J. Amer. Chem. Soc. 61: 1232–1235.
831. ——— and ———. 1939. *Erythrina* alkaloids. VI. Studies on the constitution of erythramine. J. Amer. Chem. Soc. 61: 3053–3057.
832. ——— and K. Unna. 1939. *Erythrina* alkaloids. V. Comparative curare-like potencies of species of the genus *Erythrina*. J. Amer. Pharm. Assoc. 28: 1019–1028.
833. Foote, D. E., E. L. Hill, S. Nakamura, and F. Stephens. 1972. Soil survey of the islands of Kauai, Oahu, Maui, Molokai, and Lanai, State of Hawaii. U.S.D.A., Soil Conservation Serv., 232 pp.
834. Forbes, C. N. 1909. Some new Hawaiian plants. Occas. Pap. Bernice P. Bishop Mus. 4(3): 213–223.
835. ———. 1910. New Hawaiian plants—II. Occas. Pap. Bernice P. Bishop Mus. 4(4): 296–297.
836. ———. 1911. Notes on the naturalized flora of the Hawaiian Islands. Occas. Pap. Bernice P. Bishop Mus. 4(5): 323–334.
837. ———. 1912. New Hawaiian plants—III. Occas. Pap. Bernice P. Bishop Mus. 5(1): 1–13.
838. ———. 1912. Preliminary observations concerning the plant invasion on some of the lava flows of Mauna Loa, Hawaii. Occas. Pap. Bernice P. Bishop Mus. 5(1): 15–23.
839. ———. 1913. Notes on the flora of Kahoolawe and Molokini. Occas. Pap. Bernice P. Bishop Mus. 5(3): 85–97.
840. ———. 1913. An enumeration of Niihau plants. Occas. Pap. Bernice P. Bishop Mus. 5(3): 99–111.
841. ———. 1914. Plant invasion on lava. Mid-Pacific Mag. 7: 360–365.

842. ———. 1914. New Hawaiian plants—IV. Occas. Pap. Bernice P. Bishop Mus. 6(1): 39.
843. ———. 1916. New Hawaiian plants—V. Occas. Pap. Bernice P. Bishop Mus. 6(3): 173–191.
844. ———. 1917. Botany: *in* Director's report for 1916. Occas. Pap. Bernice P. Bishop Mus. 6(4): 201–204.
845. ———. 1917. New Hawaiian plants—VI. Occas. Pap. Bernice P. Bishop Mus. 6(4): [243–246]. [Incorrect pagination of "237–240" on pages.]
846. ———. 1918. The genus *Lagenophora* in the Hawaiian Islands, with descriptions of new species. Occas. Pap. Bernice P. Bishop Mus. 6(5): 301–309.
847. ———. 1920. Botany: *in* Director's report for 1918. Occas. Pap. Bernice P. Bishop Mus. 7(1): 7–8.
848. ———. 1920. New Hawaiian plants—VII. Occas. Pap. Bernice P. Bishop Mus. 7(3): 33–39.
849. ———. 1920. Botany: *in* Director's report for 1919. Occas. Pap. Bernice P. Bishop Mus. 7(8): 175–176.
850. ———. 1921. Salient features of Hawaiian botany. Proc. 1st Pan-Pacific Sci. Conf., Honolulu, pp. 125–130.
851. ——— and G. C. Munro. 1920. A new *Cyanea* from Lanai, Hawaii. Occas. Pap. Bernice P. Bishop Mus. 7(4): 43–44.
852. Forbes, D. 1918. Report of committee on forestry. Hawaiian Pl. Rec. 18: 202–205.
853. Forehand, S. 1970. The phytosociology of an alpine tussock grassland on East Maui, Hawaii. Master's thesis, Calif. State Coll. Los Angeles, 93 pp.
854. Forman, L. L. 1962. The Menispermaceae of Malaysia: IV. *Cocculus* A. P. de Candolle. Kew Bull. 15: 479–487.
855. Fosberg, F. R. 1934. A key to the families of Monocotyledons in the Hawaiian Islands. Occas. Pap. Univ. Hawaii 18: 1–8.
856. ———. 1935. Revision of the genus *Gouldia* (Rubiaceae). Master's thesis, Univ. Hawaii, Honolulu, 166 pp.
857. ———. 1936. Plant collecting on Lanai, 1935. Mid-Pacific Mag. 49: 119–123.
858. ———. 1936. Miscellaneous Hawaiian plant notes—I. Occas. Pap. Bernice P. Bishop Mus. 12(15): 1–11.
859. ———. 1936. The Hawaiian geraniums. Occas. Pap. Bernice P. Bishop Mus. 12(16): 1–19.
860. ———. 1936. A study of the Hawaiian genus, *Gouldia* (Abstr.). Proc. Hawaiian Acad. Sci. 11: 20.
861. ———. 1937. The genus *Gouldia* (Rubiaceae). Bernice P. Bishop Mus. Bull. 147: 1–82.
862. ———. 1937. Immigrant plants in the Hawaiian Islands. I. Occas. Pap. Univ. Hawaii 32: 1–11.
863. ———. 1938. An aggressive *Lantana* mutation (Abstr.). Proc. Hawaiian Acad. Sci. 12: 18.

864. ———. 1939. Notes on Polynesian grasses. Occas. Pap. Bernice P. Bishop Mus. 15(3): 37–48.
865. ———. 1939. Taxonomy of the Hawaiian genus *Broussaisia* (Saxifragaceae). Occas. Pap. Bernice P. Bishop Mus. 15(4): 49–60.
866. ———. 1939. *Diospyros ferrea* (Ebenaceae) in Hawaii. Occas. Pap. Bernice P. Bishop Mus. 15(10): 119–131.
867. ———. 1941. Varieties of the strawberry guava. Proc. Biol. Soc. Wash. 54: 179–180.
868. ———. 1942. Miscellaneous notes on Hawaiian plants—2. Occas. Pap. Bernice P. Bishop Mus. 16(15): 337–347.
869. ———. 1943. The polynesian species of *Hedyotis* (Rubiaceae). Bernice P. Bishop Mus. Bull. 174: 1–102.
870. ———. 1943. Notes on plants of the Pacific islands—III. Bull. Torrey Bot. Club 70: 386–397.
871. ———. 1948. Immigrant plants in the Hawaiian Islands. II. Occas. Pap. Univ. Hawaii 46: 1–17.
872. ———. 1948. Derivation of the flora of the Hawaiian Islands: *in* Zimmerman, E. C., Insects of Hawaii. Vol. 1. Univ. Press Hawaii, Honolulu, pp. 107–119.
873. ———. 1949. The problem of rare and vanishing plant species. Proc. Pap. Int. Techn. Conf. Protect. Nature, pp. 502–504.
874. ———. 1950. The American element in the Hawaiian flora (Abstr.). Proc. 7th Int. Bot. Cong., pp. 866–867.
875. ———. 1951. The American element in the Hawaiian flora. Pacific Sci. 5: 204–206.
876. ———. 1955. Pacific forms of *Lepturus* R. Br. (Gramineae). Occas. Pap. Bernice P. Bishop Mus. 21(14): 285–294.
877. ———. 1956. Studies in Pacific Rubiaceae: I–IV. Brittonia 8: 165–178.
878. ———. 1956. The protection of nature in the islands of the Pacific. Proc. 8th Congr. Int. Bot. 1954, pp. 104–116.
879. ———. 1957. Tropical Pacific grasslands and savannas. Proc. 9th Pacific Sci. Congr., Thailand 4: 118–123.
880. ———. 1958. Vegetation of the islands of Oceania: *in* Study of tropical vegetation. Proc. Kandy Symp., UNESCO, pp. 54–60.
881. ———. 1959. Upper limits of vegetation on Mauna Loa, Hawaii. Ecology 40: 144–146.
882. ———. 1960. Grazing animals and the vegetation of oceanic islands: *in* Symposium on the impact of man on humid tropics vegetation. UNESCO, Goroka, New Guinea, pp. 168–169.
883. ———. 1961. *Scaevola sericea* Vahl versus *S. taccada* (Gaertn.) Roxb. Taxon 10: 225–226.
884. ———. 1961. Guide to excursion III, Tenth Pacific Science Congress. Tenth Pacific Sci. Congr., Honolulu, 207 pp.
885. ———. 1962. Miscellaneous notes on Hawaiian plants—3. Occas. Pap. Bernice P. Bishop Mus. 23(2): 29–44.
886. ———. 1962. The Indo-Pacific strand *Scaevola* again. Taxon 11: 181.

887. ———. 1963. Plant dispersal in the Pacific: *in* Gressitt, J. L. (ed.), Pacific basin biogeography. Bishop Mus. Press, Honolulu, pp. 273–281.
888. ———. 1964. Studies in Pacific Rubiaceae: V. Brittonia 16: 255–271.
889. ———. 1966. Critical notes on Pacific island plants. 1. Micronesica 2: 143–152.
890. ———. 1966. Miscellaneous notes on Hawaiian plants—4. Occas. Pap. Bernice P. Bishop Mus. 23(8): 129–138.
891. ———. 1966. Observations on vegetation patterns and dynamics on Hawaiian and Galapageian volcanoes (Abstr.): *in* Biotic communities of the volcanic areas of the Pacific. Proc. 11th Pacific Sci. Congr., Tokyo 5: 3.
892. ———. 1966. Vascular plants: *in* Doty, M. S., and D. Mueller-Dombois, Atlas for bioecology studies in Hawaii Volcanoes National Park. Univ. Hawaii, Hawaii Bot. Sci. Pap. 2: 153–238.
893. ———. 1967. Observations on vegetation patterns and dynamics on Hawaiian and Galapageian volcanoes. Micronesica 3: 129–134.
894. ———. 1968. Studies in Pacific Rubiaceae: VI–VIII. Brittonia 20: 287–294.
895. ———. 1968. Critical notes on Pacific island plants. 2. Micronesica 4: 255–259.
896. ———. 1968. Systematic notes on Micronesian plants. 3. Phytologia 15: 496–502.
897. ———. 1969. Miscellaneous notes on Hawaiian plants—5. Occas. Pap. Bernice P. Bishop Mus. 24(2): 9–24.
898. ———. 1969. Paleobotany of the oceanic Pacific islands: *in* Leopold, E. B., Miocene pollen and spore flora of Eniwetok, Marshall Islands. U.S. Geol. Surv. Profess. Pap. 260-II: 1140.
899. ———. 1971. Endangered island plants. Bull. Pacific Trop. Bot. Gard. 1(3): 1–7.
900. ———. 1971. Island faunas and floras: *in* McGraw Hill encyclopedia of science and technology. McGraw Hill Book Co., New York, pp. 309–311.
901. ———. 1972. Field guide to excursion III, Tenth Pacific Science Congress, revised edition. Dept. Bot., Univ. Hawaii, Honolulu, 249 pp.
902. ———. 1973. Vascular plants—widespread island species: *in* Costin, A. B., and R. H. Groves, Nature conservation in the Pacific. Australian Natl. Univ. Press, Canberra, pp. 167–169.
903. ———. 1973. Past, present and future conservation problems of oceanic islands: *in* Costin, A. B., and R. H. Groves, Nature conservation in the Pacific. Australian Natl. Univ. Press, Canberra, pp. 209–215.
904. ———. 1973. The name of the octopus tree. Baileya 19: 45–46.
905. ———. 1975. Revised check-list of vascular plants of Hawaii Volcanoes National Park. Coop. Natl. Park Resources Stud. Unit, Hawaii, Techn. Rep. 5: 1–19.
906. ———. 1975. The deflowering of Hawaii. Natl. Parks Conservation Mag. 49(40): 4–10.

907. ———. 1976. Coral island vegetation: *in* Jones, O. A., and R. Endean (eds.), Biology and geology of coral reefs. Vol. III: Biology 2. Academic Press, New York, pp. 255–277.
908. ———. 1976. *Ipomoea indica* taxonomy: a tangle of morning glories. Bot. Notiser 129: 35–38.
909. ———. 1976. *Bobea elatior* again. Taxon 25: 188.
910. ———. 1978. Studies in the genus *Boerhavia* L. (Nyctaginaceae), 1–5. Smithsonian Contr. Bot. 39: 1–20.
911. ———. 1983. The human factor in the biogeography of oceanic islands. Compt. Rend. Sommaire Séances Soc. Biogéogr. 59: 147–190.
912. ———. 1983. Reviews & notices. Island ecosystems: biological organization in selected Hawaiian communities. Environmental Conservation 10: 82–83.
913. ———. 1983. A possible new pathogen affecting *Metrosideros* in Hawaii. Newslett. Hawaiian Bot. Soc. 22: 13–16.
914. ———. 1984. Phytogeographic comparison of Polynesia and Micronesia: *in* Radovsky, F. J., P. H. Raven, and S. H. Sohmer (eds.), Biogeography of the tropical Pacific. Bishop Mus. Special Publ. 72: 33–44.
915. ———. 1985. Range extension of *Diaporthiopsis metrosideri* to Bird Park, Hawaii Volcanoes National Park. Newslett. Hawaiian Bot. Soc. 24: 18.
916. ———, P. Boiteau, and M.-H. Sachet. 1977. Nomenclature of the Ochrosiinae (Apocynaceae): 2. Synonymy of *Ochrosia* Juss. and *Neisosperma* Raf. Adansonia 17: 23–33.
917. ——— and D. R. Herbst. 1975. Rare and endangered species of Hawaiian vascular plants. Allertonia 1: 1–72.
918. ——— and E. Y. Hosaka. 1938. An open bog on Oahu. Occas. Pap. Bernice P. Bishop Mus. 14(1): 1–6.
919. ——— and C. H. Lamoureux. 1966. Vegetational responses to volcanic activity in the Chain-of-Craters area: *in* Doty, M. S., and D. Mueller-Dombois, Atlas for bioecology studies in Hawaii Volcanoes National Park. Univ. Hawaii, Hawaii Bot. Sci. Pap. 2: 315–329.
920. ——— and M.-H. Sachet. 1956. The Indo-Pacific strand *Scaevola*. Taxon 5: 7–10.
921. ——— and ———. 1972. *Thespesia populnea* (L.) Solander ex Correa and *Thespesia populneoides* (Roxburgh) Kosteletsky (Malvaceae). Smithsonian Contr. Bot. 7: 1–13.
922. ——— and ———. 1974. Plants of southeastern Polynesia. 3. Micronesica 10: 251–256.
923. ——— and ———. 1975. Polynesian plant studies 1–5. Smithsonian Contr. Bot. 21: 1–25.
924. ——— and ———. 1980. Systematic studies of Micronesian plants. Smithsonian Contr. Bot. 45: 1–40.
925. ——— and ———. 1981. Polynesian plant studies 6–18. Smithsonian Contr. Bot. 47: 1–37.

926. Francey, P. 1935–1936. Monographie du genre *Cestrum* L. Candollea 6: 46–398 (1935); 7: 1–132 (1936).
927. Frear, M. D. 1929. Our familiar island trees. The Gorham Press, Boston, 161 pp.
928. ———. 1938. Flowers of Hawaii. Dodd, Mead & Co., New York, 42 unnum. pp.
929. ———. 1944. Presentation: *in* Inn, H., Tropical blooms. Fong Inn's Ltd., Honolulu, 6 unnum. pp.
930. Freeman, O. W. 1929. What forests mean to Hawaii. Amer. Forests 35: 80–82, 108.
931. ——— (ed.). 1951. Geography of the Pacific. J. Wiley & Sons, New York, 573 pp.
932. Frick, D. 1855. Notes upon Hawaiian indigo. Trans. Roy. Hawaiian Agric. Soc. 1: 79–81.
933. Friend, D. J. 1980 [1981]. Effect of different photon flux densities (PAR) on seedling growth and morphology of *Metrosideros collina* (Forst.) Gray. Pacific Sci. 34: 93–100.
934. Fritsch, K. 1894. Gesneriaceae. Nat. Pflanzenfam. IV. 3b: 133–185.
935. Frodin, D. G. 1975. Studies in *Schefflera* (Araliaceae): the *Cephaloschefflera* complex. J. Arnold Arbor. 56: 427–448.
936. Frost, M. D. 1979. Savannas on the island of Hawaii. Calif. Geogr. 19: 29–47.
937. Fryxell, J. E. 1983. A revision of *Abutilon* sect. *Oligocarpae* (Malvaceae), including a new species from Mexico. Madroño 30: 84–92.
938. Fryxell, P. A. 1972. A reconsideration of the correct name for the Hawaiian *Gossypium*. Sida 5: 1–2.
939. ———. 1976. A nomenclator of *Gossypium*. The botanical names of cotton. U.S.D.A. Techn. Bull. 1491: 1–114.
940. ———. 1978. Neotropical segregates from *Sida* L. (Malvaceae). Brittonia 30: 447–462.
941. ———. 1978. The correct name for the Hawaiian *Gossypium*: *G. tomentosum*. Taxon 27: 131–132.
942. ———. 1979. The natural history of the cotton tribe (Malvaceae, tribe *Gossypieae*). Texas A & M Univ. Press, College Station, 245 pp.
943. Fujii, D. M. 1976. The Nuuanu eucalyptus planting: growth, survival, stand development after 64 years. U.S.D.A. Forest Serv. Res. Note PSW-318: 1–5.
944. Fujioka, F. M. 1976. Fine fuel moisture measured and estimated in dead *Andropogon virginicus* in Hawaii. U.S.D.A. Forest Serv. Res. Note PSW-317: 1–4.
945. Funk, E. 1978. Hawaiian fiber plants. Newslett. Hawaiian Bot. Soc. 17: 27–35.
946. ———. 1979. Anatomical variation of fibers in five genera of Hawaiian Urticaceae and its significance to ethnobotany. Master's thesis, Univ. Hawaii, Honolulu, 102 pp.

947. ———. 1982. The aboriginal use and domestication of *Touchardia latifolia* Gaud. (Urticaceae) in Hawaii. Archaeol. Oceania 17: 16–19.
948. ———. 1985. A new flowering plant record. Newslett. Hawaiian Bot. Soc. 24: 67.
949. ———. 1985. A survey of Queen's Beach, O'ahu, Hawai'i. Newslett. Hawaiian Bot. Soc. 24: 68–78.
950. Gagné, B. H. 1982. Silversword alliance in the bogs of East Maui: a continuing report. Proc. 4th Conf. Nat. Sci., Hawaii Volcanoes Natl. Park, p. 62.
951. Gagné, W. C. 1974. [Comments on 26 October 1973 Draft Environmental Statement on the Natural Resources Management Plan, Hawaii Volcanoes National Park to G. Bryan Harry, Superintendent, Hawaii Volcanoes National Park]. Elepaio 35: 126–127.
952. ———. 1974. Notes on the present status of the native Hawaiian flora. Newslett. Hawaiian Bot. Soc. 13: 19–20.
953. ———. 1975. Hawaii's tragic dismemberment. Defenders 50: 461–469.
954. ———. 1976. Canopy-associated arthropods in *Acacia koa* and *Metrosideros* tree communities along the Mauna Loa transect. U.S. IBP Island Ecosystems IRP Techn. Rep. 77: 1–32.
955. ———. 1979. Canopy-associated arthropods in *Acacia koa* and *Metrosideros* tree communities along an altitudinal transect on Hawaii Island. Pacific Insects 21: 56–82.
956. ———. 1981. Hawai'i Audubon Society stops bulldozing in Kaua'i forest. 'Elepaio 41: 92–93.
957. ———. 1981. HAS opposes conversion of Honomalino forest to macadamia nut orchard. 'Elepaio 41: 95.
958. ———. 1982. Insular evolution and speciation of the genus *Nesiomiris* in Hawaii (Heteroptera: Miridae). Entomologia Generalis 8: 87–88.
959. ——— and S. Conant. 1983. Nihoa: biological gem of the Northwest Hawaiian Islands. Ka 'Elele 10(7): 3–5.
960. Galtsoff, P. S. 1933. Pearl and Hermes Reef, Hawaii, hydrographical and biological observations. Bernice P. Bishop Mus. Bull. 107: 1–49.
961. Games, D. E., A. H. Jackson, N. A. Khan, and D. S. Millington. 1974. Alkaloids of some African, Asian, Polynesian and Australian species of *Erythrina*. Lloydia 37: 581–588.
962. Ganders, F. R., and K. M. Nagata. 1983. New taxa and new combinations in Hawaiian *Bidens* (Asteraceae). Lyonia 2: 1–16.
963. ——— and ———. 1983. Relationships and floral biology of *Bidens cosmoides* (Asteraceae). Lyonia 2: 23–31.
964. ——— and ———. 1984. The role of hybridization in the evolution of *Bidens* on the Hawaiian Islands: *in* Grant, W. F. (ed.), Biosystematics. Academic Press Canada, Ontario, pp. 179–194.
965. Gardner, D. E. 1978. *Koa* rust, caused by *Uromyces koae*, in Hawaii Volcanoes National Park. Pl. Dis. Reporter 62: 957–961.
966. ———. 1978. Evaluation of a new technique for herbicidal treatment of

Myrica faya trees. Proc. 2nd Conf. Nat. Sci., Hawaii Volcanoes Natl. Park, pp. 114–119.

967. ———. 1980. An evaluation of herbicidal methods of strawberry guava control in Kīpahulu Valley: *in* Smith, C. W. (ed.), Resources base inventory of Kīpahulu Valley below 2000 feet. Coop. Natl. Park Resources Stud. Unit, Univ. Hawaii, Honolulu, pp. 63–69.

968. ———. 1980. Apparent pathological conditions of *Vaccinium* spp. and *Dodonaea* sp. in Hawaii Volcanoes and Haleakala National Parks. Proc. 3rd Conf. Nat. Sci., Hawaii Volcanoes Natl. Park, pp. 125–128.

969. ———. 1981. Nuclear behavior and clarification of the spore stages of *Uromyces koae*. Canad. J. Bot. 59: 939–946.

970. ———. 1982. *Septoria* leaf spot on *Canavalia kauensis*, a native Hawaiian bean. Pl. Dis. 66: 263–264.

971. ———. 1984. Blackberries. Newslett. Hawaiian Bot. Soc. 23: 20.

972. ———. 1984. Current biocontrol investigations: exploration for control agents in the native habitats of firetree, and evaluation of rust diseases of *Rubus* spp. in the southeastern U.S. Proc. 5th Conf. Nat. Sci., Hawaii Volcanoes Natl. Park, pp. 53–60.

973. ———. 1985. Etiology of red leaf disease of '*ohelo*. Newslett. Hawaiian Bot. Soc. 24: 3–5.

974. ———. 1985. Observations on some unusual flowering characteristics of *Myrica faya*. Newslett. Hawaiian Bot. Soc. 24: 14–17.

975. ———. 1986. *Oxalis* rust in Hawaii. Newslett. Hawaiian Bot. Soc. 25: 31–32.

976. ——— and C. J. Davis. 1982. The prospects for biological control of nonnative plants in Hawaiian national parks. Coop. Natl. Park Resources Stud. Unit, Hawaii, Techn. Rep. 45: 1–55.

977. ——— and C. S. Hodges, Jr. 1983. Leaf rust caused by *Kuehneola uredinis* on native and nonnative *Rubus* species in Hawaii. Pl. Dis. 67: 962–963.

978. ——— and ———. 1986. Hawaiian forest fungi. VII. A new species of *Elsinoe* on native *Vaccinium*. Mycologia 78: 506–508.

979. ——— and V. A. D. Kageler. 1984. The apparent yellows disease of *Dodonaea*: symptomatology and considerations of the etiology. Newslett. Hawaiian Bot. Soc. 23: 7–16.

980. ——— and W. W. McCall. 1982. Response of broomsedge to soil fertility and lime. Proc. 4th Conf. Nat. Sci., Hawaii Volcanoes Natl. Park, pp. 76–79.

981. ———, T. Miller, and E. G. Kuhlman. 1979. *Tuberculina* and the life cycle of *Uromyces koae*. Mycologia 71: 848–852.

982. Gardner, R. C. 1976. The systematics of *Lipochaeta* (Compositae: Heliantheae) of the Hawaiian Islands. Ph.D. dissertation, Ohio State Univ., Columbus, 168 pp.

983. ———. 1976. Evolution and adaptive radiation in *Lipochaeta* (Compositae) of the Hawaiian Islands. Syst. Bot. 1: 383–391.

984. ———. 1977. Chromosome numbers and their systematic implications in *Lipochaeta* (Compositae: *Heliantheae*). Amer. J. Bot. 64: 810–813.
985. ———. 1977. Observations on tetramerous disc florets in the Compositae. Rhodora 79: 139–146.
986. ———. 1979. Revision of *Lipochaeta* (Compositae: *Heliantheae*) of the Hawaiian Islands. Rhodora 81: 291–343.
987. ——— and J. C. La Duke. 1978. Phyletic and cladistic relationships in *Lipochaeta* (Compositae). Syst. Bot. 3: 197–207.
988. Gartley, A. 1913. The breeding of hibiscus. Friend 71: 26–27, 45–46.
989. ———. 1916. Hawaii's flower—hibiscus. Mid-Pacific Mag. 12: 270–275.
990. Gaudichaud-Beaupré, C. 1824. Descriptions de quelques nouveaux genres de plantes recueillies dans le voyage autour du monde, sous les ordres du capitaine Freycinet. Ann. Sci. Nat. (Paris) 3: 507–510.
991. ———. 1826–1830. Voyage autour du monde, entrepris par ordre du roi, . . . exécuté sur les corvettes de S. M. l'*Uranie* et la Physicienne, pendant les années 1817, 1818, 1819 et 1820; publié . . . par M. Louis de Freycinet. Botanique. Pillet-ainé, Paris, 522 pp. ["Îles Sandwich" (Sandwich Islands) text published in Chapter 16, pp. 88–107, 1827.]
992. ———. 1829. Voyage autour du monde, entrepris par ordre du roi, . . . exécuté sur les corvettes de S. M. l'*Uranie* et la *Physicienne*, pendant les années 1817, 1818, 1819 et 1820; publié . . . par M. Louis de Freycinet. Botanique. Part 10. Pillet-ainé, Paris, pp. 401–432.
993. ———. 1829. Voyage autour du monde, entrepris par ordre du roi, . . . exécuté sur les corvettes de S. M. l'*Uranie* et la *Physicienne*, pendant les années 1817, 1818, 1819 et 1820; publié . . . par M. Louis de Freycinet. Botanique. Part 11. Pillet-ainé, Paris, pp. 433–464.
994. ———. 1830. Voyage autour du monde, entrepris par ordre du roi, . . . exécuté sur les corvettes de S. M. l'*Uranie* et la *Physicienne*, pendant les années 1817, 1818, 1819 et 1820; publié . . . par M. Louis de Freycinet. Botanique. Part 12. Pillet-ainé, Paris, pp. 465–522.
995. ———. 1841–1866. Voyage autour du monde exécuté pendant les années 1836 et 1837 sur la corvette la *Bonite*, commandée par M. Vaillant- . . . Botanique. Arthus Bertrand, Paris.
996. ———. 1841–?1852. Voyage autour du monde exécuté pendant les années 1836 et 1837 sur la corvette la *Bonite*, commandée par M. Vaillant . . . Botanique . . . Atlas. Arthus Bertrand, Paris, *pl. 1–150*.
997. ———. 1866. Voyage autour du monde exécuté . . . sur la corvette la *Bonite* . . . Botanique . . . Explication et description des planches de l'atlas par M. Charles d'Alleizette. Arthus Bertrand, Paris, 186 pp.
998. Gauthier, R. 1959. L'anatomie vasculaire et l'interprétation de la fleur pistillée de l'*Hillebrandia sandwicensis* Oliv. Phytomorphology 9: 72–87.
999. Gay, R. A. 1967. The effect of three soil moisture regimes on two Hawaiian dry grasses grown in artificial communities in a greenhouse. Master's thesis, Univ. Hawaii, Honolulu, 62 pp.
1000. ———. 1975. Floristic and environmental variations throughout the

Metrosideros forest belt on Hawaii (Abstr.). Proc. 13th Pacific Sci. Congr., Canada 1: 98–99.
1001. Geesink, R. 1969. An account of the genus *Portulaca* in Indo-Australia and the Pacific (Portulacaceae). Blumea 17: 275–301.
1002. Gehring, P. E. 1967. A study of variation in Hawaiian species of the genus *Pipturus* of Kauai and Oahu. Master's thesis, Univ. Hawaii, Honolulu, 56 pp.
1003. Gerrish, G. 1978. The relationship of native and exotic plant species in two rain forest communities in the Koolau Mountains, Oahu, Hawaii. Master's thesis, Univ. Hawaii, Honolulu, 158 pp.
1004. ———. 1978. Factors controlling the distribution of exotic plants in the Koʻolau Mountains, Oʻahu. Proc. 2nd Conf. Nat. Sci., Hawaii Volcanoes Natl. Park, pp. 120–124.
1005. ———. 1980. Photometric monitoring of foliage loss from a wind storm, island of Hawaiʻi. Proc. 3rd Conf. Nat. Sci., Hawaii Volcanoes Natl. Park, pp. 129–135.
1006. ——— and K. W. Bridges. 1979. Botanical summary of the terrestrial ecosystems of the Hawaiian Islands, American Samoa, and the U.S. Trust Territory of the Pacific Islands: *in* Byrne, J. E. (ed.), Literature review and synthesis of information on Pacific island ecosystems. Dept. Interior, U.S. Fish & Wildlife Serv., Washington, D.C., 2-1-2-32.
1007. ——— and ———. 1984. A thinning and fertilizing experiment in *Metrosideros* dieback stands in Hawaiʻi. Univ. Hawaii, Hawaii Bot. Sci. Pap. 43: 1–107.
1008. ——— and D. Mueller-Dombois. 1980. Behavior of native and non-native plants in two tropical rain forests on Oahu, Hawaiian Islands. Phytocoenologia 8: 237–295.
1009. Gerum, S. 1983. In search of the "wild" bananas. Notes Waimea Arbor. & Bot. Gard. 10(1): 9–11.
1010. Giffard, W. M. 1918. Some observations on Hawaiian forests and forest cover in their relation to water supply. Hawaiian Pl. Rec. 18: 515–538.
1011. Gilbert, B. 1971. Then came man and a mustard seed. Sports Ill. 35(11): 96–109.
1012. Gill, L. T. 1934. Migrations of food plants in the Pacific. J. Pan-Pacific Res. Inst. 9(2): 7–9.
1013. Gillett, G. W. 1964. The Harold L. Lyon Arboretum, University of Hawaii. Pl. Sci. Bull. 10: 3–4.
1014. ———. 1966. Hybridization and its taxonomic implications in the *Scaevola gaudichaudiana* complex of the Hawaiian Islands. Evolution 20: 506–516.
1015. ———. 1967. The genus *Cyrtandra* in Fiji. Contr. U.S. Natl. Herb. 37: 107–159.
1016. ———. 1969. The nomenclatural and taxonomic status of the Hawaiian shrub *Scaevola gaudichaudii* H. & A. Pacific Sci. 23: 125–128.
1017. ———. 1972. The role of hybridization in the evolution of the Hawaiian

flora: *in* Valentine, D. H. (ed.), Taxonomy, phytogeography, and evolution. Academic Press, New York, pp. 205–219.

1018. ———. 1972. Additional experimental crosses in Hawaiian *Bidens* (Asteraceae). Pacific Sci. 26: 415–417.

1019. ———. 1972. Genetic affinities between Hawaiian and Marquesan *Bidens* (Asteraceae). Taxon 21: 479–483.

1020. ———. 1973. Genetic relationships between *Bidens forbesii* and six species in the Hawaiian and the Marquesas Islands. Brittonia 25: 10–14.

1021. ———. 1975. The diversity and history of Polynesian *Bidens*, section *Campylotheca*. Univ. Hawaii Harold L. Lyon Arbor. Lecture 6: 1–32.

1022. ——— and E. K. S. Lim. 1970. An experimental study of the genus *Bidens* (Asteraceae) in the Hawaiian Islands. Univ. Calif. Publ. Bot. 56: 1–63.

1023. Gilmartin, A. J. 1956. Post-fertilization seed and ovary development of *Passiflora edulis* forma *flavicarpa*. Master's thesis, Univ. Hawaii, Honolulu, 32 pp.

1024. ———. 1968. Baker's Law and dioecism in the Hawaiian flora: an apparent contradiction. Pacific Sci. 22: 285–292.

1025. Gingins de la Sarraz, F.-C.-J. 1826. Description de quelques espèces nouvelles de Violacées reçues de Mr. Adelbert de Chamisso examinée en 1825 par Mr. de Gingins. Linnaea 1: 406–413.

1026. Glück, H. 1941. Kritische untersuchungen über das Indischasiatische Pfeilkraut (*Sagittaria sinensis* Sims = *S. sagittifolia* aut.). Bot. Jahrb. Syst. 72: 1–68.

1027. Gon, S. M., III. 1978. Altitudinal effects on the general diversity of endemic insect communities in a leeward Hawaiian forest system, Manuka Forest Reserve, South Kona, Hawai'i. Proc. 2nd Conf. Nat. Sci., Hawaii Volcanoes Natl. Park, pp. 134–149.

1028. Gonzalez, V. 1966. Genetic and agronomic studies on the genus *Leucaena* Benth. Master's thesis, Univ. Hawaii, Honolulu, 81 pp.

1029. ———, J. L. Brewbaker, and D. E. Hamill. 1967. *Leucaena* cytogenetics in relation to the breeding of low mimosine lines. Crop Sci. 7: 140–143.

1030. Good, R. 1953. The geography of the flowering plants. 2nd ed. Longmans, Green & Co., New York, 452 pp.

1031. Goodell, B., and D. Ward. 1979. Are you a carrier? Dept. Parks and Recreation, Honolulu, 14 pp.

1032. Gorman, M., N. Neuss, C. Djerassi, J. P. Kutney, and P. J. Scheuer. 1957. Alkaloid studies—XIX. Alkaloids of some Hawaiian *Rauwolfia* species: the structure of sandwicine and its interconversion with ajmaline and ajmalidine. Tetrahedron 1: 328–337.

1032a. Gould, F. W. 1972. A systematic treatment of *Garnotia* (Gramineae). Kew Bull. 27: 515–562.

1033. Grady, M. 1986. Stand structure of an isolated forest in Lyon Arboretum, 'O'ahu [O'ahu], Hawai'i. Newslett. Hawaiian Bot. Soc. 25: 47–59.

1034. Graham, A., and A. S. Tomb. 1974. Palynology of *Erythrina* (Leguminosae: Papilionoideae): preliminary survey of the subgenera. Lloydia 37: 465–481.

1035. Gray, A. 1849. On some plants of the Order Compositae from the Sandwich Islands. Proc. Amer. Assoc. Advancem. Sci. 2: 397–398.

1036. ———. 1852. [Account of *Argyroxiphium*, a remarkable genus of Compositae, belonging to the mountains of the Sandwich Islands]. Proc. Amer. Acad. Arts 2: 159–160.

1037. ———. 1852. [Characters of two new genera of plants of the order Violaceae, discovered by the naturalists of the United States Exploring Expedition]. Proc. Amer. Acad. Arts 2: 323–325.

1038. ———. 1853. Characters of some new genera of plants, mostly from Polynesia, in the collection of the United States Exploring Expedition, under Captain Wilkes. Proc. Amer. Acad. Arts 3: 48–54.

1039. ———. 1854. United States Exploring Expedition. During the years 1838, 1839, 1840, 1841, 1842. Under the command of Charles Wilkes, U.S.N. vol. XV. Botany. Phanerogamia. Part I. C. Sherman, Philadelphia, 777 pp.

1040. ———. 1854. Characters of new genera of plants, mostly from Polynesia, in the collection of the United States Exploring Expedition, under Captain Wilkes (continued). Proc. Amer. Acad. Arts 3: 127–129.

1041. ———. 1855. Description de cinq nouveaux genres de plantes de la Polynésie recueillies dans le voyage d'exploration du Capitaine Wilkes. Ann. Sci. Nat., Bot. Sér. 4, 4: 176–178.

1042. ———. 1858. Notes upon some Rubiaceae, collected in the United States South-Sea Exploring Expedition under Captain Wilkes, with characters of new species, &c. Proc. Amer. Acad. Arts 4: 33–50.

1043. ———. 1859. Notes upon some Rubiaceae, collected in the South-Sea Exploring Expedition under Captain Wilkes. Proc. Amer. Acad. Arts 4: 306–318.

1044. ———. 1859. Notes upon some Polynesian plants of the order Loganiaceae. Proc. Amer. Acad. Arts 4: 319–324.

1045. ———. 1859. Diagnosis of the species of sandal-wood (*Santalum*) of the Sandwich Islands. Proc. Amer. Acad. Arts 4: 326–327.

1046. ———. 1861. Characters of some Compositae in the collection of the United States South Pacific Exploring Expedition under Captain Wilkes, with observations &c. Proc. Amer. Acad. Arts 5: 114–146.

1047. ———. 1861. Notes on Lobeliaceae, Goodeniaceae, &c. of the collection of the U.S. South Pacific Exploring Expedition. Proc. Amer. Acad. Arts 5: 146–152.

1048. ———. 1862. Characters of new or obscure species of plants of monopetalous orders in the collection of the United States South Pacific Exploring Expedition under Captain Charles Wilkes, U.S.N. with occasional remarks, &c. Proc. Amer. Acad. Arts 5: 321–352.

1049. ———. 1862. Characters of some new or obscure species of plants, of monopetalous orders, in the collection of the United States South

Pacific Exploring Expedition under Captain Charles Wilkes, U.S.N. with various notes and remarks. Proc. Amer. Acad. Arts 6: 37–55.

1050. ———. 1865. New or little-known Polynesian Thymeleae. J. Bot. 3: 302–306.

1051. ———. 1865. Characters of some new plants of California and Nevada, chiefly from the collections of Professor William H. Brewer, botanist of the State Geological Survey of California, and of Dr. Charles L. Anderson, with revisions of certain genera or groups. Proc. Amer. Acad. Arts 6: 519–556.

1052. ———. 1870. Miscellaneous botanical notes and characters. Proc. Amer. Acad. Arts 8: 282–296.

1053. Greenwell, A. B. H. 1947. Taro—with special reference to its culture and uses in Hawaii. Econ. Bot. 1: 276–289.

1054. ———. 1951. Hawaiian violets. Paradise Pacific 63(9): 22–23.

1055. ———. 1957. Flora of Honaunau: *in* Bryan, E. H., Jr., C. K. Wentworth, A. B. H. Greenwell, M. C. Neal, A. Suehiro, and E. A. Kay, The natural and cultural history of Honaunau, Kona, Hawaii. Volume I. The natural history of Honaunau. Bernice P. Bishop Museum, Honolulu, pp. 39–43.

1056. Greenwell, B. H. 1933. Hawaiians made use of many plants for medical purposes. J. Pan-Pacific Res. Inst. 8(1): 14–15.

1057. Gressitt, J. L. 1978. Evolution of the endemic Hawaiian Cerambycid beetles. Pacific Insects 18: 137–167.

1058. ———. 1980. The endemic Hawaiian Cerambycid beetles. Proc. 3rd Conf. Nat. Sci., Hawaii Volcanoes Natl. Park, pp. 139–142.

1059. Grisebach, A. 1853. *Schenkia*, novum genus Gentianearum. Bonplandia 1: 226.

1060. Guillaumin, A. 1928. Les régions floristiques du Pacifique d'après leur endémisme et la répartition de quelques plants phanérogames. Proc. 3rd Pan-Pacific Sci. Congr., Tokyo 1: 920–938.

1061. ———. 1934. Les régions florales du Pacifique: *in* Berland, L., et al., Contribution à l'étude du peuplement zoologique et botanique des îles du Pacifique. Mém. Soc. Biogéogr. 4: 255–270.

1062. ———. 1937. Matériaux pour la flore de la Nouvelle-Calédonie. XLVI.—Révision des Fluviales. Bull. Soc. Bot. France 84: 255–257.

1063. ———. 1946. Le tî. 1.—Note de systématique. J. Soc. Océanistes 2: 191.

1064. Gulick, A. 1932. Biological peculiarities of oceanic islands. Quart. Rev. Biol. 7: 405–427.

1065. Guppy, H. B. 1906. Observations of a naturalist in the Pacific between 1896 and 1899. Macmillan & Co., Ltd., London, 627 pp.

1066. Gupta, S. 1967. Cytotaxonomic and chemotaxonomic studies of Hawaiian *Wikstroemia*. Master's thesis, Univ. Hawaii, Honolulu, 44 pp.

1067. ——— and G. W. Gillett. 1969. Chemotaxonomic studies of Hawaiian *Wikstroemia*. Econ. Bot. 23: 24–31.

1068. ——— and ———. 1969. Observations on Hawaiian species of *Wikstroemia* (Angiospermae: Thymelaeaceae). Pacific Sci. 23: 83–88.

1069. Gustafson, R. J. 1979. Hawaii's unique and vanishing flora. The genesis of an exhibit. Terra 18(2): 3–9.
1070. ———. 1981. Captain Cook's legacy: inadvertent consequences of an early expedition. Terra 20(2): 33–36.
1071. Gutmanis, J. 1976. *Kahuna La'au Lapa'au*. The practice of Hawaiian herbal medicine. Island Heritage Press, Honolulu, 267 pp.
1072. Haas, J. E. 1977. The Pacific species of *Pittosporum* Banks ex Gaertn. (Pittosporaceae). Allertonia 1: 73–167.
1072a. Hackel, E. 1889. Andropogoneae. Monogr. phan. 6: 1–716.
1073. ———. 1897. Gramineae. Nat. Pflanzenfam. Nachtr. II. 2: 39–47.
1074. ———. 1911. Gramineae novae. VIII. Repert. Spec. Nov. Regni Veg. 10: 165–174.
1075. ———. 1912. Gramineae novae. IX. Repert. Spec. Nov. Regni Veg. 11: 18–30.
1076. Hackler, R. E. A. 1986. Foster Botanic Garden. A history of Foster Park and Garden and Lili'uokalani Garden. Friends of Foster Garden, Honolulu, 164 pp.
1077. Hadden, F. C. 1941. Midway Islands. Hawaiian Pl. Rec. 45: 174–221.
1078. Hadley, T. H. 1966. Waimea Canyon and Kokee. A nature guide. Kauai Publ. Co., Lihue, Hawaii, 72 pp.
1079. ——— and M. S. Williams. 1960. Kauai, the garden island of Hawaii. Garden Island Publ. Co., Ltd., Lihue, Hawaii, 106 pp.
1080. Hall, E. O. 1839. Notes of a tour around Oahu. Hawaiian Spectator 2: 94–112.
1081. Hall, W. L. 1904. Hawaiian forests. A description of the island forests based on recent observations. Hawaiian Forester Agric. 1: 14–20.
1082. ———. 1904. The forests of the Hawaiian Islands. Hawaiian Forester Agric. 1: 84–102.
1083. ———. 1904. The forests of the Hawaiian Islands. Planters' Monthly 23: 355–380.
1084. ———. 1904. The forests of the Hawaiian Islands. U.S.D.A. Bur. Forest. Bull. 48: 1–29.
1085. ———. 1916. The forests of the Hawaiian Islands. Mid-Pacific Mag. 12: 456–463.
1086. Hallier, H. 1893. Versuch einer natürlichen Gliederung der Convolvulaceen auf morphologischer und anatomischer Grundlage. Bot. Jahrb. Syst. 16: 453–591.
1087. Hamlin, B. G. 1959. A revision of the genus *Uncinia* (Cyperaceae–Caricoideae) in New Zealand. Dominion Mus. Bull. 19: 1–106.
1088. Handel-Mazzetti, H. 1928. A revision of the Chinese species of *Lysimachia*, with a new system of the whole genus. Notes Roy. Bot. Gard. Edinburgh 77: 51–122.
1089. Handley, L. L. 1982. The role of a tropical grass cover in the fate of applied wastewater nitrogen. Ph.D. dissertation, Univ. Hawaii, Honolulu, 170 pp.
1090. Handy, E. G. 1955. Ka-'u, Hawai'i, in ecological and historical perspec-

tive: *in* Handy, E. S. C., and M. K. Pukui, The Polynesian family system in Ka-'u, Hawai'i. J. Polynes. Soc. 64: 56–101.
1091. Handy, E. S. C. 1940. The Hawaiian planter. Volume I. His plants, methods and areas of cultivation. Bernice P. Bishop Mus. Bull. 161: 1–227.
1092. ———, K. P. Emory, E. H. Bryan, Jr., P. H. Buck, J. H. Wise, and others. 1933. Ancient Hawaiian civilization. A series of lectures delivered at The Kamehameha Schools. The Kamehameha Schools, Honolulu, 323 pp. (Rev. ed., 1965, Charles E. Tuttle Co., Rutland, Vt., 333 pp.).
1093. ——— and E. G. Handy. 1972. Native planters in old Hawaii. Their life, lore, and environment. Bernice P. Bishop Mus. Bull. 233: 1–641.
1094. ———, M. K. Pukui, and K. Livermore. 1934. Outline of Hawaiian physical therapeutics. Bernice P. Bishop Mus. Bull. 126: 1–51.
1095. Harborne, J. B., and P. S. Green. 1980. A chemotaxonomic survey of flavonoids in leaves of the Oleaceae. J. Linn. Soc., Bot. 81: 155–167.
1096. Hargreaves, D., and B. Hargreaves. 1958. Hawaii blossoms. Hargreaves Industrial, Portland, Oreg., 62 pp.
1097. ——— and ———. 1964. Tropical trees of Hawaii. Hargreaves Industrial, Portland, Oreg., 64 pp.
1098. ——— and ———. 1970. Tropical trees of the Pacific. Hargreaves Co., Kailua, Hawaii, 64 pp.
1099. ——— and ———. 1970. Tropical blossoms of the Pacific. Hargreaves Co., Kailua, Hawaii, 64 pp.
1100. Harland, S. C. 1932. The genetics of *Gossypium*. Bibliogr. Genet. 9: 107–182.
1101. Harms, H. 1897. Zur Kenntnis der Gattungen *Aralia* und *Panax*. Bot. Jahrb. Syst. 23: 1–23.
1102. ———. 1898. Araliaceae. Nat. Pflanzenfam. III. 8: 1–62.
1103. ———. 1926. Über eine neue Gattung der Araliaceen auf Papuasien. Notizbl. Bot. Gart. Berlin-Dahlem 9: 478–484.
1104. Harriman, N. A. 1981. [Reports by Neil A. Harriman]: *in* Löve, Á., Chromosome number reports LXX. Taxon 30: 77–78.
1105. Harris, J. A. 1934. The physico-chemical properties of plant saps in relation to phytogeography. Data on native vegetation in its natural environment. Univ. Minnesota Press, Minneapolis, 339 pp.
1106. Hartog, C. den. 1959. A key to the species of *Halophila* (Hydrocharitaceae), with descriptions of the American species. Acta Bot. Neerl. 8: 484–489.
1107. ———. 1970. The sea-grasses of the world. Verh. Kon. Ned. Akad. Wetensch., Afd. Natuurk., Tweede Sect. 59: 1–275.
1108. Hartt, C. E., and W. W. G. Moir. 1964. The Pacific Tropical Botanical Garden. Pl. Sci. Bull. 10: 1–3.
1109. ——— and M. C. Neal. 1940. The plant ecology of Mauna Kea, Hawaii. Ecology 21: 237–266.

1110. Haselwood, E. L., and G. G. Motter (eds.). 1966. Handbook of Hawaiian weeds. Hawaiian Sugar Planters' Assoc., Honolulu, 479 pp.
1111. ——— and ——— (eds.). 1983. Handbook of Hawaiian weeds. 2nd ed. [revised & expanded by R. T. Hirano]. Univ. Hawaii Press, Honolulu, 491 pp.
1112. Hastings, G. T. 1941. Book reviews. Flora of the Hawaiian Islands. Torreya 41: 16–18.
1113. Hatheway, W. H. 1952. Composition of certain native dry forests: Mokuleia, Oahu, T.H. Ecol. Monogr. 22: 153–168.
1114. Hawaii Audubon Society Board of Directors. 1982. Critical habitats at Pohakuloa Training Area. 'Elepaio 42: 100–102.
1115. Hawaii Division of Forestry and Wildlife. n.d. Our official State tree . . . *kukui* tree. Hawaii, Dept. Land Nat. Resources, Honolulu, 1 p.
1116. Hayden, W. J. 1980. Systematic anatomy of Oldfieldioideae (Euphorbiaceae). Ph.D. dissertation, Univ. Maryland, College Park, 320 pp.
1117. ——— and D. S. Brandt. 1984. Wood anatomy and relationships of *Neowawraea* (Euphorbiaceae). Syst. Bot. 9: 458–466.
1118. Heed, W. B. 1968. Ecology of the Hawaiian Drosophilidae: *in* Wheeler, M. R. (ed.), Studies in genetics IV. Research reports. Univ. Texas Publ. 6818: 387–419.
1119. ———. 1971. Host plant specificity and speciation in Hawaiian *Drosophila*. Taxon 20: 115–121.
1120. Heimerl, A. 1913. Die Nyctaginaceen-Gattungen *Calpidia* und *Rockia*. Oesterr. Bot. Z. 63: 279–290.
1121. ———. 1934. Nyctaginaceae. Nat. Pflanzenfam. ed. 2, 16c: 86–134.
1122. ———. 1937. Nyctaginaceae of southeastern Polynesia and other Pacific islands. Occas. Pap. Bernice P. Bishop Mus. 13(4): 27–47.
1123. Heinicke, R. M. 1985. The pharmacologically active ingredient of *noni*. Bull. Pacific Trop. Bot. Gard. 15: 10–14.
1124. Helenurm, K. 1983. Genetic differentiation of Hawaiian *Bidens*. Master's thesis, Univ. British Columbia, Vancouver.
1125. ——— and F. R. Ganders. 1985. Adaptive radiation and genetic differentiation in Hawaiian *Bidens*. Evolution 39: 753–765.
1126. Heller, A. A. 1897. Observations on the ferns and flowering plants of the Hawaiian Islands. Minnesota Bot. Stud. 1: 760–922.
1127. Hemsley, W. B. 1885. Report on present state of knowledge of various insular floras, being an introduction to the first three parts of the botany of the *Challenger* Expedition. Rep. Voy. H.M.S. *Challenger*, Bot. 1(1): 1–75.
1128. ———. 1892. Notes. *Trematocarpus*. Ann. Bot. (London) 6: 154.
1129. ———. 1893. Notes. The genus *Trematocarpus*. Ann. Bot. (London) 7: 289–290.
1130. Hennen, J. F., and C. S. Hodges, Jr. 1981. Hawaiian forest fungi. II. Species of *Puccinia* and *Uredo* on *Euphorbia*. Mycologia 73: 1116–1122.

1131. Henrard, J. T. 1949. Monograph of the genus *Digitaria*. Universitaire Pers Leiden, Leiden, 999 pp.
1132. Henrickson, J. 1971. Vascular flora of the northeast outer slopes of Haleakala Crater, East Maui, Hawaii. Contr. Nature Conservancy 7: 1–14.
1133. Henshaw, H. W. 1918. A mid-Pacific bird reservation. Mid-Pacific Mag. 15: 282–285.
1134. Herat, R. M. 1981. A systematic study of the genus *Cheirodendron* (Araliaceae). Ph.D. dissertation, Univ. Hawaii, Honolulu, 246 pp.
1135. Herat, T. R., and R. M. Herat. 1976. Common cultivated plants of the Hawaiian Islands 1. Euphorbiaceae (spurge or poinsettia family). Bull. Pacific Trop. Bot. Gard. 6: 52–63.
1136. ———— and ————. 1976. Common cultivated plants of the Hawaiian Islands 2. Moraceae (fig family). Bull. Pacific Trop. Bot. Gard. 6: 73–80.
1137. Herbert, D. A. 1984. The growth dynamics of *Halophila hawaiiana*, an endemic Hawaiian seagrass. Master's thesis, Univ. Hawaii, Honolulu, 95 pp.
1137a. ————. 1986. The growth dynamics of *Halophila hawaiiana*. Aquatic Bot. 23: 351–360.
1137b. ————. 1986. Staminate flowers of *Halophila hawaiiana*: description and notes on its flowering ecology. Aquatic Bot. 25: 97–102.
1138. Herbst, D. R. 1971. A new *Euphorbia* (Euphorbiaceae) from Hawaii. Pacific Sci. 25: 489–490.
1139. ————. 1971. The ontogeny of the disjunct foliar veins in *Euphorbia forbesii* Sherff. Ph.D. dissertation, Univ. Hawaii, Honolulu, 162 pp.
1140. ————. 1971. Disjunct foliar veins in Hawaiian euphorbias. Science 171: 1247–1248.
1141. ————. 1972. Ontogeny of foliar venation in *Euphorbia forbesii*. Amer. J. Bot. 59: 843–850.
1142. ————. 1972. Botanical survey of the Waiehu sand dunes. Bull. Pacific Trop. Bot. Gard. 2: 6–7.
1143. ————. 1972. *Ohai*, a rare and endangered Hawaiian plant. Bull. Pacific Trop. Bot. Gard. 2: 58.
1144. ————. 1974. Addenda to the new edition: *in* Rock, J. F., The indigenous trees of the Hawaiian Islands. Charles E. Tuttle Co., Rutland, Vt., pp. 513–542.
1145. ————. 1974. [Review of] Memoir number 1. List and summary of the flowering plants in the Hawaiian Islands. Pacific Trop. Bot. Gard., Lawai, Hawaii, 2 pp.
1146. ————. 1974. Identification of plants in "Ethnobotany of the Hawaiians." Univ. Hawaii Harold L. Lyon Arbor. Lecture 5, Suppl: 1–4.
1147. ————. 1975. An introduction to the Hawaiian flora. Bull. Amer. Assoc. Bot. Gard. Arbor. 9: 49–52.
1148. ————. 1975. Knapp's plant nomenclature updated. Newslett. Hawaiian Bot. Soc. 14: 118–121.

1149. ———. 1975. Field guide to the Awaawapuhi Trail, Kokee, Kauai, Hawaii. Pacific Trop. Bot. Gard. Bot. Guide No. 1: 1–11.

1150. ———. 1976. An introduction to the Hawaiian flora. 'Elepaio 37: 31–32.

1151. ———. 1977. Vegetation: *in* Clapp, R. B., and E. Kridler, The natural history of Necker Island, northwestern Hawaiian Islands. Atoll Res. Bull. 206: 25–31.

1152. ———. 1977. Vegetation: *in* Clapp, R. B., E. Kridler, and R. R. Fleet, The natural history of Nihoa Island, Northwestern Hawaiian Islands. Atoll Res. Bull. 207: 26–38.

1153. ———. 1977. Endangered Hawaiian plants. Newslett. Hawaiian Bot. Soc. 16: 22–29.

1154. ———. 1977. Vanishing plants. Water Spectrum 9(4): 20–26.

1155. ———. 1980. Miscellaneous notes on the Hawaiian flora I. Phytologia 45: 67–81.

1156. ———. 1983. Endangered and threatened wildlife and plants; proposed endangered status and critical habitat for *Gouania hillebrandii*. Fed. Reg. 48: 40407–40411.

1157. ———. 1983. Endangered and threatened wildlife and plants; proposed endangered status and critical habitat for *Kokia drynarioides* (*hauhele'ula*). Fed. Reg. 48: 40920–40923.

1158. ———. 1984. Endangered and threatened wildlife and plants; final rule to list *Bidens cuneata* and *Schiedea adamantis* as endangered species. Fed. Reg. 49: 6099–6101.

1159. ———. 1984. Endangered and threatened wildlife and plants; final rule to list *Gouania hillebrandii* as an endangered species and to designate its critical habitat. Fed. Reg. 49: 44753–44757.

1160. ———. 1985. Endangered and threatened wildlife and plants; proposed rule to determine endangered status for *Hibiscadelphus distans* (Kauai hau kuahiwi). Fed. Reg. 50: 28873–28876.

1161. ———. 1985. Endangered and threatened wildlife and plants; proposed endangered status for *Abutilon menziesii* (*ko'oloa 'ula*). Fed. Reg. 50: 28876–28878.

1162. ———. 1985. Endangered and threatened wildlife and plants; determination of endangered status for *Gardenia brighamii* (*na'u* or Hawaiian gardenia) and withdrawal of proposed designation of critical habitat. Fed. Reg. 50: 33728–33731.

1163. ———. 1986. Endangered and threatened wildlife and plants; determination of endangered status for *Santalum freycinetianum* var. *lanaiense* (Lanai sandalwood or *'iliahi*). Fed. Reg. 51: 3182–3185.

1164. ———. 1986. Endangered and threatened wildlife and plants; determination of endangered status for *Argyroxiphium sandwicense* ssp. *sandwicense* (*'ahinahina* or Mauna Kea silversword). Fed. Reg. 51: 9814–9820.

1165. ———. 1986. Endangered and threatened wildlife and plants; determination of endangered status for *Achyranthes rotundata*. Fed. Reg. 51: 10518–10521.

1166. ———. 1986. Endangered and threatened wildlife and plants; determination of endangered status for *Hibiscadelphus distans* (Kauai *hau kuahiwi*). Fed. Reg. 51: 15903–15905.

1166a. ———. 1986. Endangered and threatened wildlife and plants; determination of endangered status for *Scaevola coriacea* (dwarf *naupaka*). Fed. Reg. 51: 17971–17974.

1167. ———. 1986. Endangered and threatened wildlife and plants; determination of endangered status for *Mezoneuron kavaiense* (*uhiuhi*). Fed. Reg. 51: 24672–24675.

1168. ———. 1986. Endangered and threatened wildlife and plants; determination of endangered status for *Abutilon menziesii* (*ko'oloa'ula*). Fed. Reg. 51: 34412–34415.

1169. ——— and J. J. Fay. 1979. Endangered and threatened wildlife and plants; determination that three Hawaiian plants are endangered species. Fed. Reg. 44: 62468–62469.

1170. ——— and ———. 1981. Proposal to list *Panicum carteri* (Carter's panicgrass) as an endangered species and determine its critical habitat. Fed. Reg. 46: 9976–9979.

1171. ——— and ———. 1983. Endangered and threatened wildlife and plants; rule to list *Panicum carteri* (Carter's panicgrass) as an endangered species and determine its critical habitat. Fed. Reg. 48: 46328–46332.

1172. ——— and B. H. Gagné. 1984. Endangered and threatened wildlife and plants; proposed endangered status and critical habitat for "*Gardenia brighamii*" Mann (*na'u* or Hawaiian gardenia). Fed. Reg. 49: 40058–40062.

1173. ——— and W. N. Takeuchi. 1982. Endangered and threatened wildlife and plants; proposed endangered status for *Bidens cuneata* and *Schiedea adamantis*. Fed. Reg. 47: 36675–36678.

1174. Heywood, V. H. 1979. The future of island floras: *in* Bramwell, D. (ed.), Plants and islands. Academic Press, New York, pp. 431–441.

1175. Hiern, W. P. 1873. A monograph of Ebenaceae. Trans. Cambridge Philos. Soc. 12: 27–300.

1176. Higa, T., and P. J. Scheuer. 1974. [Hawaiian plant studies. Part XV.] Alkaloids from *Pelea barbigera*. Phytochemistry 13: 1269–1272.

1177. ——— and ———. 1974. Hawaiian plant studies. Part XVI. Coumarins and flavones from *Pelea barbigera* (Gray) Hillebrand (Rutaceae). J. Chem. Soc. Perkin Trans. I: 1350–1352.

1178. Higashino, P. K., and L. K. Croft. 1979. An ecological survey of Pua'alu'u Stream. Part II. Catalogue of the vascular plants at Pua'alu'u, Maui. Coop. Natl. Park Resources Stud. Unit, Hawaii, Techn. Rep. 27: 25–33.

1179. ———, W. Guyer, and C. P. Stone. 1983. The Kilauea wilderness marathon and crater rim runs: sole searching experiences. Newslett. Hawaiian Bot. Soc. 22: 25–28.

1180. ——— and C. P. Stone. 1982. The fern jungle exclosure in Hawaii

Volcanoes National Park: 13 years without feral pigs in a rain forest. Proc. 4th Conf. Nat. Sci., Hawaii Volcanoes Natl. Park, p. 86.
1181. Higgins, J. E. 1904. The banana in Hawaii. Hawaii Agric. Exp. Sta. Bull. 7: 1–53.
1182. ———. 1906. The mango in Hawaii. Hawaii Agric. Exp. Sta. Bull. 12: 1–32.
1183. ——— and V. S. Holt. 1914. The papaya in Hawaii. Hawaii Agric. Exp. Sta. Bull. 32: 1–44.
1184. ———, C. J. Hunn, and V. S. Holt. 1911. The avocado in Hawaii. Hawaii Agric. Exp. Sta. Bull. 25: 1–48.
1185. Hill, W. 1882. Something about bananas. Hawaiian Almanac and Annual for 1883: 62–63.
1186. Hillebrand, W. 1888. Die Vegetationsformationen der Sandwich-Inseln. Bot. Jahrb. Syst. 9: 305–314.
1187. ———. 1888. Flora of the Hawaiian Islands: a description of their phanerogams and vascular cryptogams. Carl Winter, Heidelberg, Germany; Williams & Norgate, London; B. Westermann & Co., New York, 673 pp. (Facsimile ed., 1965, Hafner Publ. Co., New York, 673 pp.; Facsimile ed., 1981, Lubrecht & Cramer, Monticello, N.Y., 673 pp.)
1188. ———. 1920. The relation of forestry to agriculture. Hawaiian Pl. Rec. 22: 174–200.
1189. Hirano, R. T. 1967. Chromosomal and pollination studies as related to intra-specific and inter-specific compatibility in the genus *Psidium*. Master's thesis, Univ. Hawaii, Honolulu, 48 pp.
1190. ———. 1977. Propagation of *Santalum*, sandalwood tree. Pl. Propag. 23(2): 11–14.
1191. ——— and K. M. Nagata. 1972. A checklist of indigenous and endemic plants of Hawaii in cultivation at the Harold L. Lyon Arboretum. Harold L. Lyon Arbor., Univ. Hawaii, Honolulu, 22 unnum. pp.
1192. ——— and H. Y. Nakasone. 1969. Chromosome numbers of ten species and clones in the genus *Psidium*. J. Amer. Soc. Hort. Sci. 94: 83–86.
1193. ——— and ———. 1969. Pollen germination and compatibility studies of some *Psidium* species. J. Amer. Soc. Hort. Sci. 94: 287–289.
1194. Hitchcock, A. S. 1917. A botanical trip to the Hawaiian Islands. Sci. Monthly 5: 323–349.
1195. ———. 1917. A botanical trip to the Hawaiian Islands. II. Sci. Monthly 5: 419–432.
1196. ———. 1917. Botanical explorations in the Hawaiian Islands. Smithsonian Misc. Collect. 66: 59–73.
1197. ———. 1919. Floral aspects of the Hawaiian Islands. Annual Rep. Board Regents Smithsonian Inst. 1917: 449–462.
1198. ———. 1922. The grasses of Hawaii. Mem. Bernice P. Bishop Mus. 8: 101–230.
1199. ———. 1933. Remarks on type-specimens, and on a new species of grass from Hawaii. J. Bot. 71: 3–7.

1200. Hitchcock, C. L. 1932. A monographic study of the genus *Lycium* of the Western Hemisphere. Ann. Missouri Bot. Gard. 19: 179–375.
1201. ———. 1933. A taxonomic study of the genus *Nama*. I. Amer. J. Bot. 20: 415–431.
1202. ———. 1933. A taxonomic study of the genus *Nama*. II. Amer. J. Bot. 20: 518–534.
1203. Hobdy, R. W. 1982. A vegetative survey of Maui County's offshore islets. Proc. 4th Conf. Nat. Sci., Hawaii Volcanoes Natl. Park, pp. 87–92.
1204. ———. 1984. A re-evaluation of the genus *Hibiscadelphus* (Malvaceae) and the description of a new species. Occas. Pap. Bernice P. Bishop Mus. 25(11): 1–7.
1205. ———. 1986. The floral biology of the Hawaiian Malvaceae (Abstr.). Proc. 6th Conf. Nat. Sci., Hawaii Volcanoes Natl. Park, p. 24.
1206. Hochreutiner, B. P. G. 1900. Revision du genre *Hibiscus*. Annuaire Conserv. Jard. Bot. Genève 4: 23–191.
1207. ———. 1912. Plantae Hochreutineranae, étude systématique et biologique des collections faites par l'auteur au cours de son voyage aux Indes Néerlandaises et autour du monde pendant les années 1903 à 1905. Fascicule I. Annuaire Conserv. Jard. Bot. Genève 15: 145–247.
1208. ———. 1925. Plantae Hochreutineranae, étude systématique et biologique des collections faites par l'auteur au cours de son voyage aux Indes Néerlandaises et autour du monde pendant les années 1903 à 1905. Fascicule II. Candollea 2: 317–513.
1209. ———. 1928. Un *Cyrtandropsis* nouveau dans les Îles Hawaï. Arch. Sci. Phys. Nat. sér. 5, 10: 76–77.
1210. ———. 1934. Plantae Hochreutineranae, étude systématique et biologique des collections faites par l'auteur au cours de son voyage aux Indes Néerlandaises et autour du monde pendant les années 1903 à 1905. Fascicule III. Candollea 5: 175–341.
1211. ———. 1936. Plantae Hochreutineranae, étude systématique et biologique des collections faites par l'auteur au cours de son voyage aux Indes Néerlandaises et autour du monde pendant les années 1903 à 1905. Fascicule IV. Candollea 6: 399–488.
1212. ———. 1943. Plantae Hochreutineranae, étude systématique et biologique des collections faites par l'auteur au cours de son voyage aux Indes Néerlandaises et autour du monde pendant les années 1903 à 1905. Fascicule VI. Index des genres. Candollea 9: 481–493.
1213. Hodel, D. R. 1980. Notes on *Pritchardia* in Hawaii. Principes 24: 65–81.
1214. ———. 1982. Hal's last palms. Principes 26: 135–137.
1215. ———. 1985. A new *Pritchardia* from South Kona, Hawaii. Principes 29: 31–34.
1216. Hodges, C. S., Jr., K. T. Adee, J. D. Stein, H. B. Wood, and R. D. Doty. 1986. Decline of *ohia* (*Metrosideros polymorpha*) in Hawaii: a review. U.S.D.A. Forest Serv. Gen. Techn. Rep. PSW-86: 1–22.

1217. ——— and D. E. Gardner. 1982. Rusts of *Acacia* in Hawaii (Abstr.). Phytopathology 72: 965.
1218. ——— and ———. 1984. Hawaiian forest fungi. IV. Rusts on endemic *Acacia* species. Mycologia 76: 332–349.
1219. Hoffman, E. G. 1931. The flowers and trees of Hawaii. Mid-Pacific Mag. 42: 472–477.
1220. Holden, C. 1985. Hawaiian rainforest being felled. Science 228: 1073–1074.
1221. Holm, L., J. V. Rancho, J. P. Herberger, and D. L. Plucknett. 1979. A geographical atlas of world weeds. John Wiley & Sons, New York, 391 pp.
1222. Holt, R. A. 1982. The Maui Forest Disease and its impact on forestry in Hawaii (Abstr.). Proc. 4th Conf. Nat. Sci., Hawaii Volcanoes Natl. Park, p. 93.
1223. ———. 1983. Exotic species control: an island perspective. Nat. Conservancy News 33(4): 23–24.
1224. ———. 1983. The Maui forest trouble: a literature review and proposal for research. Univ. Hawaii, Hawaii Bot. Sci. Pap. 42: 1–67.
1225. Holt, W. W., and R. E. Nelson. 1959. A timber resource survey for Hawaii. U.S.D.A. Forest Serv., Forest Surv. Release 36: 1–13.
1226. Honda, N., W. H. C. Wong, Jr., and R. E. Nelson. 1967. Plantation timber on the island of Kauai—1965. U.S.D.A. Forest Serv. Resource Bull. PSW-6: 1–34.
1227. Hooker, J. D. 1865. *Railliardia ciliolata*. Ciliate-leaved *Railliardia*. Bot. Mag. 91: *pl. 5517*.
1228. ———. 1870. *Obbea timonioides*, Hook. f. Icon. pl. 11: 56.
1229. ———. 1870. *Rytidotus sandvicensis*, Hook. f. Icon. pl. 11: 56–57.
1230. ———. 1877. *Stenogyne rotundifolia*, A. Gray. Icon. pl. 13: 37–38.
1231. ———. 1887. *Hillebrandia sandwicensis*. Native of the Sandwich Islands. Bot. Mag. 113: *pl. 6953*.
1232. ———. 1894. *Osteomeles anthyllidifolia*. Native of eastern Asia and the Pacific Islands. Bot. Mag. 120: *pl. 7354*.
1233. Hooker, W. J. 1836. A brief memoir of the life of Mr. David Douglas, with extracts from his letters. Companion Bot. Mag. 2: 79–182.
1234. ———. 1837. *Argyroxiphium sandwicense*. Icon. pl. 1: *pl. 75*.
1235. ———. 1837. *Geranium cuneatum*, Hook. Icon. pl. 2: *pl. 198*.
1236. ———. 1839. A brief memoir of the life of Mr. David Douglas, with extracts from his letters. Hawaiian Spectator 2: 1–49, 131–180, 276–333, 396–437.
1237. ———. 1844. *Schiedea nuttallii*, Hook. Icon. pl. 7: *pl. 649–650*.
1238. ——— and G. A. W. Arnott. 1830–1841. The botany of Captain Beechey's Voyage; comprising an account of the plants collected by Messrs. Lay and Collie, and other officers of the expedition, during the voyage to the Pacific and Bering's Strait, performed in His Majesty's Ship *Blossom*, under the command of Captain F. W. Beechey, R.N., F.R.S. & A.S., in the years 1825, 26, 27, and 28. Parts 1–10. Henry G.

Bohn, London, 485 pp. (Facsimile ed., 1965, J. Cramer, Weinheim, Germany). ["Sandwich Islands" section published in Parts 2–3, pp. 78–111, 1832.]

1239. Horigan, D. P. 1983. An update on endangered species in Hawai'i from the Federal Register. 'Elepaio 44: 25–29.

1240. Horner, A. 1912. Report of the committee on forestry. Hawaiian Pl. Rec. 6: 60–69.

1241. Hosaka, E. Y. 1935. Floristic and ecological studies in Kipapa Gulch, Oahu. Master's thesis, Univ. Hawaii, Honolulu, 606 pp.

1242. ———. 1936. A troublesome introduced grass. Mid-Pacific Mag. 49: 126.

1243. ———. 1937. Ecological and floristic studies in Kipapa Gulch, Oahu. Occas. Pap. Bernice P. Bishop Mus. 13(17): 175–232.

1244. ———. 1937. Floristic and ecological studies in Kipapa Gulch, Oahu (Abstr.). Proc. Hawaiian Acad. Sci. 11: 6–7.

1245. ———. 1937. Phytogeography and ecology of Oahu (Abstr.). Proc. Hawaiian Acad. Sci. 11: 7–8.

1246. ———. 1939. Life-forms of the flowering plants of Kipapa Gulch, Oahu (Abstr.). Proc. Hawaiian Acad. Sci. 13: 12–13.

1247. ———. 1940. A revision of the Hawaiian species of *Myrsine* (*Suttonia*, *Rapanea*), (Myrsinaceae). Occas. Pap. Bernice P. Bishop Mus. 16(2): 25–76.

1248. ———. 1942. A new Hawaiian *Panicum* (Gramineae). Occas. Pap. Bernice P. Bishop Mus. 17(5): 67–69.

1249. ———. 1945. Kaimi Spanish clover for humid lowland pastures of Hawaii. Hawaii Agric. Exp. Sta. Circ. 22: 1–8.

1250. ———. 1945. Noxious weeds of Hawaii. Hawaii, Board Commiss. Agric. Forest., Honolulu, 26 unnum. pp.

1251. ——— and O. Degener. 1938. A new species of *Phyllostegia* and two new varieties of *Cyanea* of the Hawaiian Islands. Occas. Pap. Bernice P. Bishop Mus. 14(3): 27–30.

1252. ——— and J. C. Ripperton. 1939. Grasses of Hawaiian ranges (Abstr.). Proc. Hawaiian Acad. Sci. 13: 22–23.

1253. ——— and ———. 1944. Legumes of the Hawaiian ranges. Hawaii Agric. Exp. Sta. Bull. 93: 1–80.

1254. ——— and ———. 1953. Molasses grass on Hawaiian ranges. Univ. Hawaii Extens. Bull. 59: 1–9.

1255. ——— and A. Thistle. 1954. Noxious plants of the Hawaiian ranges. Univ. Hawaii Extens. Bull. 62: 1–39.

1256. ———, S. Uehara, and N. S. Hanson. 1957. Common weeds of Hawaii. Univ. Hawaii Agric. Extens. Serv., Honolulu, 16 pp.

1257. Hosmer, R. S. 1910. Kahoolawe forest reserve. Hawaiian Forester Agric. 7: 264–267.

1258. ———. 1912. The choice of street trees for planting in Honolulu. Hawaiian Almanac and Annual for 1913: 75–81.

1259. ———. 1912. Reclamation of Kahoolawe. Hawaiian Forester Agric. 9: 93–96.

1260. ———. 1913. Forest trails of Hawaii. Mid-Pacific Mag. 6: 561–564.
1261. ———. 1959. The beginning five decades of forestry in Hawaii. J. Forest. (Washington) 57: 83–89.
1262. Howard, R. A. 1962. Hawaii—a botanical and horticultural opportunity. Gard. J. 12: 223–227.
1263. Howarth, F. G. 1985. Impacts of alien land arthropods and mollusks on native plants and animals in Hawai'i: in Stone, C. P., and J. M. Scott (eds.), Hawai'i's terrestrial ecosystems: preservation and management. Coop. Natl. Park Resources Stud. Unit, Univ. Hawaii, Honolulu, pp. 149–179.
1264. Hsu, C.-C. 1965. The classification of *Panicum* (Gramineae) and its allies, with special reference to the characters of lodicule, style-base and lemma. J. Tokyo Univ. Fac. Sci. Sect. III 9: 43–150.
1265. Hu, S.-Y. 1966. The evolution and distribution of the species of Aquifoliaceae in the Pacific area (Abstr.): in Evolution, distribution and migration of the plant and animal in the Pacific. Proc. 11th Pacific Sci. Congr., Tokyo 5: 23.
1266. ———. 1967. The evolution and distribution of the species of Aquifoliaceae in the Pacific area (1). J. Jap. Bot. 42: 13–27.
1267. ———. 1967. The evolution and distribution of the species of Aquifoliaceae in the Pacific area (2). J. Jap. Bot. 42: 49–59.
1268. Hubbard, D. H., and V. R. Bender, Jr. 1950. Trailside plants of Hawaii National Park. Hawaii Nature Notes 4(1): 1–29.
1269. Hubbell, T. H. 1968. The biology of islands. Proc. Natl. Acad. U.S.A. 60: 22–32.
1270. Hudgins, W. R., and P. J. Scheuer. 1964. Chemotaxonomic value of some fingerprint gas chromatograms of oils from genus *Pelea*. Naturwissenschaften 51: 511–512.
1271. Hull, D. 1980. Palm questions and answers. Principes 24: 64, 81.
1272. Hume, L. H. 1986. Some plants of Hanalei. Bull. Pacific Trop. Bot. Gard. 16: 37–48.
1273. Hurov, H. R. 1972. New crops for Hawaii and the tropics. Newslett. Hawaiian Bot. Soc. 11: 43–46.
1274. Hutchinson, J. 1964. The genera of flowering plants (Angiospermae). Dicotyledones. Volume I. Clarendon Press, Oxford, 516 pp.
1275. ———. 1967. The genera of flowering plants (Angiospermae). Dicotyledones. Volume II. Clarendon Press, Oxford, 659 pp.
1276. Hutchinson, J. B., R. A. Silow, and S. G. Stephens. 1947. The evolution of *Gossypium* and the differentiation of the cultivated cottons. G. Cumberledge, Oxford Univ. Press, New York, 160 pp.
1277. Hutton, E. M. 1960. Flowering and pollination in *Indigofera spicata*, *Phaseolus lathyroides*, *Desmodium unicinatum*, and some other tropical pasture legumes. Empire J. Exp. Agric. 28: 235–243.
1278. Huynh, K.-L. 1970. Le pollen et la systématique chez le genre *Lysimachia* (Primulaceae). I. Morphologie générale du pollen et palynotaxonomie. Candollea 25: 267–296.

1279. ———. 1971. Le pollen et la systématique chez le genre *Lysimachia* (Primulaceae). II. Considérations générales. Candollea 26: 279–295.
1280. Hwang, S. C. 1976. *Phytophthora cinnamomi*: its biology in soil and relation to *ohia* decline. Ph.D. dissertation, Univ. Hawaii, Honolulu, 71 pp.
1281. ———. 1977. *Phytophthora cinnamomi*: its survival in soil and relation to *ohia* decline. Coop. Natl. Park Resources Stud. Unit, Hawaii, Techn. Rep. 12: 1–71.
1282. ——— and W. H. Ko. 1976. Quantitative studies of *Phytophthora cinnamomi* in *ohia* forests (Abstr.). Proc. Amer. Phytopathol. Soc. 3: 237.
1282a.——— and ———. 1978. Quantitative studies of *Phytophthora cinnamomi* in decline and healthy *ohia* forests. Trans. Brit. Mycol. Soc. 70: 312–315.
1283. Hyde, C. M. 1885. Helps to the study of Hawaiian botany. Hawaiian Almanac and Annual for 1886: 39–42.
1284. Ingham, J. L. 1980. Induced isoflavonoids of *Erythrina sandwicensis*. Z. Naturf. 35c: 384–386.
1285. ——— and K. R. Markham. 1980. Identification of the *Erythrina* phytoalexin cristacarpin and a note on the chirality of other 6a-hydroxypterocarpans. Phytochemistry 19: 1203–1207.
1286. Irwin, H. S., and R. C. Barneby. 1982. The American *Cassiinae*. A synoptical revision of Leguminosae tribe *Cassieae* subtribe *Cassiinae* in the New World. Mem. New York Bot. Gard. 35: 455–918.
1287. Jablonski, E. 1973. Catalogus Euphorbiarum 1973. Vol. I–a. Phytologia 25: 329–349.
1288. Jacobi, J. D. 1974. A vegetational description of the IBP small mammal trapline transects—Mauna Loa transect. U.S. IBP Island Ecosystems IRP Techn. Rep. 48: 1–19.
1289. ———. 1976. The influence of feral pigs on a native alpine grassland in Haleakala National Park. Proc. 1st Conf. Nat. Sci., Hawaii Volcanoes Natl. Park, pp. 107–112.
1290. ———. 1978. Description of a new large-scale vegetation mapping project in Hawai'i. Proc. 2nd Conf. Nat. Sci., Hawaii Volcanoes Natl. Park, pp. 165–176.
1291. ———. 1978. Vegetation map of the Kau Forest Reserve and adjacent lands, island of Hawaii. U.S.D.A. Forest Serv. Resource Bull. PSW-16, 1 sheet.
1292. ———. 1980. Problems with the long-term maintenance of *māmane* (*Sophora chrysophylla*) in the central crater area of Haleakala National Park. Proc. 3rd Conf. Nat. Sci., Hawaii Volcanoes Natl. Park, pp. 167–171.
1293. ———. 1981. Vegetation changes in a subalpine grassland in Hawai'i following disturbance by feral pigs. Coop. Natl. Park Resources Stud. Unit, Hawaii, Techn. Rep. 41: 29–52.
1294. ———. 1983. *Metrosideros* dieback in Hawai'i: a comparison of adjacent

dieback and non-dieback rain forest stands. New Zealand J. Ecol. 6: 79–97.
1295. ———, G. Gerrish, and D. Mueller-Dombois. 1983 [1984]. 'Ōhi'a dieback in Hawai'i: vegetation changes in permanent plots. Pacific Sci. 37: 327–337.
1296. ——— and J. M. Scott. 1985. An assessment of the current status of native upland habitats and associated endangered species on the island of Hawai'i: *in* Stone, C. P., and J. M. Scott (eds.), Hawai'i's terrestrial ecosystems: preservation and management. Coop. Natl. Park Resources Stud. Unit, Univ. Hawaii, Honolulu, pp. 3–22.
1297. Jacobs, M. 1965. The genus *Capparis* (Capparaceae) from the Indus to the Pacific. Blumea 12: 385–541.
1298. Jaeger, A. 1884. A word for the palms. Planters' Monthly 2: 314–320.
1299. James, M. C. 1979. The taxonomic significance of foliar sclereids in the genus *Cyrtandra* on Oahu. Master's thesis, Univ. Hawaii, Honolulu, 105 pp.
1300. Jarves, J. J. 1844. Scenes and scenery in the Sandwich Islands, and a trip through Central America: being observations from my note-book during the years 1837–1842. James Munroe & Co., Boston, 341 pp.
1301. Jedwabnick, E. 1924. Eragrostidis specierum imprimis ad herb. Berol., Hamburg., Monac., Regimont. digestarum conspectus. Bot. Arch. 5: 177–216.
1302. Jeffrey, C. 1978. Further notes on Cucurbitaceae: IV. Some New-World taxa. Kew Bull. 33: 347–380.
1303. ———. 1980. On the nomenclature of the strand *Scaevola* species (Goodeniaceae). Kew Bull. 34: 537–545.
1304. Jenkins, D. W. 1975. At last—a brighter outlook for endangered plants. Natl. Parks Conservation Mag. 49(1): 13–17.
1305. ——— and E. S. Ayensu. 1975. One-tenth of our plant species may not survive. Smithsonian 5(10): 92–96.
1306. Jepsen, S. M. 1967. Trees of Hawaii. Amer. Forests 73(12): 18–21, 36.
1307. Johnson, L. A. S. 1957. A review of the family Oleaceae. Contr. New South Wales Natl. Herb. 2: 395–418.
1308. Johnson, M. A. 1984. The chromosome number of *Brighamia citrina* var. *napaliensis*; a rare succulent endemic from Hawaii. Brit. Cact. Succ. J. 1: 106–107.
1309. ———. 1986. Plate 60. *Brighamia citrina* var. *napaliensis*. Campanulaceae. Kew Mag. 3: 68–72.
1310. Johnson, M. F. 1971. A monograph of the genus *Ageratum* L. (Compositae–*Eupatorieae*). Ann. Missouri Bot. Gard. 58: 6–88.
1311. Johnson, R. K., and B. G. Decker. 1980. Implications of the distribution of names for cotton (*Gossypium* spp.) in the Indo-Pacific. Asian Perspect. 23: 249–307.
1312. Johnston, I. M. 1923. Diagnoses and notes relating to the spermatophytes chiefly of North America. Contr. Gray Herb. 68: 80–104.

1313. ———. 1937. Studies in the Boraginaceae, XII. J. Arnold Arbor. 18: 1–25.
1314. Johnston, M. C. 1971. Revision of *Colubrina* (Rhamnaceae). Brittonia 23: 2–53.
1315. Jones, A. G. 1984. Nomenclatural notes on *Aster* (Asteraceae)—III. The status of *A. sandwicensis*. Brittonia 36: 463–466.
1316. Jordan, W., and P. J. Scheuer. 1965. Hawaiian plant studies—XIV. Alkaloids of *Ochrosia sandwicensis* A. Gray. Tetrahedron 21: 3731–3740.
1317. Jouan, H. 1865. Recherches sur l'origine et la provenance de certains végétaux phanérogames observés dans les îles du Grand-Océan. Mém. Soc. Sci. Nat. Cherbourg 11: 81–178.
1318. ———. 1873. Notes sur l'Archipel Hawaiien (Îles Sandwich). Mém. Soc. Sci. Nat. Cherbourg 17: 5–105.
1319. Judd, A. F. 1933. Trees and plants: *in* Handy, E. S. C., K. P. Emory, E. H. Bryan, Jr., P. H. Buck, J. H. Wise, and others, Ancient Hawaiian civilization. A series of lectures delivered at The Kamehameha Schools. The Kamehameha Schools, Honolulu, pp. 273–281. (Also publ. in Handy et al., rev. ed., 1965, Charles E. Tuttle Co., Rutland, Vt., pp. 277–285).
1320. Judd, C. S. 1916. Kahoolawe. Hawaiian Almanac and Annual for 1917: 117–125.
1321. ———. 1916. The first algaroba and royal palm in Hawaii. Hawaiian Forester Agric. 13: 330–335.
1322. ———. 1917. Street trees for Hawaii. Hawaiian Forester Agric. 14: 89–93.
1323. ———. 1918. Forestry as applied in Hawaii. Hawaiian Forester Agric. 15: 117–133.
1324. ———. 1918. The Hawaiian sumach. Hawaiian Forester Agric. 15: 441–442.
1325. ———. 1919. The *kukui* or candlenut tree. Hawaiian Forester Agric. 16: 222–223.
1326. ———. 1919. Forestry in Hawaii. Hawaiian Forester Agric. 16: 271–297.
1327. ———. 1919. Original algaroba tree gone. Hawaiian Forester Agric. 16: 308–310.
1328. ———. 1920. The *koa* tree. Hawaiian Forester Agric. 17: 30–35.
1329. ———. 1920. The Australian red cedar. Hawaiian Forester Agric. 17: 57–59.
1330. ———. 1920. The *wiliwili* tree. Hawaiian Forester Agric. 17: 95–97.
1331. ———. 1921. Hawaiian forests and trails. Hawaiian Forester Agric. 18: 79–82.
1332. ———. 1921. The *alahee* tree. Hawaiian Forester Agric. 18: 133–137.
1333. ———. 1921. Hilo grass and trails. Hawaiian Forester Agric. 18: 198–200.
1334. ———. 1921. Kilauea [Hawaii] National Park trees. Hawaiian Forester Agric. 18: 255–260.

1335. ———. 1922. Chaulmoogra oil plantation. Hawaiian Forester Agric. 19: 64–66.
1336. ———. 1926. Wild food plants in Hawaii. Hawaiian Forester Agric. 23: 125–127.
1337. ———. 1927. The natural resources of the Hawaiian forest regions and their conservation. Hawaiian Forester Agric. 24: 40–47.
1338. ———. 1927. Factors deleterious to the Hawaiian forest. Hawaiian Forester Agric. 24: 47–53.
1339. ———. 1927. Bamboo against staghorn fern. Hawaiian Forester Agric. 24: 54–55.
1340. ———. 1927. The story of the forests of Hawaii. Paradise Pacific 40(10): 9–18.
1341. ———. 1928. The propagation of indigenous tree seed. Hawaiian Forester Agric. 25: 38–40.
1342. ———. 1929. Memoranda for ten-year program of forest work on Oahu. Hawaiian Forester Agric. 26: 9–11.
1343. ———. 1929. Island of Oahu cover classification. Hawaiian Forester Agric. 26: 19-A [map].
1344. ———. 1929. The forests of the Hawaiian Islands. Mid-Pacific Mag. 38: 333–336.
1345. ———. 1930. Botanical discoveries: *in* Routine report, territorial forester, December, 1929. Hawaiian Forester Agric. 27: 13.
1346. ———. 1930. Chaulmoogra oil trees in Hawaii. Hawaiian Forester Agric. 27: 105–106.
1347. ———. 1931. Botanical bonanzas. Hawaiian Almanac and Annual for 1932: 61–69.
1348. ———. 1931. *Neowawraea* trees: *in* Routine report, territorial forester, November, 1931. Hawaiian Forester Agric. 28: 108.
1349. ———. 1931. Botanical bonanzas (Abstr.). Proc. Hawaiian Acad. Sci. 6: 17.
1350. ———. 1932. The parasitic habit of the sandalwood tree. Hawaiian Almanac and Annual for 1933: 81–88.
1351. ———. 1932. Niihau. Hawaiian Forester Agric. 29: 5–9.
1352. ———. 1932. Botanical discoveries: *in* Annual report, 1931, territorial forester. Hawaiian Forester Agric. 29: 15–16.
1353. ———. 1932. The parasitic habit of the sandalwood tree (Abstr.). Proc. Hawaiian Acad. Sci. 7: 5–6.
1354. ———. 1933. Preserving our botanical treasures. Friend 103: 54, 61.
1355. ———. 1933. Grove of *kauila* trees: *in* Routine report territorial forester, September, 1932. Hawaiian Forester Agric. 29: 162.
1356. ———. 1933. Forest notes. Nature Notes, Hawaii Natl. Park 3(2): 17–18.
1357. ———. 1933. Forest Notes, Hawaii. J. Forest. (Washington) 31: 600–601.
1358. ———. 1935. *Koa* reproduction after fire. J. Forest. (Washington) 33: 176.

1359. ———. 1936. Growing sandalwood in the Territory of Hawaii. J. Forest. (Washington) 34: 82–83.
1360. ———. 1936. Seed dispersal in Hawaii. Mid-Pacific Mag. 49: 111–118.
1361. ———. 1939. Forestry in Hawaii: in First progress report. An historic inventory of the physical, social and economic resources of the Territory of Hawaii. Territorial Planning Board, Honolulu, pp. 106–109.
1362. ———. 1941. Forest resources of the Territory of Hawaii, U.S.A. Proc. 6th Pacific Sci. Congr., California 4: 797–800.
1363. ———. 1950. Preserving our botanical treasures. Elepaio 10: 40–42.
1364. Judd, H. P. 1938. A week on Niihau. Paradise Pacific 50(9): 9–10.
1365. ———. 1938. A visit to Kahoolawe. Paradise Pacific 50(10): 11–12.
1366. Juvik, J. O., and S. P. Juvik. 1984. Mauna Kea and the myth of multiple use. Endangered species and mountain management in Hawaii. Mountain Res. Developm. 4: 191–202.
1367. ——— and L. Lawrence. 1982. Late Holocene vegetation history from Hawaiian peat deposits. Proc. 4th Conf. Nat. Sci., Hawaii Volcanoes Natl. Park, p. 100.
1368. Kaaiakamanu, D. M., and J. K. Akina (Transl. A. Akana). 1922. Hawaiian herbs of medicinal value. Hawaii, Territorial Board Health, Honolulu, 74 pp. (Facsimile ed., 1972, Charles E. Tuttle Co., Rutland, Vt.).
1369. Kadooka, M. M., M. Y. Chang, H. Fukami, and P. J. Scheuer. 1976. Mamanine and pohakuline, two unprecedented quinolizidine alkaloids from *Sophora chrysophylla*. Tetrahedron 32: 919–924.
1370. Kai, M. H. 1937. The Foster Garden. Paradise Pacific 49(2): 11, 30.
1371. Kaikainahaole, M. 1968. Hawaiian uses of herbs—past and present. Newslett. Hawaiian Bot. Soc. 7: 31–38.
1372. Kamakau, S. M. 1918. *Olona*, its cultivation and uses. Hawaiian Almanac and Annual for 1919: 69–74.
1373. Kamemoto, H., and W. B. Storey. 1955. Genetics of flower color in *Asystasia gangetica*, Linn. Pacific Sci. 9: 62–68.
1374. Karg, K. H. 1940. Orchids in Hawaii. Amer. Orchid Soc. Bull. 9: 84–85.
1375. Kartawinata, K. 1971. Phytosociology and ecology of the natural dry-grass communities on Oahu, Hawaii. Ph.D. dissertation, Univ. Hawaii, Honolulu, 301 pp.
1376. ——— and D. Mueller-Dombois. 1972. Phytosociology and ecology of the natural dry-grass communities on Oahu, Hawaii. Reinwardtia 8: 369–494.
1377. Kartesz, J. T., and R. Kartesz. 1980. A synonymized checklist of the vascular flora of the United States, Canada, and Greenland. Volume II. The biota of North America. Univ. North Carolina Press, Chapel Hill, 500 pp.
1378. Kaschko, M. W., and M. S. Allen. 1978. The impact of the sweet potato on prehistoric Hawaiian cultural development. Proc. 2nd Conf. Nat. Sci., Hawaii Volcanoes Natl. Park, pp. 177–183.

1379. Katahira, L. K. 1980. The effects of feral pigs on a montane rain forest in Hawaii Volcanoes National Park. Proc. 3rd Conf. Nat. Sci., Hawaii Volcanoes Natl. Park, pp. 173–178.
1380. ——— and P. M. Finnegan. 1986. Hunting as a pig control method in Hawai'i Volcanoes National Park (Abstr.). Proc. 6th Conf. Nat. Sci., Hawaii Volcanoes Natl. Park, p. 26.
1381. ——— and C. P. Stone. 1982. Status of management of feral goats in Hawaii Volcanoes National Park. Proc. 4th Conf. Nat. Sci., Hawaii Volcanoes Natl. Park, pp. 102–108.
1382. Kato, S. S. 1969. The role of plants in the *kapu* system of Hawaii. Newslett. Hawaiian Bot. Soc. 8: 1–6.
1383. Kato, T. 1963. An anatomical investigation of the arborescent Nyctaginaceae in Hawaii. Master's thesis, Univ. Hawaii, Honolulu, 120 pp.
1384. Kay, E. A. (ed.). 1972. A natural history of the Hawaiian Islands. Selected readings. Univ. Press Hawaii, Honolulu, 653 pp.
1385. ———. 1972. Hawaiian natural history: 1778–1900: *in* A natural history of the Hawaiian Islands. Selected readings. Univ. Press Hawaii, Honolulu, pp. 604–653.
1386. ———. 1980. Little worlds of the Pacific. An essay on Pacific Basin biogeography. Univ. Hawaii Harold L. Lyon Arbor. Lecture 9: 1–40.
1387. Kearney, T. H. 1957. Wild and domesticated cotton plants of the world. Leafl. Western Bot. 8: 103–109.
1388. Keck, D. D. 1936. The silverswords of Hawaii. Carnegie Inst. Wash. News Serv. Bull. 4: 75–78.
1389. ———. 1936. The Hawaiian silverswords: systematics, affinities, and phytogeographic problems of the genus *Argyroxiphium*. Occas. Pap. Bernice P. Bishop Mus. 11(19): 1–38.
1390. Keeler, K. H. 1982. Distribution of plants with extrafloral nectaries in Hawaii Volcanoes National Park. Proc. 4th Conf. Nat. Sci., Hawaii Volcanoes Natl. Park, p. 109.
1391. Kenn, C. W. 1957. *Na okika maoli o Hawaii nei* (the indigenous orchids of Hawaii). Na Pua Okika o Hawaii Nei 7: 128–133.
1392. Kennedy, J. (ed.). 1982. Native planters. *Ho'okupu kalo*. Native Planters, Honolulu, 55 pp.
1393. Kepler, A. K. 1983. Hawaiian heritage plants. Oriental Publ. Co., Honolulu, 150 pp.
1394. Kepler, C. B., A. K. Kepler, and T. R. Simons. 1984. Hawaii's seabird islands, no. 1: Moke'ehia. 'Elepaio 44: 71–74.
1395. Ker, J. B. 1823. *Edwardsia chrysophylla*. Bot. Reg. 9: *pl. 738*.
1396. Kern, J. H. 1962. On the delimitation of the genus *Gahnia* (Cyperaceae). Acta Bot. Neerl. 11: 216–224.
1397. Kerr, M. 1948. Bird walk: Pa Lehua Trail, July 11, 1948. Elepaio 9: 26.
1398. Kikuta, K., L. D. Whitney, and G. K. Parris. 1938. Seeds and seedlings of the taro, *Colocasia esculenta*. Amer. J. Bot 25: 186–188.
1399. Kim, J. Y. 1969. "*Myrica faya*" control in Hawaii. Down to Earth 25(3): 23–25.

1400. Kimura, B. Y., and K. M. Nagata. 1980. Hawaii's vanishing flora. Oriental Publ. Co., Honolulu, 88 pp.
1401. King, J. 1874. A voyage to the Pacific Ocean, undertaken, by the command of his majesty, for making discoveries in the Northern Hemisphere. Vol. III. G. Nicol & T. Cadell, London, 558 pp.
1402. Kirch, P. V. 1982. Ecology and the adaptation of Polynesian agricultural systems. Archaeol. Oceania 17: 1–6.
1403. ———. 1982. Transported landscapes. Nat. Hist. 91(12): 32–35.
1404. ———. 1982. The impact of the prehistoric Polynesians on the Hawaiian ecosystem. Pacific Sci. 36: 1–14.
1405. ———. 1983. Man's role in modifying tropical and subtropical Polynesian ecosystems. Archaeol. Oceania 18: 26–31.
1406. ——— and M. Kelly (eds.). 1975. Prehistory and ecology in a windward Hawaiian valley: Halawa Valley, Molokai. Pacific Anthropol. Rec. 24: 1–205.
1407. Kircher, H. W. 1969. Sterols in the leaves of the *Cheirodendron gaudichaudii* tree and their relationship to Hawaiian *Drosophila* ecology. J. Insect Physiol. 15: 1167–1173.
1408. ——— and W. B. Heed. 1970. Phytochemistry and host plant specificity in *Drosophila*. Recent Advances Phytochem. 3: 191–209.
1409. Kirkaldy, G. W. 1909. The entomological work of the Hawaiian Sugar Planters' Association as seen by Dr. Silvestri and Mr. Froggatt. Hawaiian Pl. Rec. 1: 179–197.
1410. Kjargaard, J. I. 1982. Population, distribution and adverse effects of feral goats at Haleakala National Park. Proc. 4th Conf. Nat. Sci., Hawaii Volcanoes Natl. Park, p. 110.
1411. Klett, W. 1924. Umfang und Inhalt der Familie der Loganiaceen. Bot. Arch. 5: 312–338.
1412. Kliejunas, J. T. 1979. Effects of *Phytophthora cinnamomi* on some endemic and exotic plant species in Hawaii in relation to soil type. Pl. Dis. Reporter 63: 602–606.
1413. ——— and W. H. Ko. 1973. Root rot of ohia (*Metrosideros collina* subsp. *polymorpha*) caused by *Phytophthora cinnamomi*. Pl. Dis. Reporter 57: 383–384.
1414. ——— and ———. 1974. Deficiency of inorganic nutrients as a contributing factor to *ohia* decline. Phytopathology 64: 891–896.
1415. ——— and ———. 1974. Role of *Phytophthora cinnamomi* in *ohia* decline (Abstr.). Proc. Amer. Phytopathol. Soc. 1: 58.
1416. ——— and ———. 1975. The occurrence of *Pythium vexans* in Hawaii and its relation to *ohia* decline. Pl. Dis. Reporter 59: 392–395.
1417. ——— and ———. 1976. Association of *Phytophthora cinnamomi* with *ohia* decline on the island of Hawaii. Phytopathology 66: 116–121.
1418. ——— and ———. 1976. Dispersal of *Phytophthora cinnamomi* on the island of Hawaii. Phytopathology 66: 457–460.
1419. ———, R. F. Scharpf, and R. S. Smith, Jr. 1977. The occurrence of

Phytophthora cinnamomi in Hawaii in relation to *ohia* forest site and edaphic factors. Pl. Dis. Reporter 61: 290–293.

1420. ———, ———, and ———. 1979. *Teichospora obducens* and a *Pleospora* species on *Korthalsella* in Hawaii. Pl. Dis. Reporter 63: 1060–1062.

1421. Klotzsch, J. F. 1843. Euphorbiaceae: *in* Meyen, F. J. F., Observationes botanicas in itinere circum terram institutas . . . Nov. Actorum Acad. Caes. Leop.-Carol. Nat. Cur. 19, Suppl. 1: 412–422.

1422. ———. 1851. Studien über die natürliche Klass Bicornes Linné. Linnaea 24: 1–88.

1423. ———. 1860. Linné's natürliche Pflanzenklasse Tricoccae des Berliner Herbarium's im Allgemeinen und die natürliche Ordnung Euphorbiaceae insbesondere. Abh. Königl. Akad. Wiss. Berlin 1859(1): 1–108; not seen.

1424. Knapp, R. 1965. Die Vegetation von Nord- und Mittelamerika und der Hawaii-Inseln: *in* Walter, H., Vegetationsmonographien der einzelnen GroBräume. Gustav Fischer Verlag, Stuttgart, 373 pp.

1425. ——— (Transl. A. Y. Yoshinaga and H. H. Iltis). 1975. Vegetation of the Hawaiian Islands. Newslett. Hawaiian Bot. Soc. 14: 95–121.

1426. Knoblauch, E. 1895. Zur Kenntniss einiger Oleaceen-Genera. Bot. Centralbl. 61: 81–87.

1427. Kobayashi, H. K. 1973. Ecology of the silversword, Haleakala Crater, Hawaii. Hawaii Nat. Hist. Assoc., Hawaii Volcanoes Natl. Park, 124 pp.

1428. ———. 1973. Present status of the *ahinahina* or silversword *Argyroxiphium sandwicense* DC. on Haleakala, Maui. Newslett. Hawaiian Bot. Soc. 12: 23–26.

1429. ———. 1973. Putative generic hybrids of Haleakala's silversword and *kupaoa* (*Argyroxiphium sandwicense* × *Dubautia menziesii*) Compositae. Pacific Sci. 27: 207–208.

1430. ———. 1973. Ecology of the silversword *Argyroxiphium sandwicense* DC. (Compositae), Haleakala Crater, Hawaii. Ph.D. dissertation, Univ. Hawaii, Honolulu, 91 pp.

1431. ———. 1974. Preliminary investigations on insects affecting the reproductive stage of the silversword (*Argyroxiphium sandwicense* DC.) Compositae, Haleakala Crater, Maui, Hawaii. Proc. Hawaiian Entomol. Soc. 21: 397–402.

1432. Kobayashi, J. 1976. Early Hawaiian uses of medicinal plants in pregnancy and childbirth. J. Trop. Pediatrics 22: 260–262.

1433. Kobuski, C. E. 1935. Studies in Theaceae, I. *Eurya* subgen. *Ternstroemiopsis*. J. Arnold Arbor. 16: 347–352.

1434. ———. 1938. Studies in Theaceae. III. *Eurya* subgenera *Euryodes* and *Penteurya*. Ann. Missouri Bot. Gard. 25: 299–359.

1435. ———. 1947. Studies in the Theaceae, XV. A review of the genus *Adinandra*. J. Arnold Arbor. 28: 1–98.

1436. Koebele, A. 1900. Hawaii's forest foes. Hawaiian Almanac and Annual for 1901: 90–97.
1437. ———. 1901. Report of Prof. Koebele on destruction of forest trees, Hawaii: *in* Taylor, W., Rep. Commiss. Agric. Forest. for 1900. Hawaiian Gaz. Co., Ltd., Honolulu, pp. 50–60.
1438. ———. 1901. Notes on insects affecting the *koa* trees at Haiku forest on Maui: *in* Taylor, W., Rep. Commiss. Agric. Forest. for 1900. Hawaiian Gaz. Co., Ltd., Honolulu, pp. 61–66.
1439. Komkris, T. 1963. Structure and ontogeny of *Euphorbia degeneri* Sherff. Master's thesis, Univ. Hawaii, Honolulu, 75 pp.
1440. Kores, P. J. 1979. Taxonomy and pollination in the wild Hawaiian orchids. Master's thesis, Univ. Hawaii, Honolulu, 182 pp.
1441. ———. 1979. A review of the literature on Hawaiian orchids. Newslett. Hawaiian Bot. Soc. 18: 34–55.
1442. ———. 1980. Pollination mechanisms as a limiting factor in the development of the orchidaceous flora of Hawai'i. Proc. 3rd Conf. Nat. Sci., Hawaii Volcanoes Natl. Park, pp. 183–191.
1443. Korte, K. H. n.d. From World War II to the present . . . the history of forestry in Hawaii. Hawaii, Dept. Land Nat. Resources, Honolulu, 3 pp.
1444. Koster, J. T. 1952. Notes on Malay Compositae III. Blumea 7: 288–291.
1445. Kostermans, A. J. G. H. 1954. A monograph of the Asiatic, Malaysian, Australian and Pacific species of Mimosaceae, formerly included in *Pithecolobium* Mart. Organ. Sci. Res. Indonesia Bull. 20: 1–122.
1446. ———. 1977. Notes on Ceylonese ebony trees (Ebenaceae). Ceylon J. Sci., Biol. Sci. 12: 89–108.
1447. Koutnik, D. L. 1982. A taxonomic revision of the Hawaiian species of the genus *Chamaesyce* (Euphorbiaceae). Ph.D. dissertation, Univ. California, Davis, 173 pp.
1448. ———. 1985. New combinations in Hawaiian *Chamaesyce* (Euphorbiaceae). Brittonia 37: 397–399.
1449. Koyama, T. 1956. Taxonomic study of Cyperaceae V. Bot. Mag. (Tokyo) 69: 59–67.
1450. ———. 1962. The genus *Scirpus* Linn. Some North American aphylloid species. Canad. J. Bot. 40: 913–937.
1451. ———. 1964. The Cyperaceae of Micronesia. Micronesica 1: 59–111.
1452. ——— and B. C. Stone. 1960. The genus *Scirpus* in the Hawaiian Islands. Bot. Mag. (Tokyo) 73: 288–294.
1453. Kraebel, C. J. 1922. Mauna Kea plant list. Hawaiian Forester Agric. 19: 2–4.
1454. Krajina, V. J. 1930. Generis Gunnerae species hawaiienses. Acta Bot. Bohem. 9: 49–52.
1455. ———. 1930. New Hawaiian species of *Pipturus*. Occas. Pap. Bernice P. Bishop Mus. 9(3): 1–6.
1456. ———. 1931. Generis Cheirodendrum species hawaiienses ex affinitate Cheirodendri platyphylli. Preslia 10: 91–100.

1457. ———. 1963. Biogeoclimatic zones on the Hawaiian Islands. Newslett. Hawaiian Bot. Soc. 2: 93–98.
1458. ———. 1966. Biogeoclimatic zones of the Hawaiian Islands and their variation by volcanic activities (Abstr.): *in* Biotic communities of the volcanic areas of the Pacific. Proc. 11th Pacific Sci. Congr., Tokyo 5: 1.
1459. ———, J. F. Rock, and H. St. John. 1962. Campus trees and plants. Univ. Hawaii, Honolulu, 28 unnum. pp.
1460. Krämer, A. 1906. Hawaii, Ostmikronesien und Samoa. Meine zweite Südseereise (1897–1899) zum Studium der Atolle und ihrer Bewohner. Strecker & Schröder, Stuttgart, 585 pp.
1461. Kränzlin, F. 1929. Beiträge zur Kenntnis der Familie der Myoporinae R. Br. mit besonderer Berücksichtigung der Myoporinous Plants of Australia. Tome II—Lithograms. Repert. Spec. Nov. Regni Veg. Beih. 54: 1–129.
1462. Krause, K. 1912. Goodeniaceae. Pflanzenr. IV. 227 (Heft 54): 1–207.
1463. Krauss, B. H. 1974. Ethnobotany of Hawaii. Dept. Botany, Univ. Hawaii, Honolulu, 248 pp.
1464. ———. 1974. Ethnobotany of the Hawaiians. Univ. Hawaii Harold L. Lyon Arbor. Lecture 5: 1–32.
1465. ———. 1980. Ethnobotanical resources in Kīpahulu Valley below 2000 feet: *in* Smith, C. W. (ed.), Resources base inventory of Kīpahulu Valley below 2000 feet. Coop. Natl. Park Resources Stud. Unit, Univ. Hawaii, Honolulu, pp. 71–82.
1466. ———. 1980. Creating a Hawaiian ethnobotanical garden. Univ. Hawaii Harold L. Lyon Arbor. Educational Ser. 1: 1–8.
1467. ———. 1981. Native plants used as medicine in Hawaii. Harold L. Lyon Arbor., Univ. Hawaii, Honolulu, 50 pp.
1468. Krauss, F. G. 1921. The pigeon pea (*Cajanus indicus*): its culture and utilization in Hawaii. Hawaii Agric. Exp. Sta. Bull. 46: 1–23.
1469. ———. 1926. Genetic analysis of *Cajanus indicus* and the creation of improved varieties through hybridization and selection (Abstr.). Proc. Hawaiian Acad. Sci. 1: 24–25.
1470. ———. 1932. The pigeon pea (*Cajanus indicus*): its improvement, culture and utilization in Hawaii. Hawaii Agric. Exp. Sta. Bull. 64: 1–46.
1471. Krauss, R. W. 1949. A taxonomic revision of the Hawaiian species of the genus *Carex*. Master's thesis, Univ. Hawaii, Honolulu, 116 pp.
1472. ———. 1950. A taxonomic revision of the Hawaiian species of the genus *Carex*. Pacific Sci. 4: 249–282.
1473. Krohn, V. F. 1978. Hawaii dye plants and dye recipes. Univ. Press Hawaii, Honolulu, 136 pp.
1474. Krugman, S. L. 1974. *Eucalyptus* L'Herit. *Eucalyptus*: *in* Schopmeyer, C. S., Seeds of woody plants in the United States. U.S.D.A. Agric. Handbook 450: 384–392.

1475. Krukoff, B. A. 1939. Preliminary notes on Asiatic-Polynesian species of *Erythrina*. J. Arnold Arbor. 20: 225–233.
1476. ———. 1969. Supplementary notes on the American species of *Erythrina*. III. Phytologia 19: 113–175.
1477. ———. 1971. Supplementary notes on the American species of *Erythrina*. V. Phytologia 22: 244–277.
1478. ———. 1972. Notes on Asiatic-Polynesian-Australian species of *Erythrina*, II. J. Arnold Arbor. 53: 128–139.
1479. ———. 1978. Notes on the species of *Erythrina*. XI. Phytologia 39: 294–306.
1480. ———. 1979. Notes on the species of *Erythrina*. XII. Ann. Missouri Bot. Gard. 66: 422–445.
1481. ———. 1980. Notes on the species of *Erythrina*. XV. Phytologia 46: 88–93.
1482. ———. 1982. Notes on the species of *Erythrina*. XVIII. Allertonia 3: 121–138.
1483. ——— and R. C. Barneby. 1973. Notes on the species of *Erythrina*. VII. Phytologia 27: 108–141.
1484. ——— and ———. 1974. Conspectus of species of the genus *Erythrina*. Lloydia 37: 332–459.
1485. Kuck, L. E., and R. C. Tongg. 1939. The tropical garden, its design, horticulture and plant materials. Macmillan Co., New York, 378 pp.
1486. ——— and ———. 1943. Hawaiian flowers. Tongg Publ. Co., Honolulu, 109 pp.
1487. ——— and ———. 1955. The modern tropical garden: its design, plant materials and horticulture. Tongg Publ. Co., Honolulu, 250 pp.
1488. ——— and ———. 1958. Hawaiian flowers and flowering trees. Charles E. Tuttle Co., Rutland, Vt., 158 pp.
1489. Kükenthal, G. 1909. Cyperaceae–Caricoideae. Pflanzenr. IV. 20 (Heft 38): 1–824.
1490. ———. 1920. Cyperaceae novae. V. Repert. Spec. Nov. Regni Veg. 16: 430–435.
1491. ———. 1925. Beiträge zur Cyperaceenflora von Mikronesien. Bot. Jahrb. Syst. 59: 2–10.
1492. ———. 1935–1936. Cyperaceae–Scirpoideae–*Cypereae*. Pflanzenr. IV. 20 (Heft 101): 1–671.
1493. ———. 1936. Cyperaceae: *in* Hochreutiner, B. P. G., Plantae Hochreutineranae, étude systématique et biologique des collections faites par l'auteur au cours de son voyage aux Indes Néerlandaises et autour du monde pendant les années 1903 à 1905. Fascicule IV. Candollea 6: 412–432.
1494. ———. 1940. Vorarbeiten zu einer Monographie der Rhynchosporoideae. VIII. Repert. Spec. Nov. Regni Veg. 48: 49–72.
1495. ———. 1942. Vorarbeiten zu einer Monographie der Rhynchosporoideae. XIII. Repert. Spec. Nov. Regni Veg. 51: 139–193.

1496. ———. 1949. Vorarbeiten zu einer Monographie der Rhynchosporideae. Bot. Jahrb. Syst. 74: 375–509.
1497. ———. 1950. Vorarbeiten zu einer Monographie der Rhynchosporideae. Bot. Jahrb. Syst. 75: 127–195.
1498. Kunth, C. S. 1829. Révision des graminées. Part 3. Librairie-Gide, Paris, pp. [579]–666.
1499. ———. 1833–1850. Enumeratio plantarum omnium hucusque cognitarum, secundum familias naturales disposita, adjectis characteribus, differentiis et synonymis . . . Vols. 1–5. J. G. Cottae, Stuttgart.
1500. Kuntze, O. 1891–1898. Revisio generum plantarum vascularium omnium atque cellularium multarum secundum leges nomenclaturae internationales cum enumeratione plantarum exoticarum in itinere mundi collectarum . . . Vols. 1–3. Arthur Felix, Leipzig.
1501. Ladd, E. J., and D. E. Yen (eds.). 1972. Makaha Valley historical project. Interim report No. 3. Pacific Anthropol. Rec. 18: 1–115.
1502. Laemmlen, F. F., and R. V. Bega. 1972. Decline of *ohia* and *koa* forests in Hawaii (Abstr.). Phytopathology 62: 770.
1503. ——— and ———. 1974. Hosts of *Armillaria mellea* in Hawaii. Pl. Dis. Reporter 58: 102–103.
1504. Lam, H. J. 1922. Notiz ueber *Vitex*. Bull. Jard. Bot. Buitenzorg, sér. III, 5: 175–178.
1505. ———. 1925. The Sapotaceae, Sarcospermaceae and Boerlagellaceae of the Dutch East Indies and surrounding countries (Malay Peninsula and Philippine Islands). Bull. Jard. Bot. Buitenzorg, sér. III, 7: 1–289.
1506. ———. 1938. Monograph of the genus *Nesoluma* (Sapotaceae). A primitive Polynesian endemic of supposed Antarctic origin. Occas. Pap. Bernice P. Bishop Mus. 14(9): 127–165.
1507. ———. 1941. Note on the Sapotaceae–Mimusopoideae in general and on the far-eastern *Manilkara*-allies in particular. Blumea 4: 323–358.
1508. ———. 1941. Some notes on the distribution of the Sapotaceae of the Pacific region. Proc. 6th Pacific Sci. Congr., California 4: 673–683.
1509. ———. 1942. A tentative list of wild Pacific Sapotaceae, except those from New Caledonia. Blumea 5: 1–46.
1510. ———. 1954. *Nesoluma* and *Planchonella* from the Hawaiian Islands (Sapotaceae). Occas. Pap. Bernice P. Bishop Mus. 21(10): 209–212.
1511. Lamb, S. H. 1936. The trees of the Kilauea–Mauna Loa section, Hawaii National Park. Hawaii Natl. Park Nat. Hist. Bull. 2: 1–32.
1512. ———. 1938. Wildlife problems in Hawaii National Park. Trans. 3rd North Amer. Wildlife Conf., Washington, D.C., pp. 597–602.
1513. ———. 1981. Native trees & shrubs of the Hawaiian Islands. Sunstone Press, Santa Fe, N. Mex., 160 pp.
1514. Lamberton, A. R. H. 1955. The anatomy of some woods utilized by the ancient Hawaiians. Master's thesis, Univ. Hawaii, Honolulu, 105 pp.
1515. Lammers, T. G., and C. E. Freeman. 1986. Ornithophily among the Hawaiian Lobelioideae (Campanulaceae) (Abstr.). Amer. J. Bot. 73: 772.

1516. ——— and ———. 1986. Ornithophily among the Hawaiian Lobelioideae (Campanulaceae): evidence from floral nectar sugar compositions. Amer. J. Bot. 73: 1613–1619.
1517. Lamoureux, C. H. 1961. Botanical observations on leeward Hawaiian atolls. Atoll Res. Bull. 79: 1–10.
1518. ———. 1961. Letter from Charles Lamoureux regarding plants in Paiko Lagoon—November 28, 1960. Elepaio 21: 51–52.
1519. ———. 1963. The flora and vegetation of Laysan Island. Atoll Res. Bull. 97: 1–14.
1520. ———. 1963. Field guide to the Mauna Kapu–Palikea Trail. Newslett. Hawaiian Bot. Soc. 2: 81–83.
1521. ———. 1963. George Campbell Munro, 1866–1963. Newslett. Hawaiian Bot. Soc. 2: 131–134.
1522. ———. 1963. Additional plants from the Midway Islands. Pacific Sci. 17: 374.
1523. ———. 1963. Vegetation of Laysan (Abstr.). Proc. Hawaiian Acad. Sci. 37: 22.
1524. ———. 1964. The Leeward Hawaiian Islands. Newslett. Hawaiian Bot. Soc. 3: 7–11.
1525. ———. 1968. The vascular plants of Kipahulu Valley, Maui: *in* Warner, R. E. (ed.), Scientific report of the Kipahulu Valley Expedition. Nature Conservancy, San Francisco, pp. 23–54.
1526. ———. 1969. Field guide for the Hawaiian Botanical Society foray—July 4, 1969. Newslett. Hawaiian Bot. Soc. 8: 26–27.
1527. ———. 1970. Plants recorded from Kahoolawe. Newslett. Hawaiian Bot. Soc. 9: 6–11.
1528. ———. 1971. Some botanical observations on *koa*. Newslett. Hawaiian Bot. Soc. 10: 1–7.
1529. ———. 1973. Conservation problems in Hawaii: *in* Costin, A. B., and R. H. Groves (eds.), Nature conservation in the Pacific. Australian Natl. Univ. Press, Canberra, pp. 315–319.
1530. ———. 1973. Phenology and growth of Hawaiian plants, a preliminary report. U.S. IBP Island Ecosystems IRP Techn. Rep. 24: 1–62.
1531. ———. 1973. Plants: *in* Armstrong, R. W. (ed.), Atlas of Hawaii. Univ. Press Hawaii, Honolulu, pp. 63–66.
1532. ———. 1975. Phenological patterns in relation to ecological zones on Hawaii (Abstr.). Proc. 13th Pacific Sci. Congr., Canada 1: 104–105.
1533. ———. 1976. Trailside plants of Hawaii's National Parks. Hawaii Nat. Hist. Assoc., Hawaii Volcanoes Natl. Park, 80 pp.
1534. ———. 1976. Endangered species in Hawaii. Dr. Lamoureux's response. Newslett. Hawaiian Bot. Soc. 15: 14–21.
1535. ———. 1976. Endangered plants in Hawaii Volcanoes National Park. Proc. 1st Conf. Nat. Sci., Hawaii Volcanoes Natl. Park, pp. 123–125.
1536. ———. 1976. Phenological studies in Hawaii Volcanoes National Park (Abstr.). Proc. 1st Conf. Nat. Sci., Hawaii Volcanoes Natl. Park, p. 127.

1537. ———. 1983. Plants: *in* Armstrong, R. W. (ed.), Atlas of Hawaii. 2nd ed. Univ. Hawaii Press, Honolulu, pp. 69–72.
1538. ———. 1985. Restoration of native ecosystems: *in* Stone, C. P., and J.M. Scott (eds.), Hawai'i's terrestrial ecosystems: preservation and management. Coop. Natl. Park Resources Stud. Unit, Univ. Hawaii, Honolulu, pp. 422–431.
1539. ——— and D. Carswell. 1963. Na Laau Hawaii Arboretum. Proc. 2nd Annual Conf., Conserv. Council for Hawaii, pp. 29–32.
1540. ——— and C. A. Corn. 1973. Progress report on genecological studies of *Metrosideros*. U.S. IBP Island Ecosystems IRP Techn. Rep. 21: 6.10.
1541. ——— and J. R. Porter. 1970. Report on establishment of 7 phenological observation stations. U.S. IBP Island Ecosystems IRP Techn. Rep. 1: 78.
1542. ——— and ———. 1972. Report of phenological and growth studies, 1971. U.S. IBP Island Ecosystems IRP Techn. Rep. 2: 61–70.
1543. ———, ———, and L. Matsunami. 1973. Progress report of phenological and growth studies, 1972. U.S. IBP Island Ecosystems IRP Techn. Rep. 21: 6.5–6.6.
1544. ——— and R. L. Stemmermann. 1976. Report of the Kī-Pahulu Bicentennial Expedition, June 26–29, 1976. Coop. Natl. Park Resources Stud. Unit, Hawaii, Techn. Rep. 11: 1–18.
1545. Landgraf, L. K. 1973. Mauna Kea and Mauna Loa silversword: alive and perpetuating. Bull. Pacific Trop. Bot. Gard. 3: 64–66.
1546. Langkavel, B. 1894. Flora und Fauna der Hawaiischen Inseln. Natur 43: 294–296.
1547. Lanner, R. M. 1964. Modifications in the growth habit of exotic trees in Hawaii. Proc. Soc. Amer. Foresters, Denver, pp. 36–37.
1548. ———. 1964. Adventitious rooting—a response to Hawaii's environment. U.S.D.A. Forest Serv. Res. Note PSW-54: 1–3.
1549. ———. 1965. Phenology of *Acacia koa* on Mauna Loa, Hawaii. U.S.D.A. Forest Serv. Res. Note PSW-89: 1–10.
1550. ———. 1966. Adventitious roots of *Eucalyptus robusta* in Hawaii. Pacific Sci. 20: 379–381.
1551. LaRosa, A. M. 1982. Response of *Passiflora mollissima* (HBK.) Bailey to experimental canopy removal: a simulation of natural gap formation in a closed-canopy *Cibotium* spp. forest in Ola'a Tract, Hawai'i. Proc. 4th Conf. Nat. Sci., Hawaii Volcanoes Natl. Park, pp. 118–134.
1552. ———. 1983. The biology and ecology of *Passiflora mollissima* in Hawai'i. Master's thesis, Univ. Hawaii, Honolulu, 270 pp.
1553. ———. 1984. The biology and ecology of *Passiflora mollissima* in Hawaii. Coop. Natl. Park Resources Stud. Unit, Hawaii, Techn. Rep. 50: 1–168.
1554. Larsen, N. P. 1946. Medical art in ancient Hawaii. Hawaiian Hist. Soc., 53rd Annual Rep. 1944: 27–44.
1555. Lassetter, J. S., and C. R. Gunn. 1979 [1980]. *Vicia menziesii* Sprengel

(Fabaceae) rediscovered: its taxonomic relationships. Pacific Sci. 33: 85–101.
1556. Lauener, L. A. 1966. Catalogue of the names published by Hector Léveillé: III. Notes Roy. Bot. Gard. Edinburgh 26: 333–346.
1557. ———. 1966. Catalogue of the names published by Hector Léveillé: IV. Notes Roy. Bot. Gard. Edinburgh 27: 1–10.
1558. ———. 1967. Catalogue of the names published by Hector Léveillé: V. Notes Roy. Bot. Gard. Edinburgh 27: 265–292.
1559. ———. 1970. Catalogue of the names published by Hector Léveillé: VI. Notes Roy. Bot. Gard. Edinburgh 30: 239–294.
1560. ———. 1972. Catalogue of the names published by Hector Léveillé: VII. Notes Roy. Bot. Gard. Edinburgh 31: 397–435.
1561. ———. 1976. Catalogue of the names published by Hector Léveillé: IX [Compositae]. Notes Roy. Bot. Gard. Edinburgh 34: 327–402.
1562. ———. 1977. Catalogue of the names published by Hector Léveillé: XI. Notes Roy. Bot. Gard. Edinburgh 35: 265–279.
1563. ———. 1978. Catalogue of the names published by Hector Léveillé: XII. Notes Roy. Bot. Gard. Edinburgh 37: 125–151.
1564. ———. 1980. Catalogue of the names published by Hector Léveillé: XIII. Notes Roy. Bot. Gard. Edinburgh 38: 453–485.
1565. ———. 1980. Faurie's Hawaiian types at the British Museum. Notes Roy. Bot. Gard. Edinburgh 38: 495–497.
1566. ———. 1982. Catalogue of the names published by Hector Léveillé: XIV. Notes Roy. Bot. Gard. Edinburgh 40: 157–203.
1567. ———. 1982. Catalogue of the names published by Hector Léveillé. Errata & Emendata: I. Notes Roy. Bot. Gard. Edinburgh 40: 204.
1568. ———. 1982. Catalogue of the names published by Hector Léveillé: XV. Notes Roy. Bot. Gard. Edinburgh 40: 345–358.
1569. ———. 1983. Catalogue of the names published by Hector Léveillé: XVI. Notes Roy. Bot. Gard. Edinburgh 40: 475–505.
1570. ———. 1983. Catalogue of the names published by Hector Léveillé. Errata & emendata: II. Notes Roy. Bot. Gard. Edinburgh 41: 181–188.
1571. ———. 1983. Catalogue of the names published by Hector Léveillé: Index. Notes Roy. Bot. Gard. Edinburgh 41: 339–393.
1572. Lay, K. K. 1949. A revision of the genus *Heliocarpus* L. Ann. Missouri Bot. Gard. 36: 507–541.
1573. LeBarron, R. K. n.d. A historical report . . . Hawaii's sandalwoods. Hawaii, Dept. Land Nat. Resources, Honolulu, 2 pp.
1574. ———. 1962. Eucalypts in Hawaii: a survey of practices and research programs. Pacific Southw. Forest Range Exp. Sta. Misc. Pap. 64: 1–22.
1575. ———. 1965. The use of eucalypts in Hawaii. Newslett. Hawaiian Bot. Soc. 4: 14–16.
1576. ———. 1979. The tree made by a committee. Amer. Forests 85(4): 38–41.

1577. Lecomte, H. 1916. Le genre *Korthalsella* et la tribu des Bifariées de van Tieghem. Bull. Mus. Hist. Nat. (Paris) 22: 260–267.
1578. Lee, B. K. H. 1971. Ecological and physiological studies of soil microfungi in Heeia mangrove swamp, Oahu, Hawaii. Ph.D. dissertation, Univ. Hawaii, Honolulu, 103 pp.
1579. ——— and G. E. Baker. 1972. Environment and the distribution of microfungi in a Hawaiian mangrove swamp. Pacific Sci. 26: 11–19.
1580. Lee, H. A. 1926. The common grasses in Hawaii in relation to mosaic or yellow stripe disease. Hawaiian Pl. Rec. 30: 270–278.
1581. Lee, M. A. B. 1979. Insect damage to plants in three successional plant communities. Ph.D. dissertation, Univ. Hawaii, Honolulu, 170 pp.
1582. ———. 1981. Insect damage to leaves of two varieties of *Metrosideros collina* subsp. *polymorpha*. Pacific Sci. 35: 89–92.
1583. Lee, R. K. 1980. Legends of the Hawaiian forest. Makapuʻu Press, Honolulu, 26 unnum. pp.
1584. Leenhouts, P. W. 1983. Notes on the extra-Australian species of *Dodonaea* (Sapindaceae). Blumea 28: 271–289.
1585. Leeper, J. R., and J. W. Beardsley. 1973. The bioecology of *Psylla uncatoides* in the Hawaii Volcanoes National Park and the *Acacia koaia* sanctuary. U.S. IBP Island Ecosystems IRP Techn. Rep. 23: 1–13.
1586. Leopold, L. B. 1951. Hawaiian climate: its relation to human and plant geography. Meteorological Monogr. 1(3): 1–6.
1587. Lessing, C. F. 1831. Synanthereae: *in* Chamisso, L. C. A. von, and D. F. L. von Schlechtendal, De plantis in expeditione speculatoria Romanzoffiana observatis. Linnaea 6: 83–170, 209–260, 501–528.
1588. ———. 1832. Synopsis generum Compositarum earumque dispositionis novae tentamen monographiis multarum capensium interjectis . . . Duncker & Humblot, Berlin, 473 pp.
1589. Léveillé, H. 1911. Plantae novae sandwicenses. Repert. Spec. Nov. Regni Veg. 10: 120–124.
1590. ———. 1911. Plantae novae sandwicenses. II. Repert. Spec. Nov. Regni Veg. 10: 149–157.
1591. ———. 1912. Decades plantarum novarum. LXXV–LXXIX. Repert. Spec. Nov. Regni Veg. 10: 369–378.
1592. ———. 1912. Decades plantarum novarum. LXXX–LXXXVI. Repert. Spec. Nov. Regni Veg. 10: 431–444.
1593. ———. 1912. Decades plantarum novarum. LXXXVII–LXXXVIII. Repert. Spec. Nov. Regni Veg. 10: 473–476.
1594. ———. 1912. Decades plantarum novarum. LXXXIX. Repert. Spec. Nov. Regni Veg. 11: 31–33.
1595. ———. 1912. Decades plantarum novarum. XC–XCII. Repert. Spec. Nov. Regni Veg. 11: 63–67.
1595a. ———. 1913. Decades plantarum novarum. CXII–CXVIII. Repert. Spec. Nov. Regni Veg. 12: 181–191.

1596. ———. 1913. Decades plantarum novarum. CXXVI. Repert. Spec. Nov. Regni Veg. 12: 505–507.
1597. ———. 1914. Revisio plantarum Hawaiensium. Repert. Spec. Nov. Regni Veg. 13: 422.
1598. Levin, G. A. 1986. Systematic foliar morphology of Phyllanthoideae (Euphorbiaceae). I. Conspectus. Ann. Missouri Bot. Gard. 73: 29–85.
1599. ———. 1986. Systematic foliar morphology of Phyllanthoideae (Euphorbiaceae). II. Phenetic analysis. Ann. Missouri Bot. Gard. 73: 86–98.
1600. ———. 1986. Systematic foliar morphology of Phyllanthoideae (Euphorbiaceae). III. Cladistic analysis. Syst. Bot. 11: 515–530.
1601. Lewis, W. H. 1974. Chromosomes and phylogeny of *Erythrina* (Fabaceae). Lloydia 37: 460–464.
1602. ——— and R. L. Oliver. 1969. *In* Chromosome numbers of phanerogams. 3. Ann. Missouri Bot. Gard. 56: 472–475.
1603. ——— and ———. 1974. Revision of *Richardia* (Rubiaceae). Brittonia 26: 271–301.
1604. Lewton, F. L. 1912. *Kokia*: a new genus of Hawaiian trees. Smithsonian Misc. Collect. 60(5): 1–4.
1605. Lewton-Brain, L. 1909. The Maui forest trouble. Hawaiian Pl. Rec. 1: 92–95.
1606. Lieth, H., J. H. Lieth, and A. Lieth. 1975. *Portulaca pilosa* ssp. *villosa*; ecotype of *P. pilosa* of great variability and adaptability. Newslett. Hawaiian Bot. Soc. 14: 23–25.
1607. Lim, E. K. S. 1967. Experimental studies on the evolution of Hawaiian species of *Bidens*. Ph.D. dissertation, Univ. Hawaii, Honolulu, 179 pp.
1608. Limpricht, W. 1928. Taccaceae. Pflanzenr. IV. 42 (Heft 92): 1–31.
1609. Lindgren, W. 1908. The water resources of Molokai. Hawaiian Forester Agric. 5: 191–195.
1610. Lindley, J. 1822. Observations on the natural group of plants called Pomaceae. Trans. Linn. Soc. London 13: 88–106.
1611. ———. 1830–1840. The genera and species of orchidaceous plants. Ridgways, Piccadilly, London, 553 pp. (Facsimile ed., 1963, A. Asher & Co., Amsterdam).
1612. ———. 1835. *Dracaena terminalis*. The Sandwich Islands tee-plant. Edwards's Bot. Reg. 21: *pl. 1749*.
1613. Linney, G. 1986. *Coccinia grandis* (L.) Voigt: a new cucurbitaceous weed in Hawai'i. Newslett. Hawaiian Bot. Soc. 25: 3–5.
1614. Little, E. L., Jr. 1969. Native trees of Hawaii. Amer. Forests 75(2): 16–17, 44–45.
1615. ———. 1969. Native trees of Hawaii. Elepaio 30: 11–15.
1616. Littlecott, L. C. 1969. Hawaii first. Amer. Forests 75(2): 12–15, 59–63.
1617. Lockerbie, L. 1950. Reviews. Selling, Olof H., "On the Late Quaternary history of the Hawaiian vegetation." J. Polynes. Soc. 59: 90–92.
1618. Loesener, T. 1897. Uber die geographische Verbreitung einiger Celastraceen. Bot. Jahrb. Syst. 24: 197–201.

1619. ———. 1897. Aquifoliaceae. Nat. Pflanzenfam. Nachtr. III. 5: 217–221.
1620. ———. 1901. Monographia Aquifoliacearum. Pars I. Nova Acta Acad. Caes. Leop.-Carol. German. Nat. Cur. 78: 1–598.
1621. ———. 1908. Monographia Aquifoliacearum. Pars II. Nova Acta Acad. Caes. Leop.-Carol. German. Nat. Cur. 89: 1–313.
1622. Loope, L. L. 1982. Population biology of woody plant species of Haleakala: a progress report. Proc. 4th Conf. Nat. Sci., Hawaii Volcanoes Natl. Park, p. 135.
1623. ——— and P. G. Scowcroft. 1985. Vegetation response within exclosures in Hawai'i: a review: in Stone, C. P., and J. M. Scott (eds.), Hawai'i's terrestrial ecosystems: preservation and management. Coop. Natl. Park Resources Stud. Unit, Univ. Hawaii, Honolulu, pp. 377–402.
1624. Löve, Á. 1966. IOPB chromosome number reports VIII. Taxon 15: 279–284.
1625. ———. 1967. IOPB chromosome number reports XIII. Taxon 16: 445–461.
1626. Low, J. S., and C. S. Judd. 1927. Hawaiian forest areas. Univ. Hawaii Agric. Stud. 1: 1–8.
1627. Lowrey, T. K. 1980. Biosystematic studies in Hawaiian *Tetramolopium* (Compositae: Asteraceae [*Astereae*]). Proc. 3rd Conf. Nat. Sci., Hawaii Volcanoes Natl. Park, pp. 229–233.
1628. ———. 1981. A biosystematic study of Hawaiian *Tetramolopium* (Compositae; *Astereae*). Ph.D. dissertation, Univ. California, Berkeley, 192 pp.
1629. ———. 1986. A biosystematic revision of Hawaiian *Tetramolopium* (Compositae: *Astereae*). Allertonia 4: 203–265.
1630. ——— and D. J. Crawford. 1983. Allozyme divergence and evolution in *Tetramolopium* (Compositae: *Astereae*) of the Hawaiian Islands (Abstr.). Amer. J. Bot. 70(5/2): 122.
1631. ——— and ———. 1985. Allozyme divergence and evolution in *Tetramolopium* (Compositae: *Astereae*) on the Hawaiian Islands. Syst. Bot. 10: 64–72.
1632. Lucas, G., and H. Synge. 1978. The IUCN plant red data book. International Union for Conservation of Nature and Natural Resources, Morges, Switzerland, 540 pp.
1633. Lucas, L. 1982. Plants of old Hawaii. Bess Press, Honolulu, 101 pp.
1634. Lucas, S. A. 1980. Consider the banana. Bull. Pacific Trop. Bot. Gard. 10: 58–62.
1635. ———. 1981. Recent introductions of ornamental value. Bull. Pacific Trop. Bot. Gard. 11: 8–13.
1636. ———. 1982. Garden collections: Palmae. Bull. Pacific Trop. Bot. Gard. 12: 79–92.
1637. ———. 1985. Garden collections: Aristolochiaceae. Bull. Pacific Trop. Bot. Gard. 15: 53–55.
1638. Luer, C. A. 1975. The native orchids of the United States and Canada

excluding Florida. The New York Botanical Garden, New York, 361 pp.

1639. Lycan, E. 1885. Fruits and their seasons in the Hawaiian Islands. Hawaiian Almanac and Annual for 1886: 49–50.
1640. Lydgate, J. M. 1881. Indigenous ornamental plants. Hawaiian Almanac and Annual for 1882: 25–28.
1641. ———. 1882–1883. Hawaiian woods and forest trees. Hawaiian Almanac and Annual for 1883: 33–35 (1882); 1884: 30–32 (1883).
1642. ———. 1910. The endemic character of the Hawaiian flora. Hawaiian Almanac and Annual for 1911: 53–58.
1643. ———. 1914. Sandalwood days. Hawaiian Almanac and Annual for 1915: 50–56.
1644. ———. 1914. A day in the Kauai forests. Mid-Pacific Mag. 7: 290–296.
1645. ———. 1918. Scientific treasure trove. Hawaiian Almanac and Annual for 1919: 60–64.
1646. ———. 1919–1921. Reminiscences of an amateur collector. Hawaiian Almanac and Annual for 1920: 120–126 (1919); 1921: 68–76 (1920); 1922: 61–67 (1921).
1647. Lyon, H. L. 1909. The forest disease on Maui. Hawaiian Pl. Rec. 1: 151–159.
1648. ———. 1910. Leguminous plants for Hawaiian fields. Hawaiian Pl. Rec. 3: 51–55.
1649. ———. 1917. The pigeon pea. An important food plant for Hawaii. Hawaiian Pl. Rec. 16: 402–410.
1650. ———. 1918. The forests of Hawaii. Hawaiian Pl. Rec. 18: 276–280.
1651. ———. 1919. A dangerous bindweed. Hawaiian Pl. Rec. 20: 248–249.
1652. ———. 1919. Some observations on the forest problems of Hawaii. Hawaiian Pl. Rec. 21: 289–300.
1653. ———. 1922. Hawaiian forests. Hawaiian Forester Agric. 19: 159–162.
1654. ———. 1922. Fig trees for Hawaiian forests. Hawaiian Pl. Rec. 26: 78–87, 148–159.
1655. ———. 1923. Forestry on Oahu. Hawaiian Pl. Rec. 27: 283–310.
1656. ———. 1926. Exotic trees in Hawaii. Hawaiian Pl. Rec. 30: 255–258.
1657. ———. 1926. Exotic trees in Hawaii. Hawaiian Pl. Rec. 30: 349–353.
1658. ———. 1927. Exotic trees in Hawaii. Hawaiian Pl. Rec. 31: 163–169.
1659. ———. 1927. Botany in Hawaii (Abstr.). Proc. Hawaiian Acad. Sci. 2: 10–11.
1660. ———. 1929. Forestry on Oahu. Hawaiian Forester Agric. 26: 11–15.
1661. ———. 1929. Ten years in Hawaiian forestry. Hawaiian Pl. Rec. 33: 55–97.
1662. ———. 1930. The flora of Moanalua 100,000 years ago (Abstr.). Proc. Hawaiian Acad. Sci. 5: 6–7.
1663. ———. 1941. Polymorphic species in Hawaii (Abstr.). Proc. 6th Pacific Sci. Congr., California 4: 657.
1664. Lyons, A. B. 1896. Native plants of the Hawaiian Islands. Hawaiian Almanac and Annual for 1897: 55–70.

1665. ———. 1899. What a botanist may see in Honolulu. Hawaiian Almanac and Annual for 1900: 93–108.
1666. ———. 1907. Plant names scientific and popular. 2nd ed. Nelson, Baker & Co., Detroit, 630 pp.
1667. Mabberley, D. J. 1974. The pachycaul lobelias of Africa and St. Helena. Kew Bull. 29: 535–584.
1668. ———. 1975. The giant lobelias: pachycauly, biogeography, ornithophily and continental drift. New Phytologist 74: 365–374.
1669. ———. 1975. The giant lobelias: toxicity, inflorescence and tree-building in the Campanulaceae. New Phytologist 75: 289–295.
1670. MacBryde, B. 1983. Endangered and threatened wildlife and plants; supplement to review of plant taxa for listing; proposed rule. Fed. Reg. 48: 53639–53670.
1671. MacCaughey, V. 1910. The mountain trail from Wahiawa to Kahana. Hawaiian Forester Agric. 7: 352–358.
1672. ———. 1912. The "air-plant," *Bryophyllum*. An interesting plant of Hawaii. Hawaiian Forester Agric. 9: 10–16.
1673. ———. 1912. The *kukui* forests of Hawaii. Paradise Pacific 25(1): 21–22.
1674. ———. 1912. The *kukui* forests of Hawaii. Paradise Pacific 25(6): 14–16.
1675. ———. 1915. Some common woody plants of the Oahu lowlands. Hawaiian Forester Agric. 12: 290–292.
1676. ———. 1916. Vegetation of the Hawaiian summit bogs. Amer. Bot. (Binghamton) 22: 45–52.
1677. ———. 1916. The wild flowers of Hawaii. Amer. Bot. (Binghamton) 22: 97–105, 131–135.
1678. ———. 1916. The economic woods of Hawaii. Forest. Quart. 14: 696–716.
1679. ———. 1916. The *hau*. An interesting tree of Hawaii. Hawaiian Almanac and Annual for 1917: 108–112.
1680. ———. 1916. An annotated reference list of the more common trees and shrubs of the Konahuanui region. Hawaiian Forester Agric. 13: 28–34.
1681. ———. 1916. Precinctive flora of the Waianae Mountains, Oahu. An annotated reference list of seventy species and varieties. Hawaiian Forester Agric. 13: 85–89.
1682. ———. 1916. Passifloras in the Hawaiian Islands. J. Bot. 54: 363–368.
1683. ———. 1916. The orchids of Hawaii. Pl. World 19: 350–355.
1684. ———. 1916. The forests of the Hawaiian Islands. Pl. World 20: 162–166.
1685. ———. 1916. The genus *Eugenia* in the Hawaiian Islands. Torreya 16: 260–267.
1686. ———. 1917. The Oahu rain forest. Amer. Forestry 23: 276–278.
1687. ———. 1917. *Gunnera petaloidea* Gaud., a remarkable plant of the Hawaiian Islands. Amer. J. Bot. 4: 33–39.
1688. ———. 1917. The phytogeography of Manoa Valley, Hawaiian Islands. Amer. J. Bot. 4: 561–603.
1689. ———. 1917. A survey of the Hawaiian land flora. Bot. Gaz. (Crawfordsville) 64: 89–114.

1690. ———. 1917. Vegetation of Hawaiian lava flows. Bot. Gaz. (Crawfordsville) 64: 386–420.
1691. ———. 1917. An annotated list of the forest trees of the Hawaiian Archipelago. Bull. Torrey Bot. Club 44: 145–157.
1692. ———. 1917. The guavas of the Hawaiian Islands. Bull. Torrey Bot. Club 44: 513–524.
1693. ———. 1917. A rare fruit tree of Hawaii. Hawaiian Forester Agric. 14: 97–98.
1694. ———. 1917. The Hawaiian taro as food. Hawaiian Forester Agric. 14: 265–268.
1695. ———. 1917. The mangrove in the Hawaiian Islands. Hawaiian Forester Agric. 14: 361–366.
1696. ———. 1917. American explorers of Hawaii. Mid-Pacific Mag. 14: 281–285.
1697. ———. 1917. The food plants of the ancient Hawaiians. Sci. Monthly 4: 75–80.
1698. ———. 1917. The black persimmon or guaya-bota. *Diospyros ebenaster* Retz. A rare fruit tree of Hawaii. Trop. Agr. (Ceylon) 49: 106–107.
1699. ———. 1918. The Hawaiian *lehua*. Amer. Forestry 24: 409–418.
1700. ———. 1918. An endemic begonia of Hawaii. Bot. Gaz. (Crawfordsville) 66: 273–275.
1701. ———. 1918. The strand flora of the Hawaiian Archipelago—I. Geographical relations, origin, and composition. Bull. Torrey Bot. Club 45: 259–277.
1702. ———. 1918. The strand flora of the Hawaiian Archipelago—II. Ecological relations. Bull. Torrey Bot. Club 45: 483–502.
1703. ———. 1918. The Hawaiian *kamani* (*Calophyllum inophyllum* L.). Hawaiian Forester Agric. 15: 69–73.
1704. ———. 1918. The native bananas of the Hawaiian Islands. Pl. World 21: 1–12.
1705. ———. 1918. The genus *Morinda* in the Hawaiian flora. Pl. World 21: 209–214.
1706. ———. 1918. The endemic palms of Hawaii: *Pritchardia*. Pl. World 21: 317–328.
1707. ———. 1918. The *olona*, Hawaii's unexcelled fiber-plant. Science 48: 236–238.
1708. ———. 1918. The Hawaiian Violaceae. Torreya 18: 1–11.
1709. ———. 1918. The Hawaiian sumach. *Neneleau*; *Rhus semialata* var. *sandwicensis* Engler. Torreya 18: 183–188.
1710. ———. 1918–1919. History of botanical exploration in Hawaii. Hawaiian Forester Agric. 15: 388–396, 417–429, 508–510 (1918); 16: 25–28, 49–54 (1919).
1711. ———. 1919. Native and alien bananas of the Hawaiian Islands. Mid-Pacific Mag. 18: 454–459.
1712. ———. 1920. Hawaii's tapestry forests. Bot. Gaz. (Crawfordsville) 70: 137–147.

1713. ——— and J. S. Emerson. 1913. The *kalo* in Hawaii. (I). Hawaiian Forester Agric. 10: 186–193.
1714. ——— and ———. 1913. The *kalo* in Hawaii (II). Hawaiian Forester Agric. 10: 225–231.
1715. ——— and ———. 1913. The *kalo* in Hawaii. III. Hawaiian Forester Agric. 10: 280–288.
1716. ——— and ———. 1913. The *kalo* in Hawaii (IV). Hawaiian Forester Agric. 10: 315–323.
1717. ——— and ———. 1913. The *kalo* in Hawaii (V). Hawaiian Forester Agric. 10: 349–358.
1718. ——— and ———. 1913. The *kalo* in Hawaii (VI). Hawaiian Forester Agric. 10: 371–375.
1719. ——— and ———. 1914. The *kalo* in Hawaii (VII). Hawaiian Forester Agric. 11: 17–23.
1720. ——— and ———. 1914. The *kalo* in Hawaii (VIII). Hawaiian Forester Agric. 11: 44–51.
1721. ——— and ———. 1914. The *kalo* in Hawaii (IX). Hawaiian Forester Agric. 11: 111–123.
1722. ——— and ———. 1914. The *kalo* in Hawaii (conclusion). Hawaiian Forester Agric. 11: 201–204.
1723. ——— and ———. 1914. A revised list of Hawaiian varietal names for *kalo*. Hawaiian Forester Agric. 11: 338–341.
1724. ——— and W. Weinrich. 1918. Sisal in the Hawaiian Islands. Hawaiian Forester Agric. 15: 42–48.
1725. ——— and ———. 1918. Sisal in the Hawaiian Islands. Trop. Agr. (Ceylon) 50: 93–98.
1726. MacDaniels, L. H. 1947. A study of the feʻi banana and its distribution with reference to Polynesian migrations. Bernice P. Bishop Mus. Bull. 190: 1–56.
1727. MacIntyre, D. 1905. Cultivation of the mango in Hawaii. Hawaiian Forester Agric. 11: 116–123.
1728. Macneil, J. D., Jr., L. K. Croft, and D. E. Hemmes. 1976. Puʻu-Kohola Heiau National Historic Site plant survey. Proc. 1st Conf. Nat. Sci., Hawaii Volcanoes Natl. Park, pp. 131–133.
1729. ——— and D. E. Hemmes. 1977. Puukohola Heiau National Historic Site plant survey. Coop. Natl. Park Resources Stud. Unit, Hawaii, Techn. Rep. 15: 1–36.
1730. Maka, J. E. 1973. A mathematical approach to defining spatially recurring species groups in a montane rain forest on Mauna Loa, Hawaii. Master's thesis, Univ. Hawaii, Honolulu, 138 pp.
1731. ———. 1973. A mathematical approach to defining spatially recurring species groups in a montane rain forest on Mauna Loa, Hawaii. U.S. IBP Island Ecosystems IRP Techn. Rep. 31: 1–112.
1732. Mäkinen, Y. 1968. Havaijin kasvistosta ja kasvillisuudesta [On the flora and vegetation of the Hawaiian Islands]. Eripainos. Luonnon Tutkija 72(3): 65–81.

1733. Makuchan, E. M. 1981. The *Gunnera/Nostoc* symbiosis: an ultrastructural comparison of aging glands and nodules. Ph.D. dissertation, Univ. Hawaii, Honolulu, 194 pp.
1734. Malcolm, F. B. 1960. Factors influencing an expanded sawmilling industry for Hawaii. U.S.D.A. Forest Serv. Forest Products Lab. Rep. 2190: 1–22.
1735. Mangelsdorf, A. J. 1950. Sugar-cane—as seen from Hawaii. Econ. Bot. 4: 150–176.
1736. Manitz, H. 1983 [1984]. Die cytologie der Convolvulaceae und Cuscutaceae I. Zusammenstellung der bekannten Chromosomenzahlen. Wiss. Z. Friedrich-Schiller-Univ. Jena 32: 915–944.
1737. Mann, H. 1866. Revision of the genus *Schiedea*, and of the Hawaiian Rutaceae. Proc. Boston Soc. Nat. Hist. 10: 309–319.
1738. ———. 1866–1871. Flora of the Hawaiian Islands. Commun. Essex Inst. 5: 113–144 (1866), 161–192 (1867), 233–248 (1868); 6: 105–112 (1871).
1739. ———. 1867–1868. Enumeration of Hawaiian plants. Proc. Amer. Acad. Arts 7: 143–184 (1867), 185–235 (1868).
1740. ———. 1869. Statistics and geographical range of Hawaiian (Sandwich Islands) plants. J. Bot. 7: 171–183.
1741. ———. 1869. Notes on *Alsinidendron*, *Platydesma*, and *Brighamia*, new genera of Hawaiian plants; with an analysis of the Hawaiian flora. Mem. Boston Soc. Nat. Hist. 1: 529–541.
1742. Manning, A. 1986. Bishop and Dole on forest management. 'Elepaio 46: 163–165.
1743. Marchant, Y. Y., F. R. Ganders, C.-K. Wat, and G. H. N. Towers. 1984. Polyacetylenes in Hawaiian *Bidens*. Biochem. Syst. Ecol. 12: 167–178.
1744. Marcuse, A. 1894. Die Hawaiischen Inseln. R. Friedländer & Sohn, Berlin, 186 pp.
1745. Margolin, L. 1911. *Eucalyptus* culture in Hawaii. Hawaii, Board Agric. Forest., Forest. Bull. 1: 1–80.
1746. Markgraf, F. 1950. Neue Diagnosen. Mitt. Bot. Staatssamml. München 1: 26–31.
1747. Markin, G. P. 1984. Biological control of the noxious weed gorse *Ulex europaeus* L. A status report (Abstr.). Proc. 5th Conf. Nat. Sci., Hawaii Volcanoes Natl. Park, p. 77.
1748. Marsh, D. H. 1966. Microorganisms of the phyllosphere, with particular reference to fungi, occurring on the dominant plants of biogeoclimatic zones of the Hawaiian Islands. Master's thesis, Univ. Hawaii, Honolulu, 52 pp.
1749. Martelli, U. 1910. Enumerazione delle "Pandanaceae." Webbia 3: 307–327.
1750. ———. 1913. Enumerazione delle "Pandanaceae." II. Webbia 5: 1–105.
1751. ———. 1921. Recensione delle palme del Vecchio Mondo. Webbia 5: 1–70.
1752. ———. 1930. Two new varieties of *Pandanus odoratissimus* Linn. in the Hawaiian group. Univ. Calif. Publ. Bot. 12: 363–368.

1753. ———. 1933. La distribuzione geografica delle Pandanaceae. Atti Soc. Tosc. Sci. Nat. Pisa Mem. 43: 190–209.
1754. Marticorena, C., and O. Parra. 1975. Morfologia de los granos de polen de *Hesperomannia* Gray y Moquinia DC. (Compositae–*Mutisieae*). Estudio comparativo con generos afines. Gayana, Bot. 29: 1–22.
1755. Massal, E., and J. Barrau. 1955. Pacific subsistence crops . . . cassava. S. Pacific Commis. Quart. Bull. 5(4): 15–18.
1756. Matsuura, M., J. Y. Shigeta, and E. Y. Hosaka. 1956. Identifying native dry forest trees of Puuwaawaa. Univ. Hawaii Extens. Serv. Club Circ. 108: 1–32.
1757. Mattoon, W. R. 1936. Forest trees and forest regions of the United States. U.S.D.A. Misc. Publ. 217: 1–55.
1758. McBride, L. R. 1975. Practical folk medicine of Hawaii. Petroglyph Press, Hilo, Hawaii, 104 pp.
1759. McClelland, C. K. 1915. Grasses and forage plants of Hawaii. Hawaii Agric. Exp. Sta. Bull. 36: 1–43.
1760. ——— and C. A. Sahr. 1912. Cotton in Hawaii. Hawaii Agric. Exp. Sta. Press Bull. 34: 1–24.
1761. McClintock, E. 1982. Erythrinas cultivated in California. Allertonia 3: 139–154.
1762. McCormick, S. P., B. A. Bohm, and F. R. Ganders. 1984. Methylated chalcones from *Bidens torta*. Phytochemistry 23: 2400–2401.
1763. McCurrach, J. C. 1960. Palms of the world. Harper & Brothers, New York, 290 pp.
1764. McDonald, M. A. 1978. *Ka Lei*. The leis of Hawaii. Topgallant Publ. Co., Honolulu, 187 pp.
1765. McEldowney, G. A. 1930. Forestry on Oahu. Hawaiian Pl. Rec. 34: 267–287.
1766. McGeorge, W., and W. A. Anderson. 1912. *Euphorbia lorifolia*, a possible source of rubber and chicle. Hawaii Agric. Exp. Sta. Press Bull. 37: 1–16.
1767. McKenna, D. J. 1979. Biochemical markers in Hawaiian and extra-Hawaiian *Acacia* species. Master's thesis, Univ. Hawaii, Honolulu, 192 pp.
1768. McManus, R. E., R. F. Altevogt, and B. MacBryde. 1978. Endangered and threatened wildlife and plants; determination that 11 plant taxa are endangered species and 2 plant taxa are threatened species. Fed. Reg. 43: 17910–17916.
1769. McMillan, C., O. Zapata, and L. Escobar. 1980. Sulphated phenolic compounds in seagrasses. Aquatic Bot. 8: 267–278.
1770. ——— and S. C. Williams. 1980. Systematic implications of isozymes in *Halophila* section *Halophila*. Aquatic Bot. 9: 21–31.
1771. Mears, J. A. 1977. The nomenclature and type collections of the widespread taxa of *Alternanthera* (Amaranthaceae). Proc. Acad. Nat. Sci. Philadelphia 129: 1–21.
1772. Medeiros, A. C., Jr., L. L. Loope, and R. W. Hobdy. 1984. Remnant

native vegetation at a lowland site near Kihei, Maui. Proc. 5th Conf. Nat. Sci., Hawaii Volcanoes Natl. Park, pp. 78–82.
1773. ———, ———, and R. A. Holt. 1986. Status of native flowering plant species on the south slope of Haleakala, East Maui, Hawaii. Coop. Natl. Park Resources Stud. Unit, Hawaii, Techn. Rep. 59: 1–230.
1774. Meijer, W. 1973. Endangered plant life. Biol. Conservation 5: 163–167.
1775. Meisner, C. F. 1836–1843. Plantarum vascularium genera secundum ordines naturalis digesta eorumque differentiae et affinitates tabulis diagnosticis expositae. Libraria Weidmannia, Leipzig.
1776. ———. 1856. Polygonaceae. Prodr. 14(1): 1–185.
1777. ———. 1857. Thymelaeaceae. Prodr. 14(2): 493–605.
1778. Melchior, H. 1925. Violaceae. Nat. Pflanzenfam. ed. 2, 21: 329–377.
1779. Melville, R. 1965. Book reviews. Flora of Hawaii. Kew Bull. 19: 206.
1780. ———. 1981. Vicarious plant distributions and paleogeography of the Pacific region: *in* Nelson, G., and D. Rosen (eds.), Vicariance biogeography: a critique. Columbia Univ. Press, New York, pp. 238–274.
1781. Mensch, J. A., and G. W. Gillett. 1972. The experimental verification of natural hybridization between two taxa of Hawaiian *Bidens* (Asteraceae). Brittonia 24: 57–70.
1782. Menzel, M. Y., P. A. Fryxell, and F. D. Wilson. 1983. Relationships among New World species of *Hibiscus* section *Furcaria* (Malvaceae). Brittonia 35: 204–221.
1783. Menzies, A. 1907. An early ascent of Mauna Loa. Extract from "A. Menzies' M.S. journal, in Vancouver's voyage, 1790–1794." Hawaiian Almanac and Annual for 1908: 99–112.
1784. ———. 1908. Excursion to the mountains of Maui. Extract from M.S.—"A. Menzies' journal in Vancouver's voyage, 1790–1794." Hawaiian Almanac and Annual for 1909: 92–97.
1785. ———. 1909. Ascent of Mount Hualalai. Extract from A. Menzies' journal of Vancouver's voyage, 1790–1794. Hawaiian Almanac and Annual for 1910: 72–89.
1786. Merlin, M. D. 1976. Hawaiian forest plants. Oriental Publ. Co., Honolulu, 68 pp.
1787. ———. 1977. Hawaiian coastal plants and scenic shorelines. Oriental Publ. Co., Honolulu, 68 pp.
1788. Merrill, E. D. 1924. Bibliography of Polynesian botany. Bernice P. Bishop Mus. Bull. 13: 1–68.
1789. ———. 1941. Man's influence on the vegetation of Polynesia, with special reference to introduced species. Proc. 6th Pacific Sci. Congr., California 4: 629–639.
1790. ———. 1943. Emergency food plants and poisonous plants of the islands of the Pacific. U.S. War Dept. Techn. Manual 10–420: 1–149.
1791. ———. 1945. On the underground parts of *Tacca pinnatifida* J. R. & G. Forst. (1776) = *Tacca leontopetaloides* (Linn.) O. Kuntze. J. Arnold Arbor. 26: 85–92.

1792. ———. 1945. Plant life of the Pacific world. Macmillan Co., New York, 295 pp.
1793. ———. 1947. A botanical bibliography of the islands of the Pacific. Contr. U.S. Natl. Herb. 30: 1–322.
1794. ———. 1954. The botany of Cook's voyages. Chron. Bot. 14: 161–384.
1795. Metcalfe, C. R. 1935. The structure of some sandalwoods and their substitutes and of some other little known scented woods. Bull. Misc. Inform. 4: 165–195.
1796. Meurisse, M. G. 1898. Étude du genre *Santalum* L. Bull. Mens. Soc. Linn. Paris, sér. 2, 129: 1025–1027.
1797. Meyen, F. J. F. 1834. Reise um die Erde ausgeführt auf dem Königlich Preussischen Seehandlungs-Schiffe Prinzess Louise, commandirt von Capitain W. Wendt, in den Jahren 1830, 1831 und 1832. Part 2. Sander'schen Buchhandlung, Berlin, 413 pp.
1798. ———. 1843. Observationes botanicas in itinere circum terram institutas. Opus posthumum, sociorum Academiae curis suppletum. [Beiträge zur Botanik gesammelt auf einer Reise um die Erde. Nach dessen tode von den mitgliedern der Akademie fortgeführt und bearbeitet]. Nov. Actorum Acad. Caes. Leop.-Carol. Nat. Cur. 19, Suppl. 1: 1–512.
1799. ——— (Transl. M. Johnston). 1846. Outlines of the geography of plants: with particular enquiries concerning the native country, the culture, and the uses of the principal cultivated plants on which the prosperity of nations is based. The Ray Society, London, 422 pp.
1800. ——— (Transl. A. Jackson). 1981. A botanist's visit to Oahu in 1831, being the journal of Dr. F. J. F. Meyen's travels and observations about the island of Oahu. [Pultz, M. A., ed., An exerpt from "Reise um die Erde ausgeführt auf dem Königlich Preussischen Seehandlungs-Schiffe Prinzess Louise, commandirt von Capitain W. Wendt, in den Jahren 1830, 1831 und 1832"]. Press Pacifica, Ltd., Honolulu, 90 pp.
1801. Meyer, C. A. 1831. Cyperaceae novae. Zap. Imp. Akad. Nauk Fiz.-Mat. Otd. 1: 195–230; not seen.
1802. Meyer, E. 1850. Hortus regiomontanus seminifer. Ann. Sci. Nat. Bot. III. 14: 349–350.
1803. Meyrat, A. K. 1982. A morphometric analysis and taxonomic appraisal of the Hawaiian silversword *Argyroxiphium sandwicense* DC. (Asteraceae). Master's thesis, Univ. Hawaii, Honolulu, 113 pp.
1804. ———. 1982. A morphometric analysis and taxonomic appraisal of the Hawaiian silversword *Argyroxiphium sandwicense* DC. (Asteraceae). Coop. Natl. Park Resources Stud. Unit, Hawaii, Techn. Rep. 46: 1–58.
1805. ———, G. D. Carr, and C. W. Smith. 1983 [1984]. A morphometric analysis and taxonomic appraisal of the Hawaiian silversword *Argyroxiphium sandwicense* DC. (Asteraceae). Pacific Sci. 37: 211–225.
1806. Mez, C. 1902. Myrsinaceae. Pflanzenr. IV. 236 (Heft 9): 1–437.

1806a. Miers, J. 1867. On the Menispermaceae. Ann. Mag. Nat. Hist. ser. 3, 19: 19–29.
1807. Mill, S. W., W. L. Wagner, and D. R. Herbst. 1985. Bibliography of Otto and Isa Degeners' Hawaiian floras. Taxon 34: 229–259.
1808. Miller, C. D. 1927. Food values of poi, taro, and *limu*. Bernice P. Bishop Mus. Bull. 37: 1–25.
1809. ———. 1929. Food values of breadfruit, taro leaves, coconut, and sugar cane. Bernice P. Bishop Mus. Bull. 64: 1–23.
1810. ——— and K. Bazore. 1945. Fruits of Hawaii. Description, nutritive value, and use. Hawaii Agric. Exp. Sta. Bull. 96: 1–129.
1811. ———, ———, and M. Bartow. 1965. Fruits of Hawaii. Description, nutritive value, and use. 4th ed. Univ. Press Hawaii, Honolulu, 229 pp.
1812. ———, ———, and R. C. Robbins. 1936. Some fruits of Hawaii. Their composition, nutritive value and use. Hawaii Agric. Exp. Sta. Bull. 77: 1–133.
1813. ———, L. Louis, and K. Yanazawa. 1947. Vitamin values of foods in Hawaii. Hawaii Agric. Exp. Sta. Univ. Hawaii Techn. Bull. 6: 1–56.
1814. ———, W. Ross, and L. Louis. 1947. Hawaiian-grown vegetables. Proximate composition: calcium, phosphorus, total iron, available iron, and oxalate content. Hawaii Agric. Exp. Sta. Univ. Hawaii Techn. Bull. 5: 1–45.
1815. Millspaugh, C. F. 1916. Contributions to North American Euphorbiaceae—VI. Publ. Field Mus. Nat. Hist., Bot. Ser. 2: 401–420.
1816. Miquel, F. A. W. 1843. Systema piperacearum. Fascs. I–II. H. A. Kramers, Rotterdam, 575 pp.
1817. ———. 1843. Piperaceae: *in* Meyen, F. J. F., Observationes botanicas in itinere circum terram institutas . . . Nov. Actorum Acad. Caes. Leop.-Carol. Nat. Cur. 19, Suppl. 1: 483–495.
1818. ———. 1845. Animadversiones in Piperaceas Herbarii Hookeriani. London J. Bot. 4: 410–470.
1819. Mitchell, F. 1981. Mouflon sheep and Kau silversword. Notes Waimea Arbor. & Bot. Gard. 8(1): 6–7.
1820. ———. 1982. *Acacia koa* fad and *Vicia menziesii*. Notes Waimea Arbor. & Bot. Gard. 9(1): 6–7.
1821. Mitchell, N. J. 1980. Why are the *ohe* trees not reseeding? 'Elepaio 41: 5.
1822. Mitchell, R. 1977. An anthropological mecca: Waimea Arboretum (3): medicinal flora of the Gods. Notes Waimea Arbor. 4(1): 9–11.
1823. Mitchell, W. C., and B. M. Brennan. 1973. Progress report on insect interference in the reproductive cycle of community structure forming plants, particularly seed feeders. U.S. IBP Island Ecosystems IRP Techn. Rep. 21: 6.41.
1824. Miyoshi, M. 1927. Vegetation and natural monuments of the Hawaiian Islands. Home Dept. Japan, Tokyo, 39 pp.

1825. Mizushima, M. 1957. A revision of *Drymaria cordata* Willd. (Critical studies on Japanese plants 3). J. Jap. Bot. 32: 69–81.
1826. Moldenke, H. N. 1942. The known geographic distribution of the members of the Verbenaceae and Avicenniaceae. Publ. privately, New York, 104 pp.
1827. ———. 1942. An alphabetic list of invalid and incorrect scientific names proposed in the Verbenaceae and Avicenniaceae. Publ. privately, New York, 59 pp.
1828. ———. 1944. The known geographic distribution of the members of the Verbenaceae and Avicenniaceae. Supplement 2. Bot. Gaz. (Crawfordsville) 106: 158–164.
1829. ———. 1945. The known geographic distribution of the members of the Verbenaceae and Avicenniaceae. Supplement 3. Castanea 10: 35–46.
1830. ———. 1945. The known geographic distribution of the members of the Verbenaceae and Avicenniaceae: supplement 4. Amer. J. Bot. 32: 609–612.
1831. ———. 1947. The known geographic distribution of the members of the Verbenaceae and Avicenniaceae. Supplement 5. Bol. Soc. Venez. Ci. Nat. 11: 37–52.
1832. ———. 1948. The known geographic distribution of the members of the Verbenaceae, Avicenniaceae, Stilbaceae, and Symphoremaceae. Supplement 8. Castanea 13: 110–121.
1833. ———. 1948. The known geographic distribution of the members of the Verbenaceae, Avicenniaceae, Stilbaceae, and Symphoremaceae. Supplement 9. Phytologia 2: 477–483.
1834. ———. 1949. The known geographic distribution of the members of the Verbenaceae, Avicenniaceae, Stilbaceae, Symphoremaceae, and Eriocaulaceae. Publ. privately, New York, 215 pp.
1835. ———. 1950. The known geographic distribution of the members of the Verbenaceae, Avicenniaceae, Stilbaceae, Symphoremaceae, and Eriocaulaceae. Supplement 3. Phytologia 3: 304–307.
1836. ———. 1950. The known geographic distribution of the members of the Verbenaceae, Avicenniaceae, Stilbaceae, Symphoremaceae, and Eriocaulaceae. Supplement 4. Phytologia 3: 374–383.
1837. ———. 1953. The known geographic distribution of the members of the Verbenaceae, Avicenniaceae, Stilbaceae, Symphoremaceae, and Eriocaulaceae. Supplement 6. Phytologia 4: 184–200.
1838. ———. 1958. Materials toward a monograph of the genus *Vitex*. X. Phytologia 6: 129–192.
1839. ———. 1958. Materials toward a monograph of the genus *Vitex*. XI. Phytologia 6: 197–231.
1840. ———. 1958. Materials toward a monograph of the genus *Citharexylum*. II. Phytologia 6: 262–320.
1841. ———. 1959. Additional notes on the genus *Citharexylum*. I. Phytologia 7: 73–77.
1842. ———. 1959. A résumé of the Verbenaceae, Avicenniaceae, Stilbaceae,

Symphoremaceae, and Eriocaulaceae of the world as to valid taxa, geographic distribution and synonymy. Publ. privately, Yonkers, N.Y., 495 pp.

1843. ———. 1959. A résumé of the Verbenaceae, Avicenniaceae, Stilbaceae, Symphoremaceae, and Eriocaulaceae of the world as to valid taxa, geographic distribution and synonymy. Supplement I. Publ. privately, Yonkers, N.Y., 26 pp.

1844. ———. 1962. Materials toward a monograph of the genus *Verbena*. IV. Phytologia 8: 230–272.

1845. ———. 1962. Materials toward a monograph of the genus *Verbena*. VII. Phytologia 8: 395–453.

1846. ———. 1966. Additional notes on the genus *Citharexylum*. II. Phytologia 13: 277–304.

1847. ———. 1966. Novelties among the American Verbenaceae. Phytologia 14: 453–480.

1848. ———. 1968. Additional notes on the genus *Verbena*. VI. Phytologia 16: 87–106.

1849. ———. 1968. Additional notes on the genus *Vitex*. IX. Phytologia 17: 114–120.

1850. ———. 1971. A fifth summary of the Verbenaceae, Avicenniaceae, Stilbaceae, Dicrastylidaceae, Symphoremaceae, Nyctanthaceae, and Eriocaulaceae of the world as to valid taxa, geographic distribution, and synonymy. Vols. 1–2. Publ. privately, 974 pp.

1851. ———. 1972. Additional notes on the genus *Verbena*. XII. Phytologia 23: 257–303.

1852. ———. 1973. A fifth summary of the Verbenaceae, Avicenniaceae, Stilbaceae, Dicrastylidaceae, Symphoremaceae, Nyctanthaceae, and Eriocaulaceae of the world as to valid taxa, geographic distribution, and synonymy. Supplement 2. Phytologia 25: 225–245.

1853. ———. 1974. A fifth summary of the Verbenaceae, Avicenniaceae, Stilbaceae, Dicrastylidaceae, Symphoremaceae, Nyctanthaceae, and Eriocaulaceae of the world as to valid taxa, geographic distribution, and synonymy. Supplement 4. Phytologia 28: 425–466.

1854. ———. 1977. Additional notes on the genus *Verbena*. XXV. Phytologia 36: 216–250.

1855. ———. 1978. Additional notes on the genus *Citharexylum*. XII. Phytologia 40: 486–492.

1856. ———. 1982. A sixth summary of the Verbenaceae, Avicenniaceae, Stilbaceae, Chloanthaceae, Symphoremaceae, Nyctanthaceae, and Eriocaulaceae of the world as to valid taxa, geographic distribution, and synonymy. Supplement 1. Phytologia 50: 233–270.

1857. ———. 1982. Additional notes on the genus *Vitex*. XXXVI. Phytologia 52: 184–211.

1858. Moomaw, J. C., M. T. Nakamura, and G. D. Sherman. 1959. Aluminum in some Hawaiian plants. Pacific Sci. 13: 335–341.

1859. ——— and M. Takahashi. 1960. Vegetation on gibbsitic soils in Hawaii. J. Arnold Arbor. 41: 391–411.
1860. Moon, J. 1971. Living with nature in Hawaii or starfish on the toast. Petroglyph Press, Hilo, Hawaii, 137 pp. (Rev. ed., 1979).
1861. Mooney, H. A., and J. A. Drake (eds.). 1986. Ecology of biological invasions of North America and Hawaii. Ecol. Stud. 58: 1–321.
1862. Moore, H. E., Jr. 1963. An annotated checklist of cultivated palms. Principes 7: 119–182.
1863. ———. 1973. The major groups of palms and their distribution. Gentes Herbarum 11: 27–141.
1864. Moquin-Tandon, C. H. B. A. 1840. Chenopodearum monographica enumeratio. P.-J. Loss, Paris, 182 pp.
1864a.———. 1849. Phytolacceae. Prodr. 13: 1–40, 459–460.
1865. ———. 1849. Amaranthaceae. Prodr. 13: 231–424, 462–463.
1866. Morales, P. 1981. The rain forests of Hawaii. Pacific Disc. 34(5): 1–11.
1867. Moriarty, D. 1975. Native Hawaiian plants for tropical seaside landscaping. Bull. Pacific Trop. Bot. Gard. 5: 41–48.
1868. ———. 1976. A culinary guide to common tropical weeds. Bull. Pacific Trop. Bot. Gard. 6: 8–12.
1869. ———. 1976. Ethnobotany of taro. Bull. Pacific Trop. Bot. Gard. 6: 81–86.
1870. Morris, D. K. 1968. Summary of native plant propagation and reintroduction in Hawaii Volcanoes National Park. Newslett. Hawaiian Bot. Soc. 7: 25–27.
1871. Morris, P. C. 1931. Early records of the introduction of trees and plants in Hawaii. Friend 101: 253–255.
1872. Morrison, G. 1903. The flora of Hawaii. Fl. Life 1903: 157–159.
1873. Moseley, H. N. 1879. Notes by a naturalist on the *"Challenger,"* being an account of various observations made during the voyage of H.M.S. *"Challenger"* round the world, in the years 1872–1876, under the commands of Capt. Sir G. S. Nares, R.N., K.C.B., F.R.S., and Capt. F. T. Thomson, R.N. Macmillan & Co., London, 620 pp.
1874. Motooka, P. S., D. L. Plucknett, and D. F. Saiki. 1969. Weed problems of pastures and ranges in Hawaii: *in* Romanowski, R. R., Jr., D. L. Plucknett, and H. F. Clay (eds.), Weed control basic to agriculture development. Inst. Techn. Interchange, Univ. Hawaii, Honolulu, pp. 95–98.
1875. ———, D. F. Saiki, D. L. Plucknett, O. R. Younge, and R. E. Daehler. 1967. Aerial herbicidal control of Hawaii jungle vegetation. Hawaii Agric. Exp. Sta. Bull. 140: 1–19.
1876. Mountainspring, S. 1985. Status, research, and management needs of the native Hawaiian biota: a summary: *in* Stone, C. P., and J. M. Scott (eds.), Hawai'i's terrestrial ecosystems: preservation and management. Coop. Natl. Park Resources Stud. Unit, Univ. Hawaii, Honolulu, pp. 142–145.
1877. Mueller-Dombois, D. 1966. Topographic vegetation profiles of Hawaiian

volcanoes (Abstr.): *in* Biotic communities of the volcanic areas of the Pacific. Proc. 11th Pacific Sci. Congr., Tokyo 5: 2.

1878. ———. 1966. The vegetation map and vegetation profiles: *in* Doty, M. S., and D. Mueller-Dombois, Atlas for bioecology studies in Hawaii Volcanoes National Park. Univ. Hawaii, Hawaii Bot. Sci. Pap. 2: 391–441.

1879. ———. 1967. Ecological relations in the alpine and subalpine vegetation on Mauna Loa, Hawaii. J. Indian Bot. Soc. 46: 403–411.

1880. ———. 1970. Procedure for sampling the *koa–ohia*–tree fern forest in Kilauea Forest Reserve, Hawaii (upper montane rain forest). U.S. IBP Island Ecosystems IRP Techn. Rep. 1: 57–77.

1881. ———. 1971. Planned utilization of the lowland tropical forests. Nature and Resources 7: 18–22.

1882. ———. 1972. A non-adapted vegetation interferes with soil water removal in a tropical rain forest area in Hawaii. U.S. IBP Island Ecosystems IRP Techn. Rep. 4: 1–25.

1883. ———. 1973. A non-adapted vegetation interferes with water removal in a tropical rain forest area in Hawaii. Trop. Ecol. 14: 1–18.

1884. ———. 1973. Some aspects of island ecosystems analysis. (A preliminary conceptual synthesis). U.S. IBP Island Ecosystems IRP Techn. Rep. 19: 1–26.

1885. ———. 1973. Spatial distribution of island biota. U.S. IBP Island Ecosystems IRP Techn. Rep. 21: 2.1–2.7.

1886. ———. 1973. Studies of plant to plant interactions: studies of distributional dynamics. U.S. IBP Island Ecosystems IRP Techn. Rep. 21: 6.4.

1887. ———. 1973. Natural area system development for the Pacific region, a concept and symposium. U.S. IBP Island Ecosystems IRP Techn. Rep. 26: 1–55.

1888. ———. 1974. The *ohia* dieback problem in Hawaii. A proposal for integrated research. Coop. Natl. Park Resources Stud. Unit, Hawaii, Techn. Rep. 3: 1–35.

1889. ———. 1974. Monographs on vegetation [Review of "The vegetation of North and Central America and of the Hawaiian Islands" by R. Knapp]. Science 149: 1083–1084.

1890. ———. 1975. *Ohia* rain forest study. Coop. Natl. Park Resources Stud. Unit, Hawaii, Progr. Rep. 1: 1–25.

1891. ———. 1975. The Mauna Loa transect study of the Hawaii IBP (Abstr.). Proc. 13th Pacific Sci. Congr., Canada 1: 107.

1892. ———. 1975. Some aspects of island ecosystem analysis: *in* Golley, F. B., and E. Medina (eds.), Tropical ecological systems. Trends in terrestrial and aquatic research. Springer-Verlag, Berlin, Ecol. Stud. Ser. 11: 353–366.

1893. ———. 1975. Integrated island ecosystem ecology in Hawaii. Introductory survey. U.S. IBP Island Ecosystems IRP Techn. Rep. 54: 1–46.

1894. ———. 1976. The major vegetation types and ecological zones in Hawaii

Volcanoes National Park and their application to park management and research. Proc. 1st Conf. Nat. Sci., Hawaii Volcanoes Natl. Park, pp. 149–161.

1895. ———. 1977. Integrierung von Tier- und Pflanzensoziologie an der Ostflanke des Mauna Loa, Insel Hawaii: *in* Tüxen, R. (ed.), Vegetation und Fauna. Proc. Int. Symp. Int. Soc. Pl. Geogr. Ecol. J. Cramer, Vaduz, pp. 451–463.

1896. ———. 1978. Hawaii IBP synthesis: 2. The Mauna Loa transect analysis. Proc. 2nd Conf. Nat. Sci., Hawaii Volcanoes Natl. Park, pp. 222–230.

1897. ———. 1978. Hawaii IBP synthesis: 8. Island ecosystems: what is unique about their ecology? Proc. 2nd Conf. Nat. Sci., Hawaii Volcanoes Natl. Park, pp. 231–234.

1898. ———. 1979. Aspekte der Sukzessionsforschung auf der Insel Hawaii: *in* Tüxen, R. (ed.), Vegetationsentwicklung (Syndynamik). Proc. Int. Symp. Int. Soc. Pl. Geogr. Ecol. J. Cramer, Vaduz, pp. 491–500.

1899. ———. 1979. Succession following goat removal in Hawaii Volcanoes National Park: *in* Linn, R. M. (ed.), Proc. 1st Conf. Sci. Res. Natl. Parks. Natl. Park Serv. Trans. Proc. Ser. 5, 2: 1149–1154.

1900. ———. 1980. The 'ōhi'a dieback phenomenon in the Hawaiian rain forest: *in* Cairns, J., Jr. (ed.), The recovery process in damaged ecosystems. Ann Arbor Science, Ann Arbor, Mich., pp. 153–161.

1901. ———. 1980. Spatial variation and vegetation dynamics in the coastal lowland ecosystem, Hawaii Volcanoes National Park. Proc. 3rd Conf. Nat. Sci., Hawaii Volcanoes Natl. Park, pp. 235–247.

1902. ———. 1981. Fire in tropical ecosystems: *in* Mooney, H. A., T. M. Bonnicksen, N. L. Christensen, J. E. Lotan, and W. A. Reiners (eds.), Fire regimes and ecosystem properties. U.S.D.A. Forest Serv. Gen. Techn. Rep. WO-26: 137–176.

1903. ———. 1981. Spatial variation and succession in tropical island rain forests: a progress report. Univ. Hawaii, Hawaii Bot. Sci. Pap. 41: 1–93.

1904. ———. 1981. Vegetation dynamics in a coastal grassland of Hawaii. Vegetatio 46: 131–140.

1905. ———. 1982. Canopy dieback in indigenous forests of Pacific Islands: Hawaii, Papua New Guinea and New Zealand. Newslett. Hawaiian Bot. Soc. 21: 2–8.

1906. ———. 1982. Island ecosystem stability and *Metrosideros* dieback. Proc. 4th Conf. Nat. Sci., Hawaii Volcanoes Natl. Park, pp. 138–146.

1907. ———. 1983. Stand-level dieback in New Zealand forests and the theory of cohort senescence. Newslett. Hawaiian Bot. Soc. 22: 33–42.

1908. ———. 1983 [1984]. Canopy dieback and dynamic processes in Pacific forests. Introductory statement. Pacific Sci. 37: 313–316.

1909. ———. 1983 [1984]. Canopy dieback and successional processes in Pacific forests. Pacific Sci. 37: 317–325.

1910. ———. 1983 [1984]. Canopy dieback and dynamic processes in Pacific forests. Concluding synthesis. Pacific Sci. 37: 483–489.
1911. ———. 1983. Population death in Hawaiian plant communities: a causal theory and its successional significance. Tuexenia 3: 117–130.
1912. ———. 1984. Zum Baumgruppensterben in pazifischen Inselwäldern. Phytocoenologia 12: 1–8.
1913. ———. 1984. 'Ōhi'a dieback in Hawai'i: 1984 synthesis and evaluation. Univ. Hawaii, Hawaii Bot. Sci. Pap. 45: 1–44.
1914. ———. 1985. The biological resource value of native forest in Hawaii with special reference to the tropical lowland rainforest at Kalapana. 'Elepaio 45: 95–101.
1915. ———. 1985. 'Ohi'a dieback and protection management of the Hawaiian rain forest: *in* Stone, C. P., and J. M. Scott (eds.), Hawai'i's terrestrial ecosystems: preservation and management. Coop. Natl. Park Resources Stud. Unit, Univ. Hawaii, Honolulu, pp. 403–421.
1916. ———. 1985. 'Ōhi'a dieback in Hawaii: 1984 synthesis and evaluation. Pacific Sci. 39: 150–170.
1917. ———. 1986. Perspectives for an etiology of stand-level dieback. Ann. Rev. Ecol. Syst. 17: 221–243.
1918. ——— and K. W. Bridges. 1975. Integrated island ecosystem ecology in Hawaii. Spatial distribution of island biota. Introduction. U.S. IBP Island Ecosystems IRP Techn. Rep. 66: 1–52.
1919. ———, ———, and H. L. Carson (eds.). 1981. Island ecosystems. Biological organization in selected Hawaiian communities. US/IBP Synthesis Series 15. Hutchinson Ross Publ. Co., Woods Hole, Mass., 583 pp.
1920. ———, J. E. Canfield, R. A. Holt, and G. P. Buelow. 1983. Tree-group death in North American and Hawaiian forests: a pathological problem or a new problem for vegetation ecology? Phytocoenologia 11: 117–137.
1921. ———, R. G. Cooray, and J. Craine. 1972. Kilauea rain forest study. U.S. IBP Island Ecosystems IRP Techn. Rep. 2: 50–60.
1922. ——— and F. R. Fosberg. 1974. Vegetation map of Hawaii Volcanoes National Park (at 1:52,000). Coop. Natl. Park Resources Stud. Unit, Hawaii, Techn. Rep. 4: 1–44.
1923. ——— and W. C. Gagné. 1975. Hawaiian Islands: identification of principal natural terrestrial ecosystems (Abstr.). Proc. 13th Pacific Sci. Congr., Canada 1: 107–108.
1924. ———, J. D. Jacobi, R. G. Cooray, and N. Balakrishnan. 1977. *Ohia* rain forest study: ecological investigations of the *ohia* dieback problem in Hawaii. Coop. Natl. Park Resources Stud. Unit, Hawaii, Techn. Rep. 20: 1–117.
1925. ———, ———, ———, and ———. 1980. 'Ōhi'a rain forest study: ecological investigations of the 'ōhi'a dieback problem in Hawaii. Revised ed. Hawaii Agric. Exp. Sta. Misc. Publ. 183: 1–64.
1926. ——— and V. J. Krajina. 1968. Comparison of east-flank vegetations on

Mauna Loa and Mauna Kea, Hawaii: *in* Misra, R., and B. Gopal, Proceedings of the symposium on recent advances in tropical ecology. Int. Soc. Trop. Ecol., Varanasi, India, pp. 508–520.

1927. ——— and C. H. Lamoureux. 1966. Kipukas and soil relations: *in* Doty, M. S., and D. Mueller-Dombois, Atlas for bioecology studies in Hawaii Volcanoes National Park. Univ. Hawaii, Hawaii Bot. Sci. Pap. 2: 285–314.

1928. ——— and ———. 1967. Soil-vegetation relationships in Hawaiian kipukas. Pacific Sci. 21: 286–299.

1929. ——— and G. A. Smathers. 1975. Sukzession nach einem Vulkanausbruch auf der Insel Hawaii: *in* Schmidt, W. (ed.), Sukzessionsforschung. Proc. Int. Symp. Int. Soc. Pl. Geogr. Ecol. J. Cramer, Vaduz, pp. 159–188.

1930. ——— and G. Spatz. 1972. Mauna Loa transect study: gradient analysis of vascular plant communities. U.S. IBP Island Ecosystems IRP Techn. Rep. 2: 31–41.

1931. ——— and ———. 1972. Study on the influence of introduced large herbivores on the vegetation in Hawaii Volcanoes National Park. U.S. IBP Island Ecosystems IRP Techn. Rep. 2: 42–49.

1932. ——— and ———. 1972. The influence of feral goats on the lowland vegetation in Hawaii Volcanoes National Park. U.S. IBP Island Ecosystems IRP Techn. Rep. 13: 1–46.

1933. ——— and ———. 1972. The influence of SO_2 fuming on the vegetation surrounding the Kahe power plant on Oahu, Hawaii. U.S. IBP Island Ecosystems IRP Techn. Rep. 14: 1–12.

1934. ——— and ———. 1973. Vegetation-environment correlation studies. U.S. IBP Island Ecosystems IRP Techn. Rep. 21: 6.1–6.3.

1935. ——— and ———. 1973. Life history studies of important plants: angiosperms. U.S. IBP Island Ecosystems IRP Techn. Rep. 21: 6.7.

1936. ——— and ———. 1975. Application of the relevé method to insular tropical vegetation for an environmental impact study. Phytocoenologia 2: 417–429.

1937. ——— and ———. 1975. The influence of feral goats on the lowland vegetation in Hawaii Volcanoes National Park. Phytocoenologia 3: 1–29.

1938. ———, P. M. Vitousek, and K. W. Bridges. 1984. Canopy dieback and ecosystem processes in Pacific forests: a progress report and research proposal. Univ. Hawaii, Hawaii Bot. Sci. Pap. 44: 1–100.

1939. Mueller, F. J. H. von. 1867. Epacrideae. Fragm. Phytogr. Australiae 6: 29–76.

1940. Mueller, K. 1857. Myrtaceae. Walpers' Ann. Bot. Syst. 4: 821–854.

1941. Muir, F. 1921. The origin of the Hawaiian flora and fauna. Proc. 1st Pan-Pacific Sci. Conf., Honolulu, pp. 143–146.

1942. Mull, M. E. 1975. Comments on silversword planting project. Elepaio 36: 45–47.

1943. ———. 1978. Additional details on Hawaiian vetch. 'Elepaio 39: 14.

1944. ———. 1979. Protection for *Vicia* urged by Hawaii Audubon. 'Elepaio 39: 87–88.
1945. Müller Argoviensis, J. 1863. Euphorbiaceae. Vorläufige Mittheilungen aus dem für De Candolle's Prodromus bestimmten Manuscript über diese Familie. Linnaea 32: 1–126.
1946. ———. 1865. Euphorbiaceae. Verläufige Mittheilungen aus dem für De Candolle's Prodromus bestimmten Manuscript über diese Familie. Linnaea 34: 1–224.
1947. ———. 1866. Euphorbiaceae. Prodr. 15(2): 189–1286.
1948. Mullins, J. G. 1974. Distance was big factor in Hawaii's ecology. Honolulu 9(5): 12, 14, 16, 18.
1949. Munro, G. C. 1922. Forest covers. Hawaiian Forester Agric. 19: 45–46.
1950. ———. 1929. Windbreaks for wind eroded lands. Hawaiian Forester Agric. 26: 124–125.
1951. ———. 1930. Myriad-nested Laysan. Asia 30: 686–689.
1952. ———. 1932. The rotation and distribution of plants (Abstr.). Proc. Hawaiian Acad. Sci. 7: 22–23.
1953. ———. 1933. Preserving the rare plants of Hawaii (Abstr.). Proc. Hawaiian Acad. Sci. 8: 26–27.
1954. ———. 1941. Birds of Hawaii and adventures in bird study. Bird islands off the coast of Oahu. Elepaio 1: 46–49.
1955. ———. 1944. My first bird walks in Hawaii. Elepaio 5: 6–7, 13–15.
1956. ———. 1944. Changes in the plant life on Hawaii. Hawaiian Naturalist 1: 7–8, 10–11.
1957. ———. 1946. Laysan Island in 1891. Elepaio 6: 51–52, 60–61, 66–69.
1958. ———. 1947. Effect of a tidal wave on some sea bird nesting islands. Elepaio 7: 43–44, 51–52.
1959. ———. 1948. Hawaiian endemic flowering plants suitable for culture in gardens. Elepaio 9: 14–15, 16–18, 23–25.
1960. ———. 1949. Some seeding plants detrimental to birds. Elepaio 9: 50–51.
1961. ———. 1950. New Zealand teatree. Elepaio 11: 14.
1962. ———. 1950. Notes on an article "Birds of Moku Manu and Manana islands off Oahu, Hawaii" by Frank Richardson and Harvey I. Fisher. Elepaio 11: 22–23.
1963. ———. 1951. Leahi Native Garden. Elepaio 11: 37–38.
1964. ———. 1951. Leahi Native Garden. Elepaio 12: 9–11.
1965. ———. 1952. Revisiting the island of Lanai in 1952. Elepaio 12: 62–64.
1966. ———. 1952. Attempts to save the shoreside and dryland plants of Hawaii. Elepaio 13: 1–5.
1967. ———. 1952. *Na Laau Hawaii*. Elepaio 13: 39–43.
1968. ———. 1953. Waahila Hawaiian Garden. Elepaio 13: 72–74.
1969. ———. 1953. *Na Laau Hawaii*. Elepaio 14: 3–6.
1970. ———. 1953. Suggestion for arboretum adjoining. *Na Laau Hawaii*. Elepaio 14: 24–26.

1971. ———. 1954. *Na Laau Hawaii* (plants belonging to Hawaii). Elepaio 14: 63–65.
1972. ———. 1954. Dry-land plants in a dry season. Elepaio 14: 69–70.
1973. ———. 1954. *Na Laau Hawaii* (plants belonging to Hawaii). Elepaio 15: 3–5.
1974. ———. 1954. *Na Laau Hawaii* in 1954. Elepaio 15: 30.
1975. ———. 1955. Preserving the rare plants of Hawaii. Elepaio 15: 57–58.
1976. ———. 1955. *Na Laau Hawaii* (plants belonging to Hawaii). Elepaio 16: 1–2.
1977. ———. 1957. Fogdrip on Lanai watershed. Elepaio 17: 49–51.
1978. ———. 1957. *Na Laau Hawaii* at the crossroads. Elepaio 18: 29–30.
1979. ———. 1958. *Na Laau Hawaii* in the last two years. Elepaio 18: 81.
1980. ———. 1958. Growth patterns of some native plants at *Na Laau Hawaii* in 1958. Elepaio 19: 18–19.
1981. ———. 1959. *Ke Kua'aina* from October 1, 1958 to September 30, 1959. Elepaio 20: 39–41.
1982. ———. 1960. History of tree form of *Hibiscus brackenridgii*. Elepaio 21: 2.
1983. ———. 1960. *Ke Kua'aina* from October 1, 1959 to September 30, 1960. Elepaio 21: 34–36.
1984. ———. 1961. *Ke Kua'aina* from October 1, 1960 to March 31, 1961. Elepaio 21: 78–79.
1985. ———. 1962. *Ke Kua'aina* from April 1, 1961 to November 30, 1961. Elepaio 22: 62–63.
1986. ———. 1962. Plants endemic to Hawaii now growing at *Ke Kua'aina*. Elepaio 22: 89–92.
1987. ———. 1963. The forests of Molokai. Elepaio 24: 29–30.
1988. Munro, H. G. 1963. Report on plants endemic to Hawaii now growing at *Ke Kua'aina*. Elepaio 24: 1–2.
1989. Munro, R. C. 1967. That is the way it is on the mountain. Elepaio 27: 61–63.
1990. Myhre, S. B. 1970. Kahoolawe. Newslett. Hawaiian Bot. Soc. 9: 21–27.
1991. Myint, T., and D. B. Ward. 1968. A taxonomic revision of the genus *Bonamia* (Convolvulaceae). Phytologia 17: 121–239.
1992. Nadeaud, J. 1873. Énumération des plantes indigènes de l'île de Tahiti. F. Savy, Libraire de la Société Botanique de France, Paris, 86 pp.
1993. Nagata, K. M. 1971. Hawaiian medicinal plants. Econ. Bot. 25: 245–254.
1994. ———. 1976. Campus plants. Univ. Hawaii, Honolulu, 22 pp.
1995. ———. 1980. The phytogeography of Pahole Gulch, Waianae Mountains, Oahu. Master's thesis, Univ. Hawaii, Honolulu, 293 pp.
1996. ———. 1981. Native strand flora (Abstr.): *in* Conserving Hawaii's coastal ecosystems. Program and presentation summaries. Univ. Hawaii Sea Grant Marine Advisory Program, pp. 4–5.
1997. ———. 1985. Early plant introductions in Hawai'i. Hawaiian J. Hist. 19: 35–61.

1998. ———— and B. Y. Kimura. 1980. Hawaiian coastal environments. Observations of native flora. Sea Grant Quart. 2(2): 1–6.
1999. Nagata, R. F., and G. P. Markin. 1986. Status of insects introduced into Hawai'i for the biological control of the wild blackberry *Rubus argutus* Link. Proc. 6th Conf. Nat. Sci., Hawaii Volcanoes Natl. Park, pp. 53–64.
2000. ———— and J. D. Stein. 1980. Attraction of the two-lined 'ōhi'a borer, *Plagithmysus bilineatus* (Coleoptera: Cerambicidae), to stressed 'ōhi'a trees (Abstr.). Proc. 3rd Conf. Nat. Sci., Hawaii Volcanoes Natl. Park, p. 251.
2001. Nakai, T. 1930. Notulae ad plantas Japoniae & Koreae XXXVIII. Bot. Mag. (Tokyo) 44: 7–40.
2002. Nakao, H. K. 1969. Biological control of weeds in Hawaii: *in* Romanowski, R. R., Jr., D. L. Plucknett, and H. F. Clay (eds.), Weed control basic to agriculture development. Inst. Techn. Interchange, Univ. Hawaii, Honolulu, pp. 93–95.
2003. Nakasone, H. Y., F. A. I. Bowers, and J. H. Beaumont. 1955. Terminal growth and flowering behavior of the pirie mango (*Mangifera indica* L.) in Hawaii. Proc. Amer. Soc. Hort. Sci. 66: 183–191.
2004. ———— and F. D. Rauch. 1980. Ornamental hibiscus—propagation and culture. Hawaii Agric. Exp. Sta. Univ. Hawaii Res. Bull. 175: 1–12.
2005. ———— and W. B. Storey. 1955. Studies on the inheritance of fruiting height of *Carica papaya* L. Proc. Amer. Soc. Hort. Sci. 66: 168–182.
2006. Na Lima Kokua. 1977. Taro (*kalo*) uses and recipes. Pacific Tropical Botanical Garden, Lawai, Hawaii, 20 pp.
2007. ————. 1980. Coconut (*niu*) uses and recipes. Pacific Tropical Botanical Garden, Lawai, Hawaii, 28 pp.
2008. ————. 1983. Sweet potato ('*uala*) uses and recipes. Pacific Tropical Botanical Garden, Lawai, Hawaii, 24 pp.
2009. Neal, M. C. 1934. Plants used medicinally: *in* Handy, E. S. C., M. K. Pukui, and K. Livermore, Outline of Hawaiian physical therapeutics. Bernice P. Bishop Mus. Bull. 126: 39–49.
2010. ————. 1937. Bean trees of Hawaii. Paradise Pacific 49(6): 21, 31.
2011. ————. 1937. South Point—island of Hawaii. Paradise Pacific 49(11): 17–18, 30.
2012. ————. 1938. Native Hawaiian orchids. Paradise Pacific 50(1): 28, 31.
2013. ————. 1938. *Maile*. Paradise Pacific 50(2): 13–14.
2014. ————. 1938. The potato family. Paradise Pacific 50(3): 23.
2015. ————. 1938. Perfumes for Hawaii. Paradise Pacific 50(5): 11–12.
2016. ————. 1938. Hawaiian dyes. Paradise Pacific 50(7): 23, 36.
2017. ————. 1938. *Puapilo*. Paradise Pacific 50(9): 8.
2018. ————. 1938. The taro family. Paradise Pacific 50(10): 13–14.
2019. ————. 1939. Hawaii the melting pot. Paradise Pacific 51(3): 6.
2020. ————. 1939. Plants in Hawaii with Old Testament associations. Paradise Pacific 51(5): 8.
2021. ————. 1939. Native Hawaiian hibiscus. Paradise Pacific 51(6): 11.

2022. ———. 1939. Vegetables of Hawaii. Paradise Pacific 51(7): 28.
2023. ———. 1939. *Kolomona*. Paradise Pacific 51(8): 4.
2024. ———. 1939. Vegetation of Lake Waiau, Hawaii. Paradise Pacific 51(10): 7, 32.
2025. ———. 1939. The trees on Iolani Palace grounds. Paradise Pacific 51(11): 13–14.
2026. ———. 1939. Hawaii has her own mistletoe. Paradise Pacific 51(12): 93.
2027. ———. 1939. A list of mosses and vascular plants collected on Mauna Kea, August 1935 (Abstr.). Proc. Hawaiian Acad. Sci. 14: 13.
2028. ———. 1940. Edible weeds in Hawaii. Paradise Pacific 52(1): 5–6.
2029. ———. 1942. Go eat grass. Paradise Pacific 54(10): 12–13.
2030. ———. 1943. Hawaii's floral emblem of 5,000 variations. Paradise Pacific 55(6): 55.
2031. ———. 1943. Hawaiian Christmas wreaths. Paradise Pacific 55(12): 24–25.
2032. ———. 1944. The Christmas cactus and its relatives. Paradise Pacific 56(12): 29–30.
2033. ———. 1945. The Christmas flower. Paradise Pacific 57(12): 25.
2034. ———. 1946. Hawaii's cosmopolitan plant life. Paradise Pacific 58(12): 64–65.
2035. ———. 1947. A *Manilkara* found on Oahu, Hawaii. Pacific Sci. 1: 243–244.
2036. ———. 1948. In gardens of Hawaii. Special Publ. Bernice P. Bishop Mus. 40: 1–805.
2037. ———. 1951. Trees of Hawaii. Paradise Pacific 63(1952 Annual): 68–69, 123.
2038. ———. 1954. Botanical guide to Iolani Palace grounds. Paradise Pacific 66(1): 16–19.
2039. ———. 1955. About plants in Hawaii. California Gard. 46(1): 9, 18.
2040. ———. 1958. Trees and plants of Iolani Palace grounds. Paradise Pacific 70(3): 25–29.
2041. ———. 1963–1964. Fruits and vegetables in Hawaii. Gard. J. New York Bot. Gard. 13: 212–215, 223–224 (1963); 14: 10–14, 29–30 (1964).
2042. ———. 1965. In gardens of Hawaii. 2nd ed. Special Publ. Bernice P. Bishop Mus. 50: 1–924.
2043. ——— and B. Metzger. 1928. In Honolulu gardens. Special Publ. Bernice P. Bishop Mus. 13: 1–327.
2044. ——— and ———. 1929. In Honolulu gardens. 2nd ed. Special Publ. Bernice P. Bishop Mus. 13: 1–336.
2045. Nees von Esenbeck, C. G. D. 1832. Genera et species Astearum. I. D. Grüson, Breslau, 309 pp. (Rep., 1833, Leonardi Schrag, Nuremberg).
2046. ———. 1834. Uebersicht der Cyperaceengattungen. Linnaea 9: 273–306.
2047. ———. 1843. Cyperaceae: *in* Meyen, F. J. F., Observationes botanicas in itinere circum institutas . . . Nov. Actorum Acad. Caes. Leop.-Carol. Nat. Cur. 19, Suppl. 1: 53–124b.
2048. ———. 1843. Gramineae: *in* Meyen, F. J. F., Observationes botanicas in

itinere circum institutas . . . Nov. Actorum Acad. Caes. Leop.-Carol. Nat. Cur. 19, Suppl. 1: 133–208.

2049. Neff, J. A., and P. A. DuMont. 1955. A partial list of the plants of the Midway Islands. Atoll Res. Bull. 45: 1–11.

2050. Nellist, G. F. 1937. Hawaiian garden of useful trees and plants. Paradise Pacific 49(3): 20.

2051. Nelmes, E. 1950. Notes on Cyperaceae: XXIV. Kew Bull. 1950: 189–208.

2052. Nelson, R. E. 1960. Silk-oak in Hawaii . . . pest or potential timber? Pacific Southw. Forest Range Exp. Sta. Misc. Pap. 47: 1–5.

2053. ———. 1965. A record of forest plantings in Hawaii. U.S.D.A. Forest Serv. Resource Bull. PSW-1: 1–18.

2054. ———. 1967. Records and maps of forest types in Hawaii. U.S.D.A. Forest Serv. Resource Bull. PSW-8: 1–22.

2055. ——— and C. J. Davis. 1972. Black twig borer . . . a tree killer in Hawaii. U.S.D.A. Forest Serv. Res. Note PSW-274: 1–2.

2056. ——— and N. Honda. 1966. Plantation timber on the island of Hawaii—1965. U.S.D.A. Forest Serv. Resource Bull. PSW-3: 1–52.

2057. ——— and T. H. Schubert. 1976. Adaptability of selected tree species planted in Hawaii forests. U.S.D.A. Forest Serv. Resource Bull. PSW-14: 1–22.

2058. ——— and P. R. Wheeler. 1963. Forest resources of Hawaii—1961. Pacific Southw. Forest Range Exp. Sta., Honolulu, 48 pp.

2059. ———, W. H. C. Wong, Jr., and H. L. Wick. 1968. Plantation timber on the island of Oahu—1966. U.S.D.A. Forest Serv. Resource Bull. PSW-10: 1–52.

2060. Neuhauss, R. 1886. Die Hawaii-Inseln. Samml. Gemeinverst. Vortr. Virchow Holtzendorff. n.s. 9: 1–48; not seen. [Review in Just's Bot. Jahresber. 14(2): 221, 1889.]

2061. Newell, C. L. 1968. A phytosociological study of the major vegetation types in Hawaii Volcanoes National Park, Hawaii. Master's thesis, Univ. Hawaii, Honolulu, 191 pp.

2062. Newell, T. K. 1968. A biosystematic study of the genus *Joinvillea* (Flagellariaceae). Master's thesis, Univ. Hawaii, Honolulu, 84 pp.

2063. ———. 1969. A study of the genus *Joinvillea* (Flagellariaceae). J. Arnold Arbor. 50: 527–555.

2064. ——— and B. C. Stone. 1967. *Flagellaria* (*Chortodes*) *plicata* Hooker fil. is a *Joinvillea*. Taxon 16: 18–20.

2065. Newman, T. S. 1971. Hawaii Island agricultural zones, circa A.D. 1823: an ethnohistorical study. Ethnohistory 18: 335–351.

2066. ———. 1972. Man in the prehistoric Hawaiian ecosystem: *in* Kay, E. A. (ed.), A natural history of the Hawaiian Islands. Selected readings. Univ. Press Hawaii, Honolulu, pp. 559–603.

2067. Nicharat, S. 1966. Anatomical and cytological evaluation of Hawaiian *Pipturus*. Master's thesis, Univ. Hawaii, Honolulu, 53 pp.

2068. ——— and G. W. Gillett. 1970. A review of the taxonomy of Hawaiian

Pipturus (Urticaceae) by anatomical and cytological evidence. Brittonia 22: 191–206.

2069. Nicol, B. 1981. North Halawa Valley: a last look? Honolulu 16(4): 74–77.
2070. Nicolson, D. H. 1980 [1981]. Summary of cytological information on *Emilia* and the taxonomy of four Pacific taxa of *Emilia* (Asteraceae: Senecioneae). Syst. Bot. 5: 391–407.
2071. Niedenzu, F. J. 1893. Myrtaceae. Nat. Pflanzenfam. III. 7: 57–105.
2072. Niimoto, D. H. 1966. Chromosome numbers of some *Hibiscus* species and other Malvaceae. Baileya 14: 29–34.
2073. Nishida, T., F. H. Haramoto, and L. Nakahara. 1970. Progress report on faunal research on *Metrosideros*. U.S. IBP Island Ecosystems IRP Techn. Rep. 1: 103.
2074. ———, ———, and ———. 1972. Progress report on faunal research on *Metrosideros*. U.S. IBP Island Ecosystems IRP Techn. Rep. 2: 141–147.
2075. ———, ———, and ———. 1973. Progress report on faunal research on *Metrosideros*. U.S. IBP Island Ecosystems IRP Techn. Rep. 21: 6.31–6.34.
2076. Nishimoto, S. K. 1969. Plants used as fish poisons. Newslett. Hawaiian Bot. Soc. 8: 20–23.
2077. Nishiyama, I. 1963. The origin of the sweet potato plant: *in* Barrau, J. (ed.), Plants and the migrations of the Pacific peoples. A symposium. Bishop Mus. Press, Honolulu, pp. 119–128.
2078. Noffsinger, T. L. 1961. Leaf and air temperature under Hawaii conditions. Pacific Sci. 15: 304–306.
2079. Nowicke, J. W. 1968 [1969]. Palynotaxonomic study of the Phytolaccaceae. Ann. Missouri Bot. Gard. 55: 294–363.
2080. Nuttall, T. 1841. Descriptions of new species and genera of plants in the natural order of the Compositae, collected in a tour across the continent to the Pacific, a residence in Oregon, and a visit to the Sandwich Islands and Upper California, during the years 1834 and 1835. Trans. Amer. Philos. Soc. n.s. 7: 283–453.
2081. ———. 1843. Description and notices of new or rare plants in the natural orders Lobeliaceae, Campanulaceae, Vaccinieae, Ericaceae, collected in a journey over the continent of North America, and during a visit to the Sandwich Islands, and Upper California. Trans. Amer. Philos. Soc. n.s. 8: 251–272.
2082. ———. 1866. On a new species of *Tacca*. J. Bot. 4: 261–263.
2083. Obata, J. K. 1967. Seed germination in native Hawaiian plants. Newslett. Hawaiian Bot. Soc. 6: 13–19.
2084. ———. 1971–1973. Propagating native Hawaiian plants. Newslett. Hawaiian Bot. Soc. 10: 47–52 (1971); 11: 4–6 (1972); 12: 2–4, 9–11 (1973).
2085. ———. 1974. Flowering and fruiting observations of *Antidesma pulvinatum* (*hame, haʻa, mehame*). Newslett. Hawaiian Bot. Soc. 13: 20–21.

2086. ———. 1976. Cultivating an "extinct" species. Newslett. Hawaiian Bot. Soc. 15: 35–37.
2087. ———. 1976. The elusive *kauila* (*Alphitonia ponderosa*) of Oʻahu. Newslett. Hawaiian Bot. Soc. 15: 52–53.
2088. ———. 1977. Native plants: *in* Palmer, D. D. (ed.), Hawaiian plants—notes and news. Newslett. Hawaiian Bot. Soc. 16: 74–75.
2089. ———. 1985. Another noxious melastome? *Oxyspora paniculata*. Newslett. Hawaiian Bot. Soc. 24: 25–26.
2090. ———. 1985. The declining forest cover of the Koʻolau summit. Newslett. Hawaiian Bot. Soc. 24: 41–42.
2091. ———. 1986. The demise of a species: *Urera kaalae*. Newslett. Hawaiian Bot. Soc. 25: 74–75.
2092. Oberdorfer, E. 1983. Einige Bemerkungen zu Vegetationsstrukturen im östlichen Nordamerika, in Oahu (Hawaii) und Mitteljapan. Andrias 2: 53–63.
2093. Ohwi, J. 1947. New or noteworthy grasses from Asia. Bull. Tokyo Sci. Mus. 18: 1–15.
2094. Oliver, D. 1866. On *Hillebrandia*, a new genus of Begoniaceae. Trans. Linn. Soc. London 25: 361–363.
2095. Oliver, W. R. B. 1935. The genus *Coprosma*. Bernice P. Bishop Mus. Bull. 132: 1–207.
2096. ———. 1942. New species of *Coprosma* from New Guinea and the Hawaiian Islands. Rec. Domin. Mus. 1: 44–47.
2097. Olson, D. F., Jr., and E. Q. P. Petteys. 1974. *Casuarina* L. Casuarina: *in* Schopmeyer, C. S., Seeds of woody plants in the United States. U.S.D.A. Agric. Handbook 450: 278–280.
2098. Olson, S. L., and H. F. James. 1982. Prodromus of the fossil avifauna of the Hawaiian Islands. Smithsonian Contr. Zool. 365: 1–59.
2099. Orchard, A. E. 1975. Taxonomic revisions in the family Haloragaceae I. The genera *Haloragis*, *Haloragodendron*, *Glischrocaryon*, *Meziella* and *Gonocarpus*. Bull. Auckland Inst. Mus. 10: 1–299.
2100. Ord, W. M. 1962. Preservation of plants and wildlife in Hawaii. Elepaio 22: 75–77.
2101. Orebamjo, T. O., G. Porteous, and G. R. Stewart. 1982. Nitrate reduction in the genus *Erythrina*. Allertonia 3: 11–18.
2102. Osborn, H. T. 1924. A preliminary study of the *pamakani* plant (*Eupatorium glandulosum* H.B.K.) in Mexico with reference to its control in Hawaii. Hawaiian Pl. Rec. 28: 546–549.
2103. Otaguro, J. 1985. The sandalwood saga. Spirit of Aloha 10(1): 26–27.
2104. Ownbey, G. B. 1961. The genus *Argemone* in South America and Hawaii. Brittonia 13: 91–109.
2105. Palmer, D. D. 1970. Plant-skin interactions. Newslett. Hawaiian Bot. Soc. 9: 33–37.
2106. ———. 1977. Report of the native plant committee. Newslett. Hawaiian Bot. Soc. 16: 60–62.
2107. Papp, R. P. 1978. The role of the Hawaiian two-lined ʻohiʻa borer,

Plagithmysus bilineatus Sharp, in the decline of *'ohi'a-lehua* forests on the island of Hawai'i (Abstr.). Proc. 2nd Conf. Nat. Sci., Hawaii Volcanoes Natl. Park, p. 236.

2108. ———, J. T. Kliejunas, R. S. Smith, Jr., and R. F. Scharpf. 1979. Association of *Plagithmysus bilineatus* (Coleoptera: Cerambycidae) and *Phytophthora cinnamomi* with the decline of *'ōhi'a-lehua* forests on the island of Hawaii. Forest Sci. 25: 187–196.

2109. Park, S. J. 1967. Inheritance of flower color in *Desmodium sandwicense* E. Mey. Master's thesis, Univ. Hawaii, Honolulu, 113 pp.

2110. ——— and P. P. Rotar. 1968. Genetic studies in Spanish clover, *Desmodium sandwicense* E. Mey. I. Inheritance of flower color, stem color, and leaflet markings. Crop. Sci. 8: 467–470.

2111. ——— and ———. 1968. Genetic studies in Spanish clover, *Desmodium sandwicense* E. Mey. II. Isolation and identification of anthocyanins in flower petals. Crop Sci. 8: 470–474.

2112. Parlatore, F. 1866. Le specie dei cotoni. Stamperia Reale, Florence, 64 pp.

2113. Parman, T. T. 1975. An autecological review of *Sophora chrysophylla* in Hawaii. Newslett. Hawaiian Bot. Soc. 14: 40–49.

2114. ———. 1976. The effects of fire upon a Hawaiian montane ecosystem. Proc. 1st Conf. Nat. Sci., Hawaii Volcanoes Natl. Park, pp. 171–178.

2115. ———. 1976. Hilina Pali fire of 1975. Proc. 1st Conf. Nat. Sci., Hawaii Volcanoes Natl. Park, pp. 179–181.

2116. ——— and K. Wampler. 1977. The Hilina Pali fire: a controlled burn exercise. Coop. Natl. Park Resources Stud. Unit, Hawaii, Techn. Rep. 18: 1–28.

2117. Parmentier, P. 1892. Histologie comparée des Ébénacées dans ses rapports avec la morphologie et l'histoire généalogique de ces plantes. Ann. Univ. Lyon 6(2): 1–155.

2118. Parris, G. K. 1940. A check list of fungi, bacteria, nematodes, and viruses occurring in Hawaii, and their hosts. Pl. Dis. Reporter, Suppl. 121: 1–91.

2119. Patterson, R. 1984. Flavonoid diversification in Hawaiian species of *Scaevola* (Goodeniaceae) (Abstr.). Amer. J. Bot. 71(5/2): 183.

2120. ———. 1984. Flavonoid uniformity in diploid species of Hawaiian *Scaevola* (Goodeniaceae). Syst. Bot. 9: 263–265.

2121. Pax, F. 1893. Über die Verbreitung südamerikanischen Caryophyllaceae und die Arten der Republica Argentina. Bot. Jahrb. Syst. 18: 1–35.

2122. ——— and K. Hoffmann. 1914. Euphorbiaceae–Acalypheae–Mercurialinae. Pflanzenr. IV. 147 (Heft 63): 1–473.

2123. ——— and ———. 1931. Euphorbiaceae. Nat. Pflanzenfam. ed. 2, 19c: 11–233.

2124. ——— and ———. 1934. Caryophyllaceae. Nat. Pflanzenfam. ed. 2, 16c: 275–364.

2125. ——— and ———. 1957. Euphorbiaceae–Phyllanthoideae–*Phyllantheae*. Pflanzenr. IV. 147 (Heft 81): 1–349.

2126. ——— and R. Knuth. 1905. Primulaceae. Pflanzenr. IV. 237 (Heft 22): 1–386.
2127. Pearcy, R. W. 1983. The light environment and growth of C_3 and C_4 tree species in the understory of a Hawaiian forest. Oecologia 58: 19–25.
2128. ——— and H. W. Calkin. 1983. Carbon dioxide exchange of C_3 and C_4 tree species in the understory of a Hawaiian forest. Oecologia 58: 26–32.
2129. ———, K. Osteryoung, and D. Randall. 1982. Carbon dioxide exchange characteristics of C_4 Hawaiian *Euphorbia* species native to diverse habitats. Oecologia 55: 333–341.
2130. ——— and J. Troughton. 1975. C_4 photosynthesis in tree form *Euphorbia* species from Hawaiian rainforest sites. Pl. Physiol. (Lancaster) 55: 1054–1056.
2131. Pearsall, G. S. 1951. Phenology of ornamental plants in the Honolulu area. Master's thesis, Univ. Hawaii, Honolulu, 318 pp.
2132. Pedley, L. 1975. Revision of the extra-Australian species of *Acacia* subg. *Heterophyllum*. Contr. Queensland Herb. 18: 1–24.
2133. ———. 1986. Derivation and dispersal of *Acacia* (Leguminosae), with particular reference to Australia, and the recognition of *Senegalia* and *Racosperma*. J. Linn. Soc., Bot. 92: 219–254.
2134. Pekelo, N. K., Jr. 1972. Some observations from West Molokai. Newslett. Hawaiian Bot. Soc. 11: 36–37.
2135. ———. 1973. Some field notes from Molokai. Elepaio 33: 98–99.
2136. Perkins, R. C. L. 1902. Enemies of lantana. Hawaiian Pl. Monthly 21: 607–612.
2137. ———. 1903. Enemies of lantana. Hawaiian Pl. Monthly 22: 159–162.
2138. ———. 1904. Later notes on lantana insects. Proc. Hawaiian Live Stock Breeders' Assoc., 2nd Ann. Meeting, Honolulu, pp. 58–61.
2139. ——— and O. H. Swezey. 1924. The introduction into Hawaii of insects that attack lantana. Bull. Exp. Sta. Hawaiian Sugar Planters' Assoc., Ent. Ser. 16: 1–83.
2140. Perlman, S. P. 1978. A rare Hawaiian orchid. Bull. Pacific Trop. Bot. Gard. 8: 19.
2141. ———. 1979. *Brighamia* in Hawaii. Bull. Pacific Trop. Bot. Gard. 9: 1–2.
2142. Perry, M. H. 1985. The effects of shading on photosynthesis and leaf morphology of *Leucaena leucocephala* (Lam) de Wit cv K8. Master's thesis, Univ. Hawaii, Honolulu, 102 pp.
2143. Petteys, E. Q. P. 1974. *Tristania conferta* R. Br. Brushbox: *in* Schopmeyer, C. S., Seeds of woody plants in the United States. U.S.D.A. Agric. Handbook 450: 817–818.
2144. ———, R. E. Burgan, and R. E. Nelson. 1975. *Ohia* forest decline: its spread and severity in Hawaii. U.S.D.A. Forest Serv. Res. Pap. PSW-105: 1–11.
2145. Pfeiffer, H. 1927. *Oreobolus* R. Br., eine merkwürdige Cyperaceengattung. Repert. Spec. Nov. Regni Veg. 23: 339–353.
2146. Pfeiffer, L. K. G. 1874. Nomenclator botanicus. Nominum ad finem anni

1858 publici juris factorum, classes, ordines, tribus, familias, divisiones, genera, subgenera vel sectiones designantium enumeratio alphabetica. Adjectis auctoribus, temporibus, locis systematicis apud varios, notis literaris atque etymologicis et synonymis. Vol. 2(2). Theodori Fischeri, Cassellis, pp. 761–1698; not seen.

2147. Philipson, W. R. 1970. A redefinition of *Gastonia* and related genera (Araliaceae). Blumea 18: 497–505.

2148. ———. 1970. Constant and variable features of the Araliaceae: *in* Robson, N. K. B., D. F. Cutler, and M. Gregory (eds.), New research in plant anatomy. J. Linn. Soc., Bot. 63, Suppl. 1: 87–101.

2149. Pichon, M. 1947. Classification des Apocynacées II, genre "*Rauvolfia*." Bull. Soc. Bot. France 94: 31–39.

2150. ———. 1947. Classification des Apocynacées. III, genre *Ochrosia*. Bull. Mus. Hist. Nat. (Paris) n.s. 19: 205–212.

2151. ———. 1947. Classification des Apocynacées. VII. Genre *Aspidosperma*. Bull. Mus. Hist. Nat. (Paris) n.s. 19: 362–369.

2152. ———. 1948. Classification des Apocynacées: IX, Rauvolfiées, Alstoniées, Allamandées et Tabernémontanoidées. Bull. Mus. Hist. Nat. (Paris) n.s. 27: 153–251.

2153. Pickford, G. D. 1962. Opportunities for timber production in Hawaii. Pacific Southw. Forest Range Exp. Sta. Misc. Pap. 67: 1–11.

2154. ——— and R. K. LeBarron. 1960. A study of forest plantations for timber production on the island of Hawaii. Pacific Southw. Forest Range Exp. Sta. Techn. Pap. 52: 1–15.

2155. Pierre, J. B. L. 1890. Notes botaniques. Sapotacées. Fasc. 1. Librairie des sciences Paul Klincksieck, Paris, 36 pp.; not seen.

2156. Pilger, R. 1922. Über die Formen von *Plantago major* L. Repert. Spec. Nov. Regni Veg. 18: 257–283.

2157. ———. 1923. Beiträge zur Kenntnis der Gattung *Plantago*. III. Repert. Spec. Nov. Regni Veg. 19: 114–119.

2158. ———. 1936. Drei neue Arten von *Plantago* aus der Verwandtschaft von *P. pachyphylla* Gray. Repert. Spec. Nov. Regni Veg. 40: 237–239.

2159. ———. 1937. Plantaginaceae. Pflanzenr. IV. 269 (Heft 102): 1–466.

2160. ———. 1940. Gramineae III. Nat. Pflanzenfam. ed. 2, 14e: 1–208.

2161. Pimm, S. L., and J. W. Pimm. 1982. Resource use, competition, and resource availability in Hawaiian honeycreepers. Ecology 63: 1468–1480.

2162. Piper, C. V., and S. T. Dunn. 1922. A revision of *Canavalia*. Bull. Misc. Inform. 4: 129–145.

2163. Plucknett, D. L., J. C. Moomaw, and C. H. Lamoureux. 1963. Root development in aluminous Hawaiian soils. Pacific Sci. 17: 398–406.

2164. ——— and B. C. Stone. 1961. The principal weedy Melastomaceae in Hawaii. Pacific Sci. 15: 301–303.

2165. ——— and W. A. Whistler. 1977. Weedy species of *Stachytarpheta* in Hawaii. Proc. 6th Asian-Pacific Weed Sci. Soc. Conf. 1: 198–203.

2166. Poellnitz, K. von. 1934. Versuch einer Monographie der Gattung *Portulaca* L. Repert. Spec. Nov. Regni Veg. 37: 240–320.
2167. ———. 1936. New Species of *Portulaca* from Southeastern Polynesia. Occas. Pap. Bernice P. Bishop Mus. 12(9): 1–6.
2168. ———. 1940. Bermerkungun zu einigen *Portulaca*-Arten. Repert. Spec. Nov. Regni Veg. 49: 224–225.
2169. Pomeroy, K. B. 1968. Letter from Hawaii. Amer. Forests 74(1): 36–37, 50, 52.
2170. Pope, W. T. 1909. The banyan and some other closely allied species. Hawaiian Forester Agric. 6: 121–129.
2171. ———. 1910. Ornamental plant-life of Honolulu. Hawaiian Almanac and Annual for 1911: 71–88.
2172. ———. 1922. Possibilities of the mango in Hawaii. Hawaiian Almanac and Annual for 1923: 53–64.
2173. ———. 1924. The Guatemalan avocado in Hawaii. Hawaii Agric. Exp. Sta. Bull. 51: 1–24.
2174. ———. 1926. Banana culture in Hawaii. Hawaii Agric. Exp. Sta. Bull. 55: 1–48.
2175. ———. 1926. Bananas of the Territory of Hawaii. Hawaiian Almanac and Annual for 1927: 106–110.
2176. ———. 1926. Unsettled variations of papaya (Abstr.). Proc. Hawaiian Acad. Sci. 1: 25.
2177. ———. 1929. Manual of wayside plants of Hawaii. Advertiser Publ. Co., Honolulu, 289 pp.
2178. ———. 1929. Mango culture in Hawaii. Hawaii Agric. Exp. Sta. Bull. 58: 1–27.
2179. ———. 1930. Papaya culture in Hawaii. Hawaii Agric. Exp. Sta. Bull. 61: 1–40.
2180. ———. 1935. The edible passion fruit in Hawaii. Hawaii Agric. Exp. Sta. Bull. 74: 1–22.
2181. Porter, D. M. 1970. The genus *Dodonaea* (Sapindacae) in the Galápagos Islands. Occas. Pap. Calif. Acad. Sci. 81: 1–4.
2182. Porter, J. R. 1972. The growth and phenology of *Metrosideros* in Hawaii. Ph.D. dissertation, Univ. Hawaii, Honolulu, 291 pp.
2183. ———. 1972. Hawaiian names for vascular plants. Hawaii Agric. Exp. Sta. Univ. Hawaii Dept. Pap. 1: 1–64.
2184. ———. 1973. The growth and phenology of *Metrosideros* in Hawaii. U.S. IBP Island Ecosystems IRP Techn. Rep. 27: 1–62.
2185. Porter, S. C. 1979. Hawaiian glacial ages. Quaternary Res. 12: 161–187.
2186. ———, K. L. Pierce, and T. D. Hamilton. 1983. Late Wisconsin mountain glaciation in the western United States: *in* Wright, H. E., Jr. (ed.), Late-Quaternary environments of the United States. Univ. Minnesota Press, Minneapolis, pp. 71–107.
2187. Potztal, E. 1953. Ein neues *Panicum* aus Hawaii. Mitt. Bot. Gart. Berlin-Dahlem 1: 128–130.

2188. Powell, A. M. 1965. Taxonomy of *Tridax* (Compositae). Brittonia 17: 47–96.
2189. ———. 1978. Systematics of *Flaveria* (*Flaveriinae*–Asteraceae). Ann. Missouri Bot. Gard. 65: 590–636.
2190. Powell, E. 1985. The Mauna Kea silversword: a species on the brink of extinction. Newslett. Hawaiian Bot. Soc. 24: 44–57.
2191. ———. 1986. Breeding biology and conservation of silverswords (Abstr.). Proc. 6th Conf. Nat. Sci., Hawaii Volcanoes Natl. Park, p. 65.
2192. ——— and F. R. Warshauer. 1985. No Na Leo 'Ole. 'Elepaio 45: 131–135.
2192a. Prain, D. 1895. An account of the genus *Argemone*. J. Bot. 33: 129–135, 176–178, 207–209, 307–312, 325–333, 363–371.
2193. Pratt, T. 1973. Plant communities and bird distribution on East Molokai. Elepaio 33: 66–70.
2194. Pray, T. R. 1959. Pattern and ontogeny of the foliar venation of *Bobea elatior* (Rubiaceae). Pacific Sci. 13: 3–13.
2195. Presl, K. B. 1836. Prodromus monographiae Lobeliacearum. Abh. Böhm. Ges. Wiss. n.s. 4(9): 1–52.
2196. Preston, F. G. 1923. *Osteomeles anthyllidifolia*. Gard. Chron. III. 73: 335.
2197. Pritzel, E. 1930. Pittosporaceae. Nat. Pflanzenfam. ed. 2, 18a: 265–286.
2198. Pukui, M. K., and S. H. Elbert. 1971. Hawaiian dictionary. Univ. Press of Hawaii, Honolulu, 402 + 188 pp.
2199. ——— and ———. 1986. Hawaiian dictionary. Rev. and enl. ed. Univ. Hawaii Press, Honolulu, 572 pp.
2200. ——— and M. C. Neal. 1941. The leis of Hawaii. An interpretation of the song of the islands. Paradise Pacific 53(12): 39–44.
2201. Pultz, M. A. (ed.). 1981. A botanist's visit to Oahu in 1831, being the journal of Dr. F. J. F. Meyen's travels and observations about the island of Oahu. [Excerpt from "Reise um die Erde ausgeführt auf dem Königlich Preussischen Seehandlungs-Schiffe Prinzess Louise, commandirt von Capitain W. Wendt, in den Jahren 1830, 1831 und 1832" by F. J. F. Meyen]. Press Pacifica, Ltd., Kailua, Hawaii.
2202. Quedado, R. M. 1974. The participation of photosynthesis in floral induction of the long-day plant *Anagallis arvensis* L. Ph.D. dissertation, Univ. Hawaii, Honolulu, 179 pp.
2203. Raabe, R. D. 1965. Check list of some parasitic phanerogams and some of their hosts found on the island of Hawaii in 1963. Pl. Dis. Reporter 49: 583–585.
2204. ———. 1966. *Armillaria* root rot in Hawaii. Hawaii Farm Sci. 15: 7–8.
2205. ———. 1966. Check list of plant diseases previously unreported in Hawaii. Pl. Dis. Reporter 50: 411–414.
2206. ——— and O. V. Holtzmann. 1965. A foliar nematode in hibiscus. Phytopathology 55: 478–479.

2207. ———— and E. E. Trujillo. 1963. *Armillaria mellea* in Hawaii. Pl. Dis. Reporter 47: 776.
2208. Rabakonandrianina, E. 1979. Experimental hybridization in *Lipochaeta* (Compositae): infrageneric relationships and possible origin from *Wedelia*. Ph.D. dissertation, Univ. Hawaii, Honolulu, 126 pp.
2209. ————. 1980 [1981]. Infrageneric relationships and the origin of the Hawaiian endemic genus *Lipochaeta* (Compositae). Pacific Sci. 34: 29–39.
2210. ———— and G. D. Carr. 1981. Intergeneric hybridization, induced polyploidy, and the origin of the Hawaiian endemic *Lipochaeta* from *Wedelia* (Compositae). Amer. J. Bot. 68: 206–215.
2211. Radlkofer, L. 1878. Ueber *Sapindus* und damit in Zusammenhang stehende Pflanzen. Sitzungsber. Math.-Phys. Cl. Königl. Bayer. Akad. Wiss. München 8: 221–408.
2212. ————. 1890. Ueber die Gliederung der Familie der Sapindaceen. Sitzungsber. Math.-Phys. Cl. Königl. Bayer. Akad. Wiss. München 20: 105–379.
2213. ————. 1895. Sapindaceae. Nat. Pflanzenfam. III. 5: 277–366.
2214. ————. 1931–1934. Sapindaceae. Tribus I–XIV. Pflanzenr. IV. 165 (Heft 98a–98h): 1–1539.
2215. ———— and J. F. Rock. 1911. New and noteworthy Hawaiian plants. Hawaii, Board Agric. Forest. Bot. Bull. 1: 1–15.
2216. Rajput, M. A. 1968. Tree stand analysis and soil characteristics of the major vegetation cover types in Hawaii Volcanoes National Park, Hawaii. Master's thesis, Univ. Hawaii, Honolulu, 236 pp.
2217. Ralph, C. J. 1978. Hawaiian plant on endangered species list. 'Elepaio 38: 142–143.
2218. ————. 1980. The phenology of some fruiting and flowering plants in a mid-elevation wet forest on Mauna Loa (Abstr.). Proc. 3rd Conf. Nat. Sci., Hawaii Volcanoes Natl. Park, p. 267.
2219. ————, A. P. Pearson, and D. C. Phillips. 1980 [1981]. Observations on the life history of the endangered Hawaiian vetch (*Vicia menziesii*) (Fabaceae) and its use by birds. Pacific Sci. 34: 83–92.
2220. Randall, J. M. 1986. Aquatic and wetland vascular plants in the Lawai Valley. Bull. Pacific Trop. Bot. Gard. 16: 1–7.
2221. ————. 1986. The Sterculiaceae. Bull. Pacific Trop. Bot. Gard. 16: 89–91.
2222. Randeria, A. J. 1960. The composite genus *Blumea*, a taxonomic revision. Blumea 10: 176–317.
2223. Rao, A. S. 1956. A revision of *Rauvolfia* with particular reference to the American species. Ann. Missouri Bot. Gard. 43: 253–354.
2224. Ratter, J. A. 1975. A survey of chromosome numbers in the Gesneriaceae of the Old World. Notes Roy. Bot. Gard. Edinburgh 33: 527–543.
2225. Rau, J. 1974. A Hawaiian alphabet of flowers. Pre-Columbian Press, Sylvan Lake, Mich., 21 unnum. pp.

2226. Rauh, W. 1981. *Brighamia insignis*. A curious succulent of the lobelia family from the Hawaiian Islands. Cact. Succ. J. (Los Angeles) 53: 219–220.
2227. Raven, P. H. 1963. The Old World species of *Ludwigia* (including *Jussiaea*), with a synopsis of the genus (Onagraceae). Reinwardtia 6: 327–427.
2228. ———. 1967. The genus *Epilobium* in Malesia (Onagraceae). Blumea 15: 269–282.
2229. ———. 1974. *Erythrina* (Fabaceae): achievements and opportunities. Lloydia 37: 321–331.
2230. ———. 1976. Ethics and attitudes: *in* Simmons, J. B., R. I. Beyer, P. E. Brandham, G. L. Lucas, and V. T. H. Parry, Conservation of threatened plants. Plenum Press, New York, pp. 155–179.
2231. ———. 1980. Hybridization and the nature of species in higher plants. Canad. Bot. Assoc. Bull. 13(1), Suppl.: 3–10.
2232. ———. 1982. *Erythrina* (Fabaceae: Faboideae): introduction to symposium IV. Allertonia 3: 1–6.
2233. ——— and T. E. Raven. 1976. The genus *Epilobium* (Onagraceae) in Australasia: a systematic and evolutionary study. New Zealand Dept. Sci. Industr. Res. Bull. 216: 1–321.
2234. ——— and W. Tai. 1979. Observations of chromosomes in *Ludwigia* (Onagraceae). Ann. Missouri Bot. Gard. 66: 862–879.
2235. Read, R. W. 1965. Chromosome numbers in the Coryphoideae. Cytologia 30: 385–391.
2236. Reeser, D. 1976. Reestablishment of native flora and fauna in Hawaii Volcanoes National Park. Proc. 1st Conf. Nat. Sci., Hawaii Volcanoes Natl. Park, pp. 183–186.
2237. ———. 1978. Planting, a tool for native ecosystem restoration. Proc. 2nd Conf. Nat. Sci., Hawaii Volcanoes Natl. Park, pp. 239–242.
2238. Reichardt, H. W. 1878. Beitrag zur Phanerogamenflora der hawaiischen Inseln. Sitzungsber. Kaiserl. Akad. Wiss., Math.-Naturwiss. Cl., Abt. 1, 76: 721–734.
2239. Rémy, J. 1862. *Ka Mooolelo [Moʻolelo] Hawaii*. Histoire de L'Archipel Havaiien (Îles Sandwich), texte et traduction précédés d'une introduction sur l'état physique, moral et politique du pays. Librairie A. Franck, Paris, 254 pp.
2240. Renear, C. W. 1911. The most valuable tree in the world. Mid-Pacific Mag. 2: 247–252.
2241. Reynolds, S. 1850. Reminiscences of Hawaiian agriculture. Trans. Roy. Hawaiian Agric. Soc. 1: 49–53.
2242. Richardson, F. 1963. Birds of Lehua Island off Niihau, Hawaii. Elepaio 23: 43–45.
2243. Richmond, G. B. 1963. Species trials at the Waiakea Arboretum, Hilo, Hawaii. U.S.D.A. Forest Serv. Res. Pap. PSW-4: 1–21.
2244. Richmond, T. de A., and D. Mueller-Dombois. 1972. Coastline ecosystems on Oahu, Hawaii. Vegetatio 25: 367–400.

2245. Ridley, H. N. 1886. A monograph of the genus *Liparis*. J. Linn. Soc., Bot. 22: 244–297.
2246. ———. 1930. The dispersal of plants throughout the world. L. Reeve & Co., Ashford, Kent, 744 pp.
2247. Riper, C. van, III. 1974. Linnet breeding biology on Hawaii. U.S. IBP Island Ecosystems IRP Techn. Rep. 46: 1–19.
2248. ———. 1975. Composition and phenology of the dry forest on Mauna Kea, Hawaii, as related to the annual cycle of the *Amakihi* (*Loxops virens*) and *Palila* (*Psittirostra bailleui*). U.S. IBP Island Ecosystems IRP Techn. Rep. 51: 1–37.
2249. ———. 1980. The phenology of the dryland forest of Mauna Kea, Hawaii, and the impact of recent environmental perturbations. Biotropica 12: 282–291.
2250. Ripperton, J. C. 1948. Grasslands in Hawaii. U.S.D.A. Yearb. Agric. 1948: 617–628.
2251. ———, R. A. Goff, D. W. Edwards, and W. C. Davis. 1933. Range grasses of Hawaii. Hawaii Agric. Exp. Sta. Bull. 65: 1–58.
2252. ——— and E. Y. Hosaka. 1942. Vegetation zones of Hawaii. Hawaii Agric. Exp. Sta. Bull. 89: 1–60.
2253. ———, ———, and P. A. Gantt. 1939. A few legumes found on Hawaiian ranges. Univ. Hawaii Agric. Extens. Serv. Circ. 10: 1–11.
2254. Ritland, K., and F. R. Ganders. 1985. Variation in the mating system of *Bidens menziesii* (Asteraceae) in relation to population substructure. Heredity 55: 235–244.
2255. Roane, M. K., and F. R. Fosberg. 1983. A new pyrenomycete associated with *Metrosideros collina* subspecies *polymorpha* (Myrtaceae). Mycologia 75: 163–166.
2256. Robertson, F. W., M. Shook, G. Takei, and H. Gaines. 1968. Observations on the biology and nutrition of *Drosophila disticha*, Hardy, an indigenous Hawaiian species: *in* Wheeler, M. R. (ed.), Studies in genetics IV. Research reports. Univ. Texas Publ. 6818: 279–299.
2257. Robertson, K. R. 1971. A revision of the genus *Jacquemontia* (Convolvulaceae) in North and Central America and the West Indies. Ph.D. dissertation, Washington Univ., St. Louis, 296 pp.
2258. ———. 1974. *Jacquemontia ovalifolia* (Convolvulaceae) in Africa, North America, and the Hawaiian Islands. Ann. Missouri Bot. Gard. 61: 502–513.
2259. Roberty, G. 1942. Gossypiorum revisionis tentamen. Candollea 9: 19–103.
2260. ———. 1952. Gossypiorum revisionis tentamen (suite et fin). Candollea 13: 9–165.
2261. Robichaux, R. H. 1980. Comparative photosynthesis of Hawaiian *Euphorbia* and *Scaevola* species from diverse habitats. Ph.D. dissertation, Univ. California, Davis, 96 pp.
2262. ———. 1984. Variation in the tissue water relations of two sympatric

Hawaiian *Dubautia* species and their natural hybrid. Oecologia 65: 75–81.

2263. ———. 1985. Tissue elastic properties of a mesic forest Hawaiian *Dubautia* species with 13 pairs of chromosomes. Pacific Sci. 39: 191–194.

2264. ——— and J. E. Canfield. 1985. Tissue elastic properties of eight Hawaiian *Dubautia* species that differ in habitat and diploid chromosome number. Oecologia 66: 77–80.

2265. ——— and R. W. Pearcy. 1980. Environmental characteristics, field water relations, and photosynthetic responses of C_4 Hawaiian *Euphorbia* species from contrasting habitats. Oecologia 47: 99–105.

2266. ——— and ———. 1980. Photosynthetic responses of C_3 and C_4 species from cool shaded habitats in Hawaii. Oecologia 47: 106–109.

2267. ——— and ———. 1984. Evolution of C_3 and C_4 plants along an environmental moisture gradient: patterns of photosynthetic differentiation in Hawaiian *Scaevola* and *Euphorbia* species. Amer. J. Bot. 71: 121–129.

2268. ———, P. W. Rundel, R. L. Stemmermann, J. E. Canfield, S. R. Morse, and W. E. Friedman. 1984. Tissue water deficits and plant growth in wet tropical environments: *in* Medina, E., H. A. Mooney, and C. Vázquez-yánes (eds.), Physiological ecology of plants of the wet tropics. W. Junk Publ., Boston, pp. 99–112.

2269. Robinson, B. L. 1913. Revisions of *Alomia*, *Ageratum*, and *Oxylobus*. Proc. Amer. Acad. Arts 49: 438–491. [Rep. in Contr. Gray Herb. 42 without change in pagination.]

2270. Robinson, H., A. M. Powell, R. M. King, and J. F. Weedin. 1981. Chromosome numbers in Compositae, XII: *Heliantheae*. Smithsonian Contr. Bot. 52: 1–28.

2271. Robson, N. K. B. 1972. Notes on Malesian species of *Hypericum* (Guttiferae). Florae Malesianae Praecursores LII. Blumea 20: 251–274.

2272. Robyns, W. 1938. A naturalist in the Hawaiian Islands. Bull. Cercle Alumni Fond. Univ. (Brussels) 9: 126–139.

2273. ——— and S. H. Lamb. 1939. Preliminary ecological survey of the island of Hawaii. Bull. Jard. Bot. État 15: 241–293.

2274. Rock, J. F. 1909. A new Hawaiian *Scaevola*. Bull. Torrey Bot. Club 36: 645–646.

2275. ———. 1909. A new Hawaiian shrub. Hawaiian Forester Agric. 6: 503.

2276. ———. 1910. Some new Hawaiian plants. Bull. Torrey Bot. Club 37: 297–304.

2277. ———. 1910. A synopsis of the Hawaiian flora. Hawaiian Almanac and Annual for 1911: 82–91.

2278. ———. 1911. Notes upon Hawaiian plants with descriptions of new species and varieties. Coll. Hawaii Publ. Bull. 1: 1–20.

2279. ———. 1911. Report of the Botanical Assistant. Hawaii, Board Commiss. Agric. Forest. Bienn. Rep. 1910: 67–88.

2280. ———. 1913. Descriptions of new species of Hawaiian plants. Coll. Hawaii Publ. Bull. 2: 39–47.

2281. ———. 1913. Remarks on certain Hawaiian plants described by H. Léveillé in Fedde Repertorium X. 10/14 (1911) 156–157. Coll. Hawaii Publ. Bull. 2: 48–49.

2282. ———. 1913. List of Hawaiian names of plants. Hawaii, Board Agric. Forest. Bot. Bull. 2: 1–20.

2283. ———. 1913. Report of the Consulting Botanist. Hawaii, Board Commiss. Agric. Forest. Bienn. Rep. 1912: 95–99.

2284. ———. 1913. The indigenous trees of the Hawaiian Islands. Publ. privately, Honolulu, 512 pp. (Rep., with introduction by S. Carlquist and addendum by D. R. Herbst, 1974, Charles E. Tuttle Co., Rutland, Vt., 548 pp.).

2285. ———. 1914. Revisio plantarum Hawaiiensium a Léveillé descriptarum. Repert. Spec. Nov. Regni Veg. 13: 352–361.

2286. ———. 1915. Vegetation der Hawaii-Inseln. Bot. Jahrb. Syst. 53: 275–311.

2287. ———. 1915. A new Hawaiian *Cyanea*. Bull. Torrey Bot. Club 42: 77–78.

2288. ———. 1916. A new species of *Pritchardia*. Bull. Torrey Bot. Club 43: 385–387.

2289. ———. 1916. The sandalwoods of Hawaii. A revision of the Hawaiian species of the genus *Santalum*. Hawaii, Board Agric. Forest. Bot. Bull. 3: 1–43.

2290. ———. 1916. Some plants of Hawaii. Mid-Pacific Mag. 11: 578–583.

2291. ———. 1917. Revision of the Hawaiian species of the genus *Cyrtandra*, section *Cylindrocalyces* Hillebr. Amer. J. Bot. 4: 604–623.

2292. ———. 1917. Notes on Hawaiian Lobelioideae, with descriptions of new species and varieties. Bull. Torrey Bot. Club 44: 229–239.

2293. ———. 1917. Hawaiian trees—a criticism. Bull. Torrey Bot. Club 44: 545–546.

2294. ———. 1917. The *ohia lehua* trees of Hawaii. A revision of the Hawaiian species of the genus *Metrosideros* Banks, with special reference to the varieties and forms of *Metrosideros collina* (Forster) A. Gray subspecies *polymorpha* (Gaud.) Rock. Hawaii, Board Agric. Forest. Bot. Bull. 4: 1–76.

2295. ———. 1917. Trees recommended for planting. Hawaiian Forester Agric. 14: 331–337.

2296. ———. 1917. Sandalwood in Hawaii. Mid-Pacific Mag. 13: 356–359.

2297. ———. 1917. The ornamental trees of Hawaii. Publ. privately, Honolulu, 210 pp.

2298. ———. 1918. Cyrtandreae Hawaiienses, sect. *Crotonocalyces* Hillebr. Amer. J. Bot. 5: 259–277.

2299. ———. 1918. *Pelea* and *Platydesma*. Bot. Gaz. (Crawfordsville) 65: 261–267.

2300. ———. 1918. New species of Hawaiian plants. Bull. Torrey Bot. Club 45: 133–139.
2301. ———. 1918. Trees recommended for planting. Hawaiian Pl. Rec. 18: 414–421.
2302. ———. 1919. Cyrtandreae Hawaiienses, sections *Schizocalyces* Hillebr. and *Chaetocalyces* Hillebr. Amer. J. Bot. 6: 47–68.
2303. ———. 1919. Cyrtandreae Hawaiienses, sect. *Microcalyces* Hillebr. Amer. J. Bot. 6: 203–216.
2304. ———. 1919. The arborescent indigenous legumes of Hawaii. Hawaii, Board Agric. Forest. Bot. Bull. 5: 1–53.
2305. ———. 1919. The Hawaiian genus *Kokia*. A relative of the cotton. Hawaii, Board Agric. Forest. Bot. Bull. 6: 1–22.
2306. ———. 1919. Report of the Consulting Botanist. Hawaii, Board Commiss. Agric. Forest. Bienn. Rep. 1918: 51–53.
2307. ———. 1919. One government forest. Reserve lands at Kulani, Hawaii, described. Hawaiian Forester Agric. 16: 39–40.
2308. ———. 1919. A monographic study of the Hawaiian species of the tribe *Lobelioideae*, family Campanulaceae. Mem. Bernice P. Bishop Mus. 7(2): 1–395.
2309. ———. 1920. The genus *Plantago* in Hawaii. Amer. J. Bot. 7: 195–210.
2310. ———. 1920. The poisonous plants of Hawaii. Hawaiian Forester Agric. 17: 59–62.
2311. ———. 1920. The poisonous plants of Hawaii (concluded). Hawaiian Forester Agric. 17: 97–101.
2312. ———. 1920. The leguminous plants of Hawaii. Hawaiian Sugar Pl. Assoc. Exp. Sta., Honolulu, 234 pp.
2313. ———. 1921. The *akala* berry of Hawaii. Asa Gray's *Rubus macraei*, an endemic Hawaiian raspberry. J. Heredity 12: 146–150.
2314. ———. 1957. Some new Hawaiian lobelioids. Occas. Pap. Bernice P. Bishop Mus. 22(5): 35–66.
2315. ———. 1962. A new Hawaiian *Pritchardia*. Occas. Pap. Bernice P. Bishop Mus. 23(4): 61–63.
2316. ———. 1962. Hawaiian lobelioids. Occas. Pap. Bernice P. Bishop Mus. 23(5): 65–75.
2317. ———. 1963. A herbarium. Newslett. Hawaiian Bot. Soc. 2: 14–15.
2318. ——— and M. C. Neal. 1957. A new variety of silversword. Occas. Pap. Bernice P. Bishop Mus. 22(4): 31–33.
2319. Roddis, L. H. 1938. Official floral emblems of Hawaii. Paradise Pacific 50(1): 14; 50(2): 7.
2320. Roe, M. J. 1959. A taxonomic study of the indigenous Hawaiian species of the genus *Hibiscus*. Master's thesis, Univ. Hawaii, Honolulu, 79 pp.
2321. ———. 1961. A taxonomic study of the indigenous Hawaiian species of the genus *Hibiscus* (Malvaceae). Pacific Sci. 15: 3–32.
2322. Roelofs, F. M. 1978. The reproductive biology of some Oahu *Cyrtandra*. Master's thesis, Univ. Hawaii, Honolulu, 105 pp.

2323. ———. 1979 [1980]. The reproductive biology of *Cyrtandra grandiflora* (Gesneriaceae) on Oahu. Pacific Sci. 33: 223–231.
2324. ———. 1983. The natural history of the Makiki Valley Loop Trail. Makiki Environmental Education Center, Honolulu, 17 pp.
2325. Rolfe, R. A. 1920. *Metrosideros collina*. Bot. Mag. 146: *pl. 8846*.
2326. Rollins, R. C. 1986. Alien species of *Lepidium* (Cruciferae) in Hawaii. J. Arnold Arbor. 67: 137–141.
2327. Romeo, J. T., and E. A. Bell. 1974. Distribution of amino acids and certain alkaloids in *Erythrina* species. Lloydia 37: 543–568.
2328. Rooke, T. C. B., and J. Montgomery. 1855. Report of select committee on the subject of indigo. Trans. Roy. Hawaiian Agric. Soc. 2: 82–83.
2329. Rosendahl, P. H. 1972. Aboriginal agriculture and residence patterns in upland Lapakahi, island of Hawaii. Ph.D. dissertation, Univ. Hawaii, Honolulu, 558 pp.
2330. ——— and D. E. Yen. 1971. Fossil sweet potato remains from Hawaii. J. Polynes. Soc. 80: 379–385.
2331. Rotar, P. P. 1968. Grasses of Hawaii. Univ. Hawaii Press, Honolulu, 355 pp.
2332. ——— and B. K. Bird. 1981. Bibliography of sweet potato (*Ipomea* [*Ipomoea*] *batatas*). Hawaii Inst. Trop. Agric. Human Resources Res. Ser. 1: 1–2 + microfiche.
2333. ——— and K. H. Chow. 1971. Morphological variation and interspecific hybridization among *Desmodium intortum*, *Desmodium sandwicense*, and *Desmodium uncinatum*. Hawaii Agric. Exp. Sta. Univ. Hawaii Techn. Bull. 82: 1–25.
2334. ———, S. J. Park, A. Bromdep, and U. Urata. 1967. Crossing and flowering behavior in Spanish clover, *Desmodium sandwicense* E. Mey., and other *Desmodium* species. Hawaii Agric. Exp. Sta. Techn. Progr. Rep. 164: 1–13.
2335. ———, D. L. Plucknett, and B. K. Bird. 1978. Bibliography of taro and edible aroids. Hawaii Inst. Trop. Agric. Human Resources Misc. Publ. 158: 1–245.
2336. ——— and U. Urata. 1966. Some agronomic observations in *Desmodium* species: seed weights. Hawaii Agric. Exp. Sta. Techn. Progr. Rep. 147: 1–11.
2337. ——— and ———. 1967. Cytological studies in the genus *Desmodium*; some chromosome counts. Amer. J. Bot. 54: 1–4.
2338. Rowley, G. D. 1983. *Brighamia*: succulent endemic of Hawaii. Brit. Cact. Succ. J. 1: 9–11.
2339. Ruhle, G. C. 1959. A guide to the crater area of Haleakala National Park, island of Maui, Hawaii. Hawaii Nat. Hist. Assoc., Hawaii Natl. Park, Hawaii, 90 pp.
2340. ——— and E. J. Barton. 1957. A guide for the Haleakala Section, island of Maui, Hawaii. Hawaii Natl. Park, Hawaii Nat. Hist. Assoc., 12 pp.
2341. Rundel, P. W. 1980. The ecological distribution of C_4 and C_3 grasses in the Hawaiian Islands. Oecologia 45: 354–359.

2342. Russ, G. W. 1929. A study of natural regeneration in some introduced species of trees. Hawaiian Forester Agric. 26: 117–124.
2343. ———. 1932. Notes on the distribution of *Neowawraea* (Abstr.). Proc. Hawaiian Acad. Sci. 7: 6–7.
2344. Russell, C. A. 1980. Food habits of the roof rat (*Rattus rattus*) in two areas of Hawaii Volcanoes National Park. Proc. 3rd Conf. Nat. Sci., Hawaii Volcanoes Natl. Park, pp. 269–272.
2345. Sachet, M.-H. 1971. Letter to the editor. Newslett. Hawaiian Bot. Soc. 10: 8–9.
2346. ——— and F. R. Fosberg. 1955. Island bibliographies. Natl. Res. Council Publ. 335: 1–577.
2347. ——— and ———. 1971. Island bibliographies supplement. Natl. Acad. Sci., Washington, 427 pp.
2348. ——— and ———. 1973. Remarks on *Halophila* (Hydrocharitaceae). Taxon 22: 439–443.
2349. Safford, W. E. 1921. Cultivated plants of Polynesia and their vernacular names, an index to the origin and migration of the Polynesians. Proc. 1st Pan-Pacific Sci. Conf., Honolulu, pp. 183–187.
2350. ———. 1921. Dispersal of plants by ocean currents (Abstr.). Proc. 1st Pan-Pacific Sci. Conf., Honolulu, pp. 535–536.
2351. Saiki, D. F., D. L. Plucknett, and P. S. Motooka. 1969. A checklist of important weeds in the Asian-Pacific region: *in* Romanowski, R. R., Jr., D. L. Plucknett, and H. F. Clay (eds.), Weed control basic to agriculture development. Inst. Techn. Interchange, Univ. Hawaii, Honolulu, pp. 131–133.
2352. St. John, H. 1931. Additions to the flora of Niihau. Occas. Pap. Bernice P. Bishop Mus. 9(14): 1–11.
2353. ———. 1932. Notes on *Pritchardia*. Occas. Pap. Bernice P. Bishop Mus. 9(19): 1–5.
2354. ———. 1933. *Lysimachia*, *Labordia*, *Scaevola*, and *Pluchea*. Hawaiian plant studies I. Occas. Pap. Bernice P. Bishop Mus. 10(4): 1–10.
2355. ———. 1934. *Panicum*, *Zanthoxylum*, *Psychotria*, and *Sicyos*. Hawaiian plant studies 2. Occas. Pap. Bernice P. Bishop Mus. 10(12): 1–7.
2356. ———. 1935. Additions to the flora of Midway Islands. Hawaiian plant studies III. Occas. Pap. Bernice P. Bishop Mus. 11(14): 1–4.
2357. ———. 1936. A revision of the Hawaiian species of *Labordia* described by H. Baillon. Hawaiian plant studies 4. Occas. Pap. Bernice P. Bishop Mus. 12(8): 1–11.
2358. ———. 1939. New Hawaiian species of *Clermontia*, including a revision of the *Clermontia grandiflora* group. Hawaiian plant studies 6. Occas. Pap. Bernice P. Bishop Mus. 15(1): 1–19.
2359. ———. 1939. New Hawaiian Lobeliaceae. Hawaiian plant studies 7. Occas. Pap. Bernice P. Bishop Mus. 15(2): 21–35.
2360. ———. 1940. Hawaiian plants named by Endlicher in 1836. Hawaiian plant studies 8. Occas. Pap. Bernice P. Bishop Mus. 15(22): 229–238.

2361. ———. 1940. *Ophioglossum*, *Rollandia*, and *Scaevola*. Hawaiian plant studies 9. Occas. Pap. Bernice P. Bishop Mus. 15(28): 351–359.

2362. ———. 1942. Later travels and botanical studies of William Hillebrand. Chron. Bot. 7: 69–70.

2363. ———. 1942. New combinations in the Gleicheniaceae and in *Styphelia* (Epacridaceae). Pacific plant studies 1. Occas. Pap. Bernice P. Bishop Mus. 17(7): 79–84.

2364. ———. 1943. New Hawaiian species of. *Peperomia*. Hawaiian plant studies 10. Occas. Pap. Bernice P. Bishop Mus. 17(12): 171–175.

2365. ———. 1944. Diagnoses of Hawaiian species of *Pelea* (Rutaceae). Hawaiian plant studies 13. Lloydia 7: 265–274.

2366. ———. 1945. *Dryopteris*, *Deschampsia*, *Portulaca*, *Lupinus*, *Fagara*, *Stenogyne*, and *Dubautia*. Hawaiian plant studies 12. Bull. Torrey Bot. Club 72: 22–30.

2367. ———. 1945. Revision of *Cardamine* and related Cruciferae in Hawaii, and *Nasturtium* in Polynesia. Pacific plant studies 3. Occas. Pap. Bernice P. Bishop Mus. 18(5): 77–93.

2368. ———. 1945. *Cyrtandra* as an indicator of the former extent of forest on Oahu (Abstr.). Proc. Hawaiian Acad. Sci. 16–18: 13.

2369. ———. 1946. Endemism in the Hawaiian flora, and a revision of the Hawaiian species of *Gunnera* (Haloragidaceae). Hawaiian plant studies 11. Proc. Calif. Acad. Sci. 25: 377–420.

2370. ———. 1947. *Pleomele fernaldii* (Liliaceae), a new species from the Hawaiian Islands. Hawaiian plant studies 16. Contr. Gray Herb. 165: 39–42.

2371. ———. 1947. The history, present distribution, and abundance of sandalwood on Oahu, Hawaiian Islands: Hawaiian plant studies 14. Pacific Sci. 1: 5–20.

2372. ———. 1950. Discussion [Review of "Processes of erosion on steep slopes of Oahu, Hawaii" by S. E. White]. Amer. J. Sci. 248: 508–510.

2373. ———. 1950. The subgenera of *Dubautia* (Compositae): Hawaiian plant studies 18. Pacific Sci. 4: 339–345.

2374. ———. 1952. Notes on Hawaiian species of *Scaevola* (Goodeniaceae). Hawaiian plant studies 19. Pacific Sci. 6: 30–34.

2375. ———. 1952. Monograph of the genus *Isodendrion* (Violaceae). Hawaiian plant studies 21. Pacific Sci. 6: 213–255.

2376. ———. 1953. Vegetational provinces of the Pacific,—Hawaiian (Abstr.). Abstr. Pap., 8th Pacific Sci. Congr., Philippines, pp. 159–160.

2377. ———. 1953. Narrow endemism in Oceania, as evidenced by the genus *Cyrtandra*. Abstr. Pap., 8th Pacific Sci. Congr., Philippines, p. 214.

2378. ———. 1954. The Hawaiian variety of *Dioscorea pentaphylla*, an edible yam. Hawaiian plant studies 22. J. Polynes. Soc. 63: 27–34.

2379. ———. 1954. Review of Mrs. Sinclair's "Indigenous flowers of the Hawaiian Islands." Hawaiian plant studies 23. Pacific Sci. 8: 140–146.

2380. ———. 1955. *Cyrtandra nutans* (Gesneriaceae) from the island of Maui.

Hawaiian plant studies 24. Occas. Pap. Bernice P. Bishop Mus. 21(15): 295–298.

2381. ———. 1955. The relationship between the species of *Erythrina* (Leguminosae) native to Hawaii and Tahiti. Pacific plant studies 13. Webbia 11: 293–299.

2382. ———. 1957. Discovery of *Alternanthera* (Amaranthaceae) in the native Hawaiian flora. Hawaiian plant studies 27. Bull. Jard. Bot. État 27: 49–54.

2383. ———. 1957. The campus trees. Univ. Hawaii, Honolulu, 8 unnum. pp.

2384. ———. 1958. *Brighamia citrina* (C. N. Forbes & Lydgate) St. John, comb. nov. Pacific Sci. 12: 182.

2385. ———. 1958. The status of "*Gahnia affinis*" and "*G. gahniaeformis*" (Cyperaceae) of Polynesia. Pacific plant studies 16. Webbia 13: 331–342.

2386. ———. 1959. Botanical novelties on the island of Niihau, Hawaiian Islands. Hawaiian plant studies 25. Pacific Sci. 13: 156–190.

2387. ———. 1960. Revision of the genus *Pandanus* Stickman, Part 1. Key to the sections. Pacific Sci. 14: 224–241.

2388. ———. 1960. The name of the Indo-Pacific strand *Scaevola*. Pacific plant studies 19. Taxon 9: 200–208.

2389. ———. 1965. La distribution mondiale du genre *Pandanus*. Adansonia 5: 307–308.

2390. ———. 1965. The identity of *Senecio capillaris* Gaudichaud. Hawaiian plant studies 26 [i.e., 122]. Ann. Missouri Bot. Gard. 52: 432–433.

2391. ———. 1965. Revision of *Capparis spinosa* and its African, Asiatic, and Pacific relatives. Micronesica 2: 25–45.

2392. ———. 1966. Monograph of *Cyrtandra* (Gesneriaceae) on Oahu, Hawaiian Islands. Bernice P. Bishop Mus. Bull. 229: 1–465.

2393. ———. 1966. Dispersal of a littoral species of *Ipomoea*, and speciation in Hawaii of the montane genus *Cyrtandra* (Abstr.): in Evolution, distribution and migration of the plant and animal in the Pacific. Proc. 11th Pacific Sci. Congr., Tokyo 5: 24.

2394. ———. 1968. *Cyrtandra megistocalyx* (Gesneriaceae), a new species from Oahu, Hawaiian Islands. Hawaiian plant studies 28. Pacific Sci. 22: 422–424.

2395. ———. 1969. Monograph of the genus *Brighamia* (Lobeliaceae). Hawaiian plant studies 29. J. Linn. Soc., Bot. 62: 187–204.

2396. ———. 1969. Hawaiian novelties in the genus *Solanum* (Solanaceae). Hawaiian plant studies 30. Pacific Sci. 23: 350–354.

2397. ———. 1969. Monograph of the Hawaiian species of *Gouania* (Rhamnaceae). Hawaiian plant studies 34. Pacific Sci. 23: 507–543.

2398. ———. 1969. Types of sections in *Clermontia*, *Cyanea*, and *Delissea* (Lobeliaceae). Taxon 18: 483.

2399. ———. 1970. Classification and distribution of the *Ipomoea pes-caprae* group (Convolvulaceae). Bot. Jahrb. Syst. 89: 563–583.

2400. ———. 1970. Revision of the Hawaiian species of *Canavalia* (Leguminosae). Hawaiian plant studies. 32. Israel J. Bot. 19: 161–219.
2401. ———. 1970. The "staminodia" of the genus *Schiedea* (Caryophyllaceae) and three new Hawaiian species. Hawaiian plant studies 32 [i.e., 33]. Pacific Sci. 24: 245–254.
2402. ———. 1970. The genus *Sicyos* (Cucurbitaceae) on the Hawaiian Leeward Islands. Hawaiian plant studies 35. Pacific Sci. 24: 439–456.
2403. ———. 1970. Typification of *Charpentiera* (Amaranth.). Taxon 19: 302.
2404. ———. 1970. Typification of *Nothocestrum* Gray (Solanaceae). Taxon 19: 304.
2405. ———. 1971. The status of the genus *Wilkesia* (Compositae), and discovery of a second Hawaiian species. Hawaiian plant studies 34 [i.e., 38]. Occas. Pap. Bernice P. Bishop Mus. 24(8): 127–137.
2406. ———. 1971. Endemic plants of Kipahulu Valley, Maui, Hawaiian Islands. Hawaiian plant studies 36. Pacific Sci. 25: 39–79.
2407. ———. 1971. The identity of *Arabis o-waihiensis* Cham. & Schlecht. Hawaiian plant studies 37. Willdenowia 6: 283–284.
2408. ———. 1972. Plantae Hobdyanae Kauaienses. Hawaiian plant studies 31. Pacific Sci. 26: 275–295.
2409. ———. 1972. *Canavalia kauensis* (Leguminosae), a new species from the island of Hawaii. Hawaiian plant studies 39. Pacific Sci. 26: 409–414.
2410. ———. 1973. A new living species of *Gouania* (Rhamnaceae) on Oahu, Hawaiian Islands. Hawaiian plant studies 40. Pacific Sci. 27: 269–273.
2411. ———. 1973. List and summary of the flowering plants in the Hawaiian Islands. Pacific Trop. Bot. Gard. Mem. 1: 1–519.
2412. ———. 1974. *Luteidiscus*, new genus (Compositae). Pacific plant studies 25. Bot. Jahrb. Syst. 94: 549–555.
2413. ———. 1974 [1975]. *Skottsbergiliana* new genus (Cucurbitaceae) of Hawaii Island. Hawaiian plant studies 41. Pacific Sci. 28: 457–462.
2414. ———. 1975. Plantae Hobdyanae Kauaienses II. Hawaiian plant studies 45. Bot. Mag. (Tokyo) 88: 59–64.
2415. ———. 1975. Ethnobotany and flora of Nualolo, Kauai. [Hawaiian plant studies 48]. Bull. Pacific Trop. Bot. Gard. 5: 24–28.
2416. ———. 1975. *Cenchrus laysanensis* (Gramineae) of the Leeward Islands. Hawaiian plant studies 47. Phytologia 31: 22–24.
2417. ———. 1975. The variability of the Hawaiian *maile* (*Alyxia olivaeformis*), Apocynaceae. Hawaiian plant studies 49. Phytologia 32: 377–386.
2418. ———. 1975. More variants of *Scaevola taccada* (Goodeniaceae). Hawaiian plant studies 43. Proc. Biol. Soc. Wash. 88: 73–76.
2419. ———. 1976. List of plants introduced to Hawaii by the ancestors of the Hawaiian people. Hui Kokua No Nā Mea-Kanu Māoli O Hawai'i Publ. 1: 1.
2420. ———. 1976. Evaluation of *Waltheria indica* L. and *W. americana* L. (Sterculiaceae). Pacific plant studies 28. Phytologia 33: 89–92.

2421. ———. 1976. The status of *Jacquemontia sandwicensis* (Convolvulaceae). Hawaiian plant studies 57. Phytologia 33: 423–428.
2422. ———. 1976. Miscellaneous taxonomic notes. Hawaiian plant studies 58. Phytologia 34: 147–148.
2423. ———. 1976. A new *Peperomia* (Piperaceae) from Maui. Hawaiian plant studies 48 [i.e., 72]. Phytologia 34: 362–364.
2424. ———. 1976. A new form of *Alyxia olivaeformis* Gaud. (Apocynaceae). Hawaiian plant studies 50. Phytologia 34: 388–389.
2425. ———. 1976. Two lectotypes in *Charpentiera* (Amaranthaceae). Hawaiian plant studies 51. Phytologia 35: 132.
2426. ———. 1976. New combinations in *Zanthoxylum* (Rutaceae). Hawaiian plant studies 44. Rhodora 78: 73–74.
2427. ———. 1976. A new species of *Panicum* (Gramineae) from Molokai. Hawaiian plant studies 42. Rhodora 78: 542–545.
2428. ———. 1976 [1977]. Biography of David Nelson, and an account of his botanizing in Hawaii. Pacific Sci. 30: 1–5.
2429. ———. 1976 [1977]. New species of Hawaiian plants collected by David Nelson in 1779. Hawaiian plant studies 52. Pacific Sci. 30: 7–44.
2430. ———. 1977. The variations of *Alphitonia ponderosa* (Rhamnaceae). Hawaiian plant studies 59. Phytologia 35: 177–182.
2431. ———. 1977. Observations on Hawaiian *Panicum* and *Sapindus*. Hawaiian plant studies 61. Phytologia 36: 465–467.
2432. ———. 1977. The variations of *Delissea subcordata* Gaud., Lobeliaceae. Hawaiian plant studies 62. Phytologia 37: 417–419.
2433. ———. 1977. The native Hawaiian *Alternanthera* (Amaranthaceae). Hawaiian plant studies 63. Phytologia 37: 476–478.
2434. ———. 1977. Plants of the Sandwich Islands collected by Archibald Menzies. Phytologia 38: 1–6.
2435. ———. 1977. Revision of the genus *Pittosporum* in Hawaii. Hawaiian plant studies 64. Phytologia 38: 75–98.
2436. ———. 1978. *Ochrosia* (Apocynaceae) of the Hawaiian Islands, Hawaiian plant studies 60. Adansonia 18: 199–220.
2437. ———. 1978. Hawaiian plant studies 67. The Cucurbitaceae of Hawaii I. Generic revision. Bot. Jahrb. Syst. 99: 490–497.
2438. ———. 1978. Hawaiian plant studies 75. The Cucurbitaceae of Hawaii II. *Sicyocarya* (Gray) St. John, new genus. Bot. Jahrb. Syst. 100: 246–248.
2439. ———. 1978. St. John's Hawaiian plant novelties since 1972. Newslett. Hawaiian Bot. Soc. 17: 56–62.
2440. ———. 1978. Notes on Hawaiian *Psychotria* (Rubiaceae). Hawaiian plant studies 65. Phytologia 38: 225–226.
2441. ———. 1978. New combinations in Hawaiian Cucurbitaceae. Hawaiian plant studies 66. Phytologia 38: 407–408.
2442. ———. 1978. The new binomial *Myrsine helleri* (Myrsinaceae). Hawaiian plant studies 68. Phytologia 39: 107.

2443. ———. 1978. *Gardenia weissichii* of Oahu Island (Rubiaceae). Hawaiian plant studies 70. Phytologia 39: 108–111.
2444. ———. 1978. Plants of the Sandwich Islands collected by James Macrae. Hawaiian plant studies 77. Phytologia 39: 307–319.
2445. ———. 1978. Novelties in *Clermontia* and *Cyanea* (Lobeliaceae). Hawaiian plant studies 79. Phytologia 40: 97–98.
2446. ———. 1978. Notes on *Hesperomannia* (Compositae). Hawaiian plant studies 80. Phytologia 40: 241–242.
2447. ———. 1978. Revision of *Joinvillea* (Joinvilleaceae). Pacific plant studies 37. Phytologia 40: 369–374.
2448. ———. 1978. *Isodendrion christensenii* (Violaceae) of Kauai. Hawaiian plant studies 81. Phytologia 40: 375–378.
2449. ———. 1978 [1979]. The first collection of Hawaiian plants by David Nelson in 1779. Hawaiian plant studies 55. Pacific Sci. 32: 315–324.
2450. ———. 1979. The new *Dubautia nagatae* (Compositae) of Kauai. Hawaiian plant studies 78. Bull. Torrey Bot. Club 106: 1–3.
2451. ———. 1979. The fruit of *Gardenia weissichii* (Rubiaceae). Hawaiian plant studies 83. Phytologia 41: 144.
2452. ———. 1979. A new *Stenogyne* (Labiatae). Hawaiian plant studies 84. Phytologia 41: 305–308.
2453. ———. 1979. Plants collected in the Sandwich Islands by Thomas Nuttall. Hawaiian plant studies 85. Phytologia 41: 441–446.
2454. ———. 1979. Revision of *Nototrichium sandwicense* (Amaranthaceae). Hawaiian plant studies 87. Phytologia 42: 25–28.
2455. ———. 1979. *Metrosideros polymorpha* (Myrtaceae) and its variations. Hawaiian plant studies 88. Phytologia 42: 215–218.
2456. ———. 1979. Plants collected on the Sandwich Islands by George Barclay. Hawaiian plant studies 89. Phytologia 43: 281–286.
2457. ———. 1979. Resurrection of *Viola lanaiensis* Becker. Hawaiian plant studies 90. Phytologia 44: 323–324.
2458. ———. 1979 [1980]. The vegetation of Hawaii as seen on Captain Cook's voyage in 1779. Pacific Sci. 33: 79–83.
2459. ———. 1979 [1980]. Novelties in the genus *Pelea* (Rutaceae). Hawaiian plant studies 50 [i.e., 118]. Pacific Sci. 33: 165–171.
2460. ———. 1979 [1980]. The botany of Kauai Island, Hawaii, as seen on Captain Cook's voyage, 1778. Pacific Sci. 33: 325–326.
2461. ———. 1979 [1980]. The botany of Niihau Island, Hawaii, as seen on Captain Cook's voyage, 1778–1779. Pacific Sci. 33: 327.
2462. ———. 1979 [1980]. The native species of *Senecio* (Compositae) in Hawaii. Hawaiian plant studies 53. Pacific Sci. 33: 329–332.
2463. ———. 1979 [1980]. Monograph of the Hawaiian species of *Achyranthes* (Amaranthaceae). Hawaiian plant studies 56. Pacific Sci. 33: 333–350.
2464. ———. 1979 [1980]. New species of *Cyrtandra* (Gesneriaceae) from Kauai. Hawaiian plant studies 74. Pacific Sci. 33: 351–355.
2465. ———. 1979 [1980]. Classification of *Acacia koa* and relatives (Leguminosae). Hawaiian plant studies 93. Pacific Sci. 33: 357–367.

2466. ———. 1979 [1980]. The native Hawaiian species of *Morinda* (Rubiaceae). Hawaiian plant studies 94. Pacific Sci. 33: 369–379.

2467. ———. 1979 [1980]. A variety of *Colubrina oppositifolia* Brongn. ex Mann (Rhamnaceae). Hawaiian plant studies 95. Pacific Sci. 33: 381–383.

2468. ———. 1979 [1980]. *Cyrtandra* (Gesneriaceae) from Kauai and Maui. Hawaiian plant studies 96. Pacific Sci. 33: 385–393.

2469. ———. 1979 [1980]. Revision of the genus *Pandanus* Stickman. Part 42. *Pandanus tectorius* Parkins. ex Z and *Pandanus odoratissimus* L.f. Pacific Sci. 33: 395–401.

2470. ———. 1979 [1980]. Harold St. John—career synopsis and bibliography. Pacific Sci. 33: 435–447.

2471. ———. 1980. A new *Scaevola* (Goodeniaceae) from Kauai. Hawaiian plant studies 92. Bot. Mag. (Tokyo) 93: 181–184.

2472. ———. 1980. Lectotypes in the Lobeliaceae. Hawaiian plant studies 91. Phytologia 45: 30.

2473. ———. 1980. Evaluation of H. Léveillé's new Hawaiian species. Hawaiian plant studies 76. Phytologia 45: 289–294.

2474. ———. 1980. A new variety of *Nototrichium viride* (Amaranthaceae). Hawaiian plant studies 98. Phytologia 45: 295.

2475. ———. 1980. Key to Hawaiian species of *Rauvolfia* (Apocynaceae). Hawaiian plant studies 99. Phytologia 45: 354–355.

2476. ———. 1980. Variations of *Reynoldsia* (Araliaceae) in the Hawaiian Is. Hawaiian plant studies 100. Phytologia 46: 154.

2477. ———. 1980. Habitats of the endangered plants of the island of Hawai'i (Abstr.). Proc. 3rd Conf. Nat. Sci., Hawaii Volcanoes Natl. Park, p. 273.

2478. ———. 1980. Plants introduced to Hawai'i by ancestors of the Hawaiian people: *in* Lindo, C. K., and N. A. Mower (eds.), Polynesian seafaring heritage. Publ. privately, Honolulu, pp. 122–123.

2479. ——— (C. A. Corn, ed.). 1981. Rare endemic plants of the Hawaiian Islands. Hawaii, Dept. Land Nat. Resources, Honolulu, 74 unnum. pp.

2480. ———. 1981. New species from Kauai. Hawaiian plant studies 97. Pacific Sci. 35: 97–103.

2481. ———. 1981. *Lepidium orbiculare* (Cruciferae) of Kauai. Hawaiian plant studies 105. Phytologia 47: 371–373.

2482. ———. 1981. Novelties in *Panicum* (Gramineae) from Kahoolawe. Hawaiian plant studies 102. Phytologia 47: 374–378.

2483. ———. 1981. Additions to *Cyanea* (Lobeliaceae) of Oahu and Maui. Hawaiian plant studies 106. Phytologia 48: 143–145.

2484. ———. 1981. New combinations in *Railliardia* (Compositae). Hawaiian plant studies 111. Phytologia 49: 291.

2485. ———. 1981. A variety of *Zanthoxylum hawaiiense* Hbd. (Rutaceae). Hawaiian plant studies 110. Phytologia 49: 292.

2486. ———. 1982. Five new species in *Pelea* (Rutaceae). Hawaiian plant studies 108. Bot. Mag. (Tokyo) 95: 139–146.
2487. ———. 1982. Vernacular plant names used on Niʻihau Island. Hawaiian plant studies 69. Occas. Pap. Bernice P. Bishop Mus. 25(3): 1–10.
2488. ———. 1982. A new species of *Labordia* (Loganiaceae). Hawaiian plant studies 115. Phytologia 52: 145–147.
2489. ———. 1982 [1983]. Monograph of *Trematolobelia* (Lobeliaceae). Hawaiian plant studies 107. Pacific Sci. 36: 483–506.
2490. ———. 1983. A new *Hesperomannia* (Compositae) from Maui Island: Hawaiian plant studies 116. Ann. Missouri Bot. Gard. 70: 198–200.
2491. ———. 1983. Novelties in Hawaiian *Pittosporum* (Pittosporaceae). Hawaiian plant studies 117. Bot. Mag. (Tokyo) 96: 313–317.
2492. ———. 1983. The early plants of Hawaii. Kukui Leaf 9(2): 3–4.
2493. ———. 1983. Hawaiian novelties in *Clermontia* (Lobeliaceae). Hawaiian plant studies 114. Nord. J. Bot. 3: 543–545.
2494. ———. 1983. A new *Carex* (Cyperaceae) of the section *Stellulatae*. Hawaiian plant studies 113. Pacific Sci. 37: 25–26.
2495. ———. 1984. Novelties among the Phanerogamae, Hawaiian plant studies 123. Bull. Torrey Bot. Club 111: 479–482.
2496. ———. 1984. Revision of the Hawaiian species of *Santalum* (Santalaceae). Hawaiian plant studies 109. Phytologia 55: 217–226.
2497. ———. 1984 [1985]. Novelties in *Lipochaeta* (Compositae). Hawaiian plant studies 119. Pacific Sci. 38: 253–282.
2498. ———. 1985. Monograph of the Hawaiian species of *Pleomele* (Liliaceae). Hawaiian plant studies 103. Pacific Sci. 39: 171–190.
2499. ———. 1985. *Touchardia angusta* (Urticaceae). Hawaiian plant studies 128. Phytologia 58: 231–232.
2500. ———. 1985. Typification of the Hawaiian plants described by Asa Gray from the Wilkes Expedition collections, and an enumeration of the other Hawaiian collections. Hawaiian plant studies 54. Rhodora 87: 565–595.
2501. ———. 1985. Earlier dates of valid publication of some genera and species in Gaudichaud's Botany of the Uranie Voyage. Taxon 34: 663–665.
2502. ———. 1986. Hawaiian plant studies 104. Revision of *Sanicula* (Umbelliferae) in Hawaii. Bot. Jahrb. Syst. 108: 51–62.
2503. ———. 1986. *Delissea konaensis* sp. nov. (Lobeliaceae). Hawaiian plant studies 131. Phytologia 59: 315–316.
2504. ———. 1986. Revision of the Hawaiian *Diopyros* [*Diospyros*] (Ebenaceae). Hawaiian plant studies 120. Phytologia 59: 389–405.
2505. ———. 1986. *Nothocestrum inconcinnum* sp. nov. (Solanaceae). Hawaiian plant studies 132. Phytologia 61: 343–345.
2506. ——— and G. D. Carr. 1981. Two new species of *Dubautia* (Compositae) from Kauai. Bull. Torrey Bot. Club 108: 198–204.
2507. ——— and R. E. Daehler. 1980. The flowering plants of Kaʻula Island, Hawaiian Islands. Bull. Pacific Trop. Bot. Gard. 10: 3–7.

2508. ——— and F. R. Fosberg. 1938. Identification of Hawaiian plants: a key to the families of dicotyledons of the Hawaiian Islands, descriptions of the families, and list of the genera. Occas. Pap. Univ. Hawaii 36: 1–53.

2509. ——— and ———. 1939. A new variety of *Ruppia maritima* (Ruppiaceae) from the tropical Pacific. Occas. Pap. Bernice P. Bishop Mus. 15(16): 175–178.

2510. ——— and ———. 1940. Identification of Hawaiian plants: part 2. A key to the families and genera of the gymnosperms and of the Monocotyledons of the Hawaiian Islands, with description of the families. Occas. Pap. Univ. Hawaii 41: 1–47.

2511. ——— and L. Frederick. 1949. A second Hawaiian species of *Alectryon* (Sapindaceae). Hawaiian plant studies 17. Pacific Sci. 3: 296–301.

2512. ——— and D. R. Herbst. 1975. An earlier name for *Bobea elatior* (Rubiaceae). Hawaiian plant studies 46. Phytologia 30: 7–8.

2513. ——— and E. Y. Hosaka. 1932. Noxious weeds of the Hawaiian pineapple fields (Abstr.). Proc. Hawaiian Acad. Sci. 7: 7.

2514. ——— and ———. 1932. Weeds of the pineapple fields of the Hawaiian Islands. Univ. Hawaii Res. Publ. 6: 1–196.

2515. ——— and ———. 1935. Hawaiian *Panicum*, *Metrosideros*, *Sanicula*, *Lobelia*, and *Rollandia*. Occas. Pap. Bernice P. Bishop Mus. 11(13): 1–18.

2516. ——— and ———. 1938. Notes on Hawaiian species of *Lobelia*. Hawaiian plant studies 5. Occas. Pap. Bernice P. Bishop Mus. 14(8): 117–126.

2517. ——— and J. R. Kuykendall. 1949. Revision of the native Hawaiian species of *Gardenia* (Rubiaceae). Hawaiian plant studies 15. Brittonia 6: 431–449.

2518. ———, K. M. Nagata, and F. R. Ganders. 1983. A new subspecies of *Bidens* (Asteraceae) from Maui. Lyonia 2: 17–21.

2519. ——— and W. R. Philipson. 1962. An account of the flora of Henderson Island, South Pacific Ocean. Trans. Roy. Soc. New Zealand, Bot. 1: 175–194.

2520. ——— and W. B. Storey. 1950. Diagnoses of new species of *Cyrtandra* (Gesneriaceae) from Oahu, Hawaiian Islands. Hawaiian plant studies 20. Occas. Pap. Bernice P. Bishop Mus. 20(6): 77–88.

2521. ——— and M. Titcomb. 1983. The vegetation of the Sandwich Islands as seen by Charles Gaudichaud in 1819. Occas. Pap. Bernice P. Bishop Mus. 25(9): 1–16.

2522. Saint-Yves, A. 1929. Contribution a l'étude des *Festuca* (subgen. *Eu-Festuca*) de l'Afrique australe et de l'Océanie. Candollea 4: 65–119.

2523. ———. 1930. Apercu sur la distribution géographique des *Festuca* (subgen. *Eu-Festuca*). Candollea 4: 146–165.

2524. Sakai, A. K., and S. G. Weller. 1986. Breeding systems in *Schiedea* (Caryophyllaceae), an endemic Hawaiian genus (Abstr.). Amer. J. Bot. 73: 675–676.

2525. Sakai, W. S., S. S. Shiroma, and M. A. Nagao. 1984. A study of raphide

microstructure in relation to irritation. Scanning Electron Microscop. 2: 979–986.
2526. Sakimura, K. 1947. Virus transmission by *Cuscuta sandwichiana*. Phytopathology 37: 66–67.
2527. Saleh, N. A. M., and G. H. N. Towers. 1972. Flavonoids of *Metrosideros polymorpha* (Myrtaceae). Experientia 28: 787.
2528. Salisbury, R. A. 1808. Some remarks of the plants now referred to *Sophora*, with characters of the genus *Edwardsia*. Trans. Linn. Soc. London 9: 296–300.
2529. Salvoza, F. M. 1936. *Rhizophora*. Nat. Appl. Sci. Bull. Univ. Philipp. 5: 179–256.
2529a. Santos, J. V. 1950. A revision of the grass genus *Garnotia*. Nat. Appl. Sci. Bull. Univ. Philipp. 10: 1–179.
2530. Sastrapradja, D. S. 1965. A study of the variations in wood anatomy of Hawaiian *Metrosideros* (Myrtaceae). Ph.D. dissertation, Univ. Hawaii, Honolulu, 196 pp.
2531. ——— and C. H. Lamoureux. 1969. Variations in wood anatomy of Hawaiian *Metrosideros* (Myrtaceae). Ann. Bogor. 5: 1–83.
2532. Sastrapradja, S. N. 1967. Comparative anatomy of Hawaiian *Peperomia* (Piperaceae) species. Ph.D. dissertation, Univ. Hawaii, Honolulu, 166 pp.
2533. ———. 1968. On the morphology of the flower in *Peperomia* (Piperaceae) species. Ann. Bogor. 4: 235–244.
2534. ———. 1968. Chromosome study of Hawaiian *Peperomia* (Piperaceae) species. Ann. Bogor. 4: 245–251.
2535. Sato, D. J., S. Gemeno, Jr., and W. S. Suzuki. n.d. Plants of Kaena. Camp H. R. Erdman, YMCA of Honolulu, 33 pp.
2536. Sauer, J. 1964. Revision of *Canavalia*. Brittonia 16: 106–181.
2537. Scanlan, G. M. 1942. A study of the genus *Cyperus* in the Hawaiian Islands. Catholic Univ. Amer., Biol. Ser. 41: 1–62.
2538. Schauer, J. C. 1843. Apocyneae R. Br.: *in* Meyen, F. J. F., Observationes botanicas in itinere circum terras institutas . . . Nov. Actorum Acad. Caes. Leop.-Carol. Nat. Cur. 19, Suppl. 1: 361–363.
2539. ———. 1843. Monocotyledoneae (exclusis Glumaceis): *in* Meyen, F. J. F., Observationes botanicas in itinere circum terras institutas . . . Nov. Actorum Acad. Caes. Leop.-Carol. Nat. Cur. 19, Suppl. 1: 425–450.
2540. Schauinsland, H. 1899. Drei Monate einer Koralleninsel. (Laysan). M. Nössler, Bremen, 104 pp.
2541. ———. 1900. Ein Besuch auf Molokai, der Insel der Aussätzigen. Abh. Naturwiss. Vereine Bremen 16: 513–543.
2542. Scheuer, P. J. 1955. The constituents of *mokihana* (*Pelea anisata* Mann). Chem. & Industr. 1955: 1257–1258.
2543. ———. 1961. Natural products from Hawaiian plants (Abstr.). Proc. Hawaiian Acad. Sci. 36: 18.

2544. ———, M. Y. Chang, and H. Fukami. 1963. Hawaiian plant studies. X. The structure of mauiensine. J. Org. Chem. 28: 2641–2643.
2545. ———, ———, and C. E. Swanholm. 1962. Hawaiian plant studies. VIII. Isolation of chelerythrine and dihydrochelerythrine from *Fagara semiarticulata*. J. Org. Chem. 27: 1472–1473.
2546. ———, L. P. Horigan, and W. R. Hudgins. 1962. A survey for alkaloids in Hawaiian plants, III. Pacific Sci. 16: 63–69.
2547. ——— and T. J. Horigan. 1959. A new carbonyl compound from *Piper methysticum* Forst. Nature 184: 979.
2548. ——— and W. R. Hudgins. 1964. Major constituents of the essential oil of *Pelea christophersenii* St. John & Hume. Perfumery & Essential Oil Rec. 55: 723–724.
2549. ——— and J. T. H. Metzger. 1961. Hawaiian plant studies. VI. The structure of holeinine. J. Org. Chem. 26: 3069–3071.
2550. ——— and T. R. Pattabhiraman. 1965. Hawaiian plant studies. XIII. Isolation of a canthinone from a member of the family Amaranthaceae. Lloydia 28: 95–100.
2551. ———, C. E. Swanholm, L. A. Madamba, and W. R. Hudgins. 1963. The constituents of *Tacca leontopetaloides*. Lloydia 26: 133–140.
2552. ——— and F. Werny. 1964. Alkaloids of Hawaiian Rutaceae. Symp. Phytochem., Univ. Hong Kong, 1961, pp. 35–36.
2553. Schindler, A. K. 1905. Halorrhagaceae. Pflanzenr. IV. 225 (Heft 23): 1–133.
2554. ———. 1926. Desmodii generumque affinium species et combinationes novae. II. Repert. Spec. Nov. Regni Veg. 22: 250–288.
2555. ———. 1928. Die Desmodiinen in der botanischen Literatur nach Linné. Repert. Spec. Nov. Regni Veg. Beih. 49: 1–371.
2556. Schinz, H. 1931. Beiträge zur Kenntnis der afrikanischen Flora (XXXVI). Vierteljahrsschr. Naturf. Ges. Zürich 76: 133–146.
2557. Schlechtendal, D. F. L. von. 1851. Hortorum botanicorum plantae novae et adnotationes in seminum indicibus depositae. Linnaea 24: 154–238.
2558. Schlittler, J. 1940. Monographie der Liliaceengattung *Dianella* Lam. Buchdruckerei Fluntern, Zurich, 284 pp.
2559. Schmid, R. 1972. Floral anatomy of Myrtaceae I. *Syzygium*. Bot. Jahrb. Syst. 92: 433–489.
2560. Schönland, S. 1889. Campanulaceae. Nat. Pflanzenfam. IV. 5: 40–70.
2561. ———. 1894. Campanulaceae. Nat. Pflanzenfam. IV. 5: 394.
2562. Schopmeyer, C. S. 1974. Seeds of woody plants in the United States. U.S.D.A. Forest Serv., Agric. Handbook 450: 1–883.
2563. Schroeder, L. J. 1926. Data of certain American Piperaceae. Candollea 3: 121–140.
2564. Schröter, C. 1922–1923. Naturwissenschaftliche Skizzen von einer Reise um die Welt. Merkbl. Volkshochschule Zürich 1922–1923, 19 pp.
2565. Schultz, C. H. 1856. Verzeichniss der Cassiniaceen, welche Herr Edelstan Jardin in den Jahren 1853–55 auf den Inseln des stillen Oceans gesammelt hat. Flora 39: 353–362.

2566. Schulz, O. E. 1903. Monographie der Gattung *Cardamine*. Bot. Jahrb. Syst. 32: 280–623.
2567. Schumann, K. 1895. Apocynaceae. Nat. Pflanzenfam. IV. 2: 109–189.
2568. ——— and K. Lauterbach. 1900–1901. Die Flora der deutschen Schutzgebiete in der Südsee. Gebrüder Borntraeger, Leipzig, 613 pp.
2569. Scott, G. A. J. 1969. Relationships between vegetation and soil avalanching in the high rainfall areas of Oahu, Hawaii. Master's thesis, Univ. Hawaii, Honolulu, 98 pp.
2570. Scott, J. M., J. D. Jacobi, and F. L. Ramsey. 1981. Avian surveys of large geographical areas: a systematic approach. Wildlife Soc. Bull. 9: 190–200.
2570a. ———, S. Mountainspring, F. L. Ramsey, and C. B. Kepler. 1986. Forest bird communities of the Hawaiian Islands: their dynamics, ecology, and conservation. Stud. Avian Biol. 9: 1–431.
2571. Scowcroft, P. G. 1970. Natural regeneration of *koa*. Newslett. Hawaiian Bot. Soc. 9: 40–41.
2572. ———. 1971. *Koa*—monarch of Hawaiian forests. Newslett. Hawaiian Bot. Soc. 10: 23–26.
2573. ———. 1976. *Māmane* forest decline on Mauna Kea: a reality or myth. Proc. 1st Conf. Nat. Sci., Hawaii Volcanoes Natl. Park, pp. 187–198.
2574. ———. 1978. Direct sowing of treated *mamane* seeds: an ineffective regeneration technique. Proc. 2nd Conf. Nat. Sci., Hawaii Volcanoes Natl. Park, pp. 247–255.
2575. ———. 1978. Germination of *Sophora chrysophylla* increased by presowing treatment. U.S.D.A. Forest Serv. Res. Note PSW-327: 1–6.
2576. ———. 1981. Regeneration of *mamane*: effects of seedcoat treatment and sowing depth. Forest Sci. 27: 771–779.
2577. ———. 1983 [1984]. Tree cover changes in *māmane* (*Sophora chrysophylla*) forests grazed by sheep and cattle. Pacific Sci. 37: 109–119.
2578. ———. 1984. Vegetation response to protection from goats in a montane *koa* parkland ecosystem (Abstr.). Proc. 5th Conf. Nat. Sci., Hawaii Volcanoes Natl. Park, p. 85.
2579. ———. 1986. Fine litterfall and leaf decomposition in a montane *koa–ohia* rain forest. Proc. 6th Conf. Nat. Sci., Hawaii Volcanoes Natl. Park, pp. 66–82.
2580. ——— and J. G. Giffin. 1983. Feral herbivores suppress *mamane* and other browse species on Mauna Kea, Hawaii. J. Range Managem. 36: 638–645.
2581. ——— and R. E. Nelson. 1976. Disturbance during logging stimulates regeneration of *koa*. U.S.D.A. Forest Serv. Res. Note PSW-306: 1–7.
2582. ——— and H. F. Sakai. 1983. Impact of feral herbivores on *mamane* forests of Mauna Kea, Hawaii: bark stripping and diameter class structure. J. Range Managem. 36: 495–498.
2583. ——— and ———. 1984. Stripping of *Acacia koa* bark by rats on Hawaii and Maui. Pacific Sci. 38: 80–86.

2584. ——— and H. B. Wood. 1976 [1977]. Reproduction of *Acacia koa* after fire. Pacific Sci. 30: 177–186.
2585. Sedgwick, T. F. 1912. The story of taro. Mid-Pacific Mag. 3: 266–273.
2586. Seemann, B. 1852. Notes on the Sandwich Islands. J. Bot. (Hooker) 4: 335–341.
2587. ———. 1853. Die Flora von Oahu. Bonplandia 1: 30–32.
2588. ———. 1853. Narrative of the voyage of H.M.S. *Herald* during the years 1845–51, under the command of captain Henry Kellett, R.N., C.B.; being a circumnavigation of the globe, and three cruizes to the arctic regions in search of Sir John Franklin. Vol. II. Reeve and Co., London, 302 pp.
2589. ———. 1862. Notizen über Südsee-Pflanzen. Bonplandia 10: 153–155.
2590. ———. 1863. The Solana of tropical Polynesia. J. Bot. 1: 206–211.
2591. ———. 1864–1868. Revision of the natural order Hederaceae. J. Bot. 2: 235–250, 289–309 (1864); 3: 73–81, 173–181, 265–276, 361–363 (1865); 4: 293–299, 352–353 (1866); 5: 236–239, 285–286 (1867); 6: 52–58, 129–142, 161–165 (1868).
2592. ———. 1865–1873. Flora vitiensis: a description of the plants of the Viti or Fiji Islands with an account of their history, uses, and properties. L. Reeve and Co., London, 453 pp.
2593. ———. 1868. On two new genera of Smilacineae. J. Bot. 6: 193–194, 257–258.
2594. Selling, O. H. 1941. Hawaii-öarnas lobeliacéer och drepanider. Popular Biologisk Revy 1: 1–6.
2595. ———. 1942. The post-glacial vegetation history of the Hawaiian Islands. Acta Horti Gothob. 15: 31–34.
2596. ———. 1947. Studies in Hawaiian pollen statistics. Part II. The pollens of the Hawaiian phanerogams. Special Publ. Bernice P. Bishop Mus. 38: 1–430.
2597. ———. 1948. Studies in Hawaiian pollen statistics, Part III. On the late quaternary history of the Hawaiian vegetation. Special Publ. Bernice P. Bishop Mus. 39: 1–154.
2598. ———. 1951. A contribution to the history of the Hawaiian vegetation. Svensk Bot. Tidskr. 45: 12–41.
2599. ———. 1952. Pollenanalys: *in* Sjöstedt, E. (ed.), Ny Kunskap. Natur Och Kultur, Stockholm, pp. 362–387.
2600. ———. 1953. Late Quaternary history of the Hawaiian vegetation (Abstr.). Proc. 7th Pacific Sci. Congr., New Zealand 5: 99.
2601. Senn, H. A. 1939. The North American species of *Crotalaria*. Rhodora 41: 317–367.
2602. Setchell, W. A. 1928. Migration and endemism with reference to Pacific insular floras. Proc. 3rd Pan-Pacific Sci. Congr., Tokyo 1: 869–875.
2603. ———. 1935. Pacific insular floras and Pacific paleogeography. Amer. Naturalist 69: 289–310.
2604. Shan, R. H., and L. Constance. 1951. The genus *Sanicula* (Umbelliferae) in the Old World and the New. Univ. Calif. Publ. Bot. 25: 1–78.

2605. Sher, S. A. 1954. Observations on plant-parasitic nematodes in Hawaii. Pl. Dis. Reporter 38: 687–689.
2606. Sherff, E. E. 1920. Studies in the genus *Bidens*. V. Bot. Gaz. (Crawfordsville) 70: 89–109.
2607. ———. 1923. Studies in the genus *Bidens*. VI. Bot. Gaz. (Crawfordsville) 76: 144–166.
2608. ———. 1925. New or otherwise noteworthy Compositae. II. Bot. Gaz. (Crawfordsville) 80: 367–389.
2609. ———. 1926. Studies in the genus *Bidens*. VII. Bot. Gaz. (Crawfordsville) 81: 25–54.
2610. ———. 1928. Studies in the genus *Bidens*. VIII. Bot. Gaz. (Crawfordsville) 85: 1–29.
2611. ———. 1928. Studies in the genus *Bidens*. IX. Bot. Gaz. (Crawfordsville) 86: 435–447.
2612. ———. 1929. New or otherwise noteworthy Compositae. III. Bot. Gaz. (Crawfordsville) 88: 285–309.
2613. ———. 1930. New or otherwise noteworthy Compositae. IV. Bot. Gaz. (Crawfordsville) 89: 362–373.
2614. ———. 1931. New or otherwise noteworthy Compositae. VI. Bot. Gaz. (Crawfordsville) 91: 308–319.
2615. ———. 1931. New or otherwise noteworthy Compositae. VII. Bot. Gaz. (Crawfordsville) 92: 202–209.
2616. ———. 1932. Studies in the genus *Bidens*. X. Family Compositae. Bot. Gaz. (Crawfordsville) 93: 213–220.
2617. ———. 1933. Some new or otherwise important Compositae of the Hawaiian Islands. Amer. J. Bot. 20: 616–619.
2618. ———. 1933. New or otherwise noteworthy Compositae. VIII. Bot. Gaz. (Crawfordsville) 94: 589–597.
2619. ———. 1933. New or noteworthy Compositae. IX. Bot. Gaz. (Crawfordsville) 95: 78–103.
2620. ———. 1934. Some new or otherwise important Labiatae of the Hawaiian Islands. Amer. J. Bot. 21: 698–701.
2621. ———. 1934. A study in the genus *Tetramolopium* Nees (family: Compositae). Bot. Gaz. (Crawfordsville) 95: 498–502.
2622. ———. 1934. Some new or otherwise noteworthy members of the families Labiatae and Compositae. Bot. Gaz. (Crawfordsville) 96: 136–153.
2623. ———. 1935. New or otherwise noteworthy Compositae. X. Amer. J. Bot. 22: 705–710.
2624. ———. 1935. Revision of *Tetramolopium*, *Lipochaeta*, *Dubautia* and *Railliardia*. Bernice P. Bishop Mus. Bull. 135: 1–136.
2625. ———. 1935. Revision of *Haplostachys*, *Phyllostegia*, and *Stenogyne*. Bernice P. Bishop Mus. Bull. 136: 1–101.
2626. ———. 1936. Additions to the genus *Euphorbia* L. and to certain genera of the Compositae. Bot. Gaz. (Crawfordsville) 97: 580–609.

2627. ———. 1936. Revision of the genus *Coreopsis*. Field Mus. Nat. Hist., Bot. Ser. 11: 277–475.

2628. ———. 1937. Certain new plants from Hawaii and Mexico. Amer. J. Bot. 24: 88–90.

2629. ———. 1937. The genus *Bidens*. Part 1. Field Mus. Nat. Hist., Bot Ser. 16: 1–346.

2630. ———. 1938 [1937]. Revision of the Hawaiian species of *Euphorbia* L. Ann. Missouri Bot. Gard. 25: 1–94.

2631. ———. 1938. Studies in the genus *Labordia* Gaud., with a new variety in *Megalodonta* E. L. Greene. Amer. J. Bot. 25: 579–589.

2632. ———. 1939. Genus *Labordia*. Field Mus. Nat. Hist., Bot. Ser. 17: 445–546.

2633. ———. 1939. Additional studies of the Hawaiian Euphorbiaceae. Field Mus. Nat. Hist., Bot. Ser. 17: 547–576.

2634. ———. 1939. Some new or otherwise noteworthy Labiatae and Compositae. Field Mus. Nat. Hist., Bot. Ser. 17: 577–612.

2635. ———. 1941. New or otherwise noteworthy plants from the Hawaiian Islands. Amer. J. Bot. 28: 18–31.

2636. ———. 1941. Additions to our knowledge of the American and Hawaiian floras. Field Mus. Nat. Hist., Bot. Ser. 22: 407–441.

2637. ———. 1942. Some recently collected specimens of *Schiedea* (Caryophyllaceae) and of Mexican Compositae. Amer. J. Bot. 29: 332–333.

2638. ———. 1942. Revision of the Hawaiian members of the genus *Pittosporum* Banks. Field Mus. Nat. Hist., Bot. Ser. 22: 467–566, 574–580.

2639. ———. 1942. Some new or otherwise noteworthy Mexican Coreopsideae (genera *Heterosperma* Cav. and *Bidens* L.) and a note on *Xylosma hawaiiense* Seem. Field Mus. Nat. Hist., Bot. Ser. 22: 567–573.

2640. ———. 1943. Some additions to our knowledge of the genus *Schiedea* Cham. & Schlecht. Amer. J. Bot. 30: 606–608.

2641. ———. 1944. Some additions to our knowledge of the flora of the Hawaiian Islands. Amer. J. Bot. 31: 151–161.

2642. ———. 1944. New or otherwise noteworthy American and Hawaiian Coreopsideae. Amer. J. Bot. 31: 277–281.

2643. ———. 1945. Some additions to the genus *Dodonaea* L. (fam. Sapindaceae). Amer. J. Bot. 32: 202–214.

2644. ———. 1945. Revision of the genus *Schiedea* Cham. & Schlecht. Brittonia 5: 308–335.

2645. ———. 1946. Some new or otherwise noteworthy dicotyledonous plants. Amer. J. Bot. 33: 499–510.

2646. ———. 1946. Remarks upon certain Hawaiian Labiatae and Compositae. Bull. Torrey Bot. Club 73: 184–193.

2647. ———. 1947. Further studies in the genus *Dodonaea*. Field Mus. Nat. Hist., Bot. Ser. 23: 269–317.

2648. ———. 1947. A preliminary study of the Hawaiian species of the genus *Rauvolfia* (Plum.) L. (family Apocynaceae). Field Mus. Nat. Hist., Bot. Ser. 23: 321–331.

2649. ———. 1947. Additions to the genera *Scalesia* Arn. and *Hidalgoa* Llave and Lex. (family Compositae). Field Mus. Nat. Hist., Bot. Ser. 23: 333–336.
2650. ———. 1948. A name for the "alpha" variety or forma of miscellaneous dicotyledonous plants. Brittonia 6: 332–342.
2651. ———. 1948. A new variety of *Gnaphalium sandwicensium* Gaud. in the Hawaiian Islands. Lloydia 11: 309.
2652. ———. 1949. Miscellaneous notes on dicotyledonous plants. Amer. J. Bot. 36: 499–511.
2653. ———. 1949. Some new or otherwise noteworthy dicotyledonous plants from the Hawaiian Islands. Occas. Pap. Bernice P. Bishop Mus. 20(1): 1–25.
2654. ———. 1950. A preliminary paper on the genus *Nototrichium* (A. Gray) Hillebr. (fam. Amaranthaceae). Bot. Leafl. 1: 2–4.
2655. ———. 1950. Notes on certain members of the Amaranthaceae, Caryophyllaceae, Euphorbiaceae and Compositae. Bot. Leafl. 2: 2–6.
2656. ———. 1951. Miscellaneous notes on new or otherwise noteworthy dicotyledonous plants. Amer. J. Bot. 38: 54–73.
2657. ———. 1951. Notes upon certain new or otherwise interesting plants of the Hawaiian Islands and Colombia. Bot. Leafl. 3: 2–8.
2658. ———. 1951. A revision of the Hawaiian Island genus *Nototrichium* Hillebr. (fam. Amaranthaceae). Bot. Leafl. 4: 2–21.
2659. ———. 1951. New entities in the genus *Cheirodendron* Nutt. ex Seem. (fam. Araliaceae) from the Hawaiian Islands. Bot. Leafl. 5: 2–14.
2660. ———. 1951. Two Hawaiian species of the genus *Sophora* L. (fam. Leguminosae). Bot. Leafl. 5: 24–25.
2661. ———. 1952. Further studies of Hawaiian Araliaceae: additions to *Cheirodendron helleri* Sherff and a preliminary treatment of the endemic species of *Reynoldsia* A. Gray. Bot. Leafl. 6: 6–19.
2662. ———. 1952. Additions to our knowledge of the genus *Tetraplasandra* A. Gray (fam. Araliaceae). Bot. Leafl. 6: 19–41.
2663. ———. 1952. Some new or otherwise noteworthy Compositae from the Hawaiian Islands. Bot. Leafl. 7: 2–6.
2664. ———. 1952. Notes on *Schiedea* Cham. & Schlecht. (fam. Caryophyllaceae) and *Phyllostegia* Benth. (fam. Labiatae) in the Hawaiian Islands. Bot. Leafl. 7: 6–7.
2665. ———. 1952. Contributions to our knowledge of the genera *Tetraplasandra* A. Gray and *Reynoldsia* A. Gray (fam. Araliaceae) in the Hawaiian Islands. Bot. Leafl. 7: 7–17.
2666. ———. 1952. *Munroidendron*, a new genus of araliaceous trees from the island of Kauai. Bot. Leafl. 7: 21–24.
2667. ———. 1953. Further notes on the genus *Tetraplasandra* A. Gray (fam. Araliaceae) in the Hawaiian Islands. Bot. Leafl. 8: 2–13.
2668. ———. 1953. Notes on miscellaneous dicotyledonous plants. Bot. Leafl. 8: 13–26.

2669. ———. 1954. Further notes upon the flora of the Hawaiian Islands. Bot. Leafl. 9: 2–10.
2670. ———. 1954. Revision of the genus *Cheirodendron* Nutt. ex Seem. for the Hawaiian Islands. Fieldiana, Bot. 29: 1–45.
2671. ———. 1955. Revision of the Hawaiian members of the genus *Tetraplasandra* A. Gray. Fieldiana, Bot. 29: 49–142.
2672. ———. 1956. Some recently collected dicotyledonous Hawaiian Island and Peruvian plants. Amer. J. Bot. 43: 475–478.
2673. ———. 1957. Further notes on Compositae (*Bidens* L., *Coreopsis* L., and *Dubautia* Gaud.), mostly in the herbarium of the British Museum of Natural History. Ann. Mag. Nat. Hist., Ser. 12. 10: 42–46.
2674. ———. 1958. Some notes upon the Hawaiian species of *Fagara* L. Amer. J. Bot. 45: 461–463.
2675. ———. 1960. Some dicotyledonous plants recently collected in the Hawaiian Islands. Brittonia 12: 170–175.
2676. ———. 1962. Miscellaneous notes on some American and Hawaiian dicotyledons. Occas. Pap. Bernice P. Bishop Mus. 22(12): 207–214.
2677. ———. 1964. Some recently collected dicotyledonous plants from the Hawaiian Islands and Mexico. Occas. Pap. Bernice P. Bishop Mus. 23(7): 121–127.
2678. ———. 1964. An annotated list of my botanical writings. Publ. privately, Chicago, 48 pp.
2679. Sherman, G. D. 1950. The genesis and morphology of Hawaiian ferruginous laterite crusts. Pacific Sci. 4: 315–322.
2680. Shigeura, G. T., and W. W. McCall. 1979. Trees and shrubs for windbreaks in Hawaii. Univ. Hawaii Coop. Extens. Serv. Circ. 447: 1–56.
2681. Shinbara, B. H. 1966. Noxious weed seeds of Hawaii. Dept. Agric., Div. Pl. Industr., Honolulu, 54 pp.
2682. Shirota, F. N. 1960. A minor alkaloid of *Rauvolfia sandwicensis* A. DC. Master's thesis, Univ. Hawaii, Honolulu, 30 pp.
2683. Shukis, C. H. 1951. The anatomy and megagametophyte development of *Scaevola gaudichaudiana* Chamisso. Master's thesis, Univ. Hawaii, Honolulu, 64 pp.
2684. Siegel, S. M., P. Carroll, C. Corn, and T. Speitel. 1970. Experimental studies on the Hawaiian silverswords (*Argyroxiphium* spp.): some preliminary notes on germination. Bot. Gaz. (Crawfordsville) 131: 277–280.
2685. Silow, R. A. 1953. The problems of trans-Pacific migration involved in the origin of the cultivated cottons of the New World. Proc. 7th Pacific Sci. Congr., New Zealand 5: 112–118.
2686. Simmonds, N. W. 1954. Notes on banana varieties in Hawaii. Pacific Sci. 8: 226–229.
2687. ———. 1962. The evolution of the bananas. Longmans, London, 170 pp.
2688. Simon, C. M., W. C. Gagné, F. G. Howarth, and F. J. Radovsky. 1984.

Hawaiʻi: a natural entomological laboratory. Bull. Entomol. Soc. Amer. 30: 8–17.

2689. Simons, T. R., C. B. Kepler, P. M. Simons, and A. K. Kepler. 1985. Hawaii's seabird islands, no. 2: Hulu Island and vicinity, Maui. ʻElepaio 45: 111–113.

2690. Sims, J. 1814. *Jacquinia aurantiaca*. Orange-flowered Jacquinia. Bot. Mag. 40: *pl. 1639*.

2691. Sinclair, I. 1885. Indigenous flowers of the Hawaiian Islands. Sampson, Low, Marston, Searle, and Rivington, London, 44 pls.

2692. Skolmen, R. G. 1963. Robusta eucalyptus wood: its properties and uses. U.S.D.A. Forest Serv. Res. Pap. PSW-9: 1–12.

2693. ———. 1968. Wood of *koa* and of black walnut similar in most properties. U.S.D.A. Forest Serv. Res. Note PSW-164: 1–4.

2694. ———. 1968. Natural durability of some woods used in Hawaii . . . preliminary findings. U.S.D.A. Forest Serv. Res. Note PSW-167: 1–7.

2695. ———. 1972. Specific gravity variation in robusta eucalyptus grown in Hawaii. U.S.D.A. Forest Serv. Res. Pap. PSW-78: 1–7.

2696. ———. 1974. Some woods of Hawaii . . . properties and uses of 16 commercial species. U.S.D.A. Forest Serv. Gen. Techn. Rep. PSW-8: 1–30.

2697. ———. 1974. Natural durability of some woods used in Hawaii . . . results of 9 1/2 years' exposure. U.S.D.A. Forest Serv. Res. Note PSW-292: 1–6.

2698. ———. 1975. Shrinkage and specific gravity variation in robusta eucalyptus wood grown in Hawaii. U.S.D.A. Forest Serv. Res. Note PSW-298: 1–6.

2699. ———. 1978. Vegetative propagation of *Acacia koa* Gray. Proc. 2nd Conf. Nat. Sci., Hawaii Volcanoes Natl. Park, pp. 260–273.

2700. ——— and D. M. Fujii. 1980. Growth and development of a pure stand of koa (*Acacia koa*) at Keauhou–Kilauea. Proc. 3rd Conf. Nat. Sci., Hawaii Volcanoes Natl. Park, pp. 301–310.

2701. ——— and M. O. Mapes. 1976. *Acacia koa* Gray plantlets from somatic callus tissue. J. Heredity 67: 114–115.

2702. ——— and ———. 1978. Aftercare procedures required for field survival of tissue culture propagated *Acacia koa*. Proc. Intl. Pl. Propagators' Soc. 28: 156–164.

2703. Skottsberg, C. 1925. Juan Fernandez and Hawaii. Bernice P. Bishop Mus. Bull. 16: 1–47.

2704. ———. 1926. Vascular plants from the Hawaiian Islands. I. Acta Horti Gothob. 2: 185–284.

2705. ———. 1927. Iakttagelser över blomningen hos *Cyanea hirtella* (H. Mann) Rock. Acta Horti Gothob. 3: 43–55.

2706. ———. 1927. *Artemisia*, *Scaevola*, *Santalum*, and *Vaccinium* of Hawaii. Bernice P. Bishop Mus. Bull. 43: 1–89.

2707. ———. 1928. On some arborescent species of *Lobelia* from tropical Asia. Acta Horti Gothob. 4: 1–26.

2708. ———. 1928. Remarks on the relative independency of Pacific floras. Proc. 3rd Pan-Pacific Sci. Congr., Tokyo 1: 914–920.
2709. ———. 1930. Further notes on Pacific sandalwoods. Acta Horti Gothob. 5: 135–145.
2710. ———. 1930. The geographical distribution of the sandalwoods and its significance. Proc. 4th Pacific Sci. Congr., Java 3: 435–442.
2711. ———. 1931. Remarks on the flora of the high Hawaiian volcanoes. Acta Horti Gothob. 6: 47–65.
2712. ———. 1931. Pipturi species Hawaiienses novae. Acta Horti Gothob. 7: 1–5.
2713. ———. 1931. The flora of the high Hawaiian volcanoes. Proc. 5th Int. Bot. Congr., Cambridge, pp. 91–97.
2714. ———. 1932. *Pipturus "albidus"* outside the Hawaiian Islands. Acta Horti Gothob. 7: 23–29.
2715. ———. 1933. *Vaccinium cereum* (L. fil.) Forst. and related species. Acta Horti Gothob. 8: 83–102.
2716. ———. 1934. Additional notes on *Santalum* and *Vaccinium* from the Pacific. Acta Horti Gothob. 9: 185–192.
2717. ———. 1934. *Astelia* and *Pipturus* of Hawaii. Bernice P. Bishop Mus. Bull. 117: 1–77.
2718. ———. 1934. Studies in the genus *Astelia* Banks et Solander. Kongl. Svenska Vetenskapsakad. Handl. 14(2): 1–106.
2719. ———. 1935. *Astelia*, an Antarctic-Pacific genus of Liliaceae. Proc. 5th Pacific Sci. Congr., Canada, pp. 3317–3327.
2720. ———. 1936. Vascular plants from the Hawaiian Islands. II. Acta Horti Gothob. 10: 97–193.
2721. ———. 1936. The arboreous Nyctaginaceae of Hawaii. Svensk Bot. Tidskr. 30: 722–743.
2722. ———. 1936. Antarctic plants in Polynesia: *in* Goodspeed, T. H. (ed.), Essays on geobotany in honor of William Albert Setchell. Univ. Calif. Press, Berkeley, pp. 291–311.
2723. ———. 1937. Further notes on *Vaccinium* of Hawaii. Acta Horti Gothob. 12: 145–151.
2724. ———. 1937. Further remarks on Hawaiian Artemisiae. Bot. Mag. (Tokyo) 51: 196–202.
2725. ———. 1937. Liliaceae of southeastern Polynesia. Occas. Pap. Bernice P. Bishop Mus. 13(18): 233–244.
2726. ———. 1937. Recent researches in *Astelia* B. and S. Trans. & Proc. Roy. Soc. New Zealand 67: 218–226.
2727. ———. 1938. Geographical isolation as a factor in species formation, and its relation to certain insular floras. Proc. Linn. Soc. London 150: 286–293.
2728. ———. 1939. A hybrid violet from the Hawaiian Islands. Bot. Not. 1939: 805–812.
2729. ———. 1939. Remarks on the Hawaiian flora. Proc. Linn. Soc. London 151: 181–186.

2730. ———. 1940. Observations on Hawaiian violets. Acta Horti Gothob. 13: 451–528.
2731. ———. 1940. En exkursion till Hawaii-öarna sommaren 1938. Ymer 1940: 1–22.
2732. ———. 1941. Plant succession on recent lava flows in the island of Hawaii. Kongl. Götheborgska Wetensk. Sam. Handl. Ser. B, 1(8): 1–32.
2733. ———. 1941. Report of the standing committee for the protection of nature in and around the Pacific for the years 1933–1938. Proc. 6th Pacific Sci. Congr., California 4: 499–546.
2734. ———. 1941. Report on Hawaiian bogs. Proc. 6th Pacific Sci. Congr., California 4: 659–661.
2735. ———. 1941. The flora of the Hawaiian Islands and the history of the Pacific Basin. Proc. 6th Pacific Sci. Congr., California 4: 685–701.
2736. ———. 1941. *Heimerliodendron*. Svensk Bot. Tidskr. 35: 364.
2737. ———. 1943. Dr. Sven Berggren's collection of Hawaiian vascular plants. Bot. Not. 1943: 358–372.
2738. ———. 1944. Vascular plants from the Hawaiian Islands. IV. Phanerogams collected during the Hawaiian Bog Survey 1938. Acta Horti Gothob. 15: 275–531.
2739. ———. 1944. On the flower dimorphism in Hawaiian Rubiaceae. Ark. Bot. 31A(4): 1–28.
2740. ———. 1945. The flower of *Canthium*. Ark. Bot. 32A(5): 1–12.
2741. ———. 1953. Chromosome numbers in Hawaiian flowering plants. Ark. Bot. 3: 63–70.
2742. ———. 1953. Report of the standing committee for the protection of nature in and around the Pacific for the years 1939–1948. Proc. 7th Pacific Sci. Congr., New Zealand 4: 586–611.
2743. ———. 1953. Vegetation of the Hawaiian Islands (Abstr.). Proc. 7th Pacific Sci. Congr., New Zealand 5: 92.
2744. ———. 1955. Notes on *Oreobolus* in the Hawaiian Islands. Arch. Soc. Zool. Bot. Fenn. "Vanamo" 9(Suppl.): 335–337.
2745. ———. 1964. Wikstroemiae novae Hawaiienses. Svensk Bot. Tidskr. 58: 177–183.
2746. ———. 1972. The genus *Wikstroemia* Endl. in the Hawaiian Islands. Acta Regiae Soc. Sci. Litt. Gothob., Bot. 1: 1–166.
2747. Sleumer, H. 1938. Die malesisch-pacifischen *Xylosma*-Arten. Notizbl. Bot. Gart. Berlin-Dahlem 14: 288–297.
2748. ———. 1963. Florae Malesianae precursores XXXVII. Materials towards the knowledge of the Epacridaceae mainly in Asia, Malaysia, and the Pacific. Blumea 12: 145–171.
2749. Smathers, G. A. 1966. Recovery of the devastated area: *in* Doty, M. S., and D. Mueller-Dombois, Atlas for bioecology studies in Hawaii Volcanoes National Park. Univ. Hawaii, Hawaii Bot. Sci. Pap. 2: 334–340.
2750. ———. 1966. Succession on new surfaces: *in* Doty, M. S., and D.

Mueller-Dombois, Atlas for bioecology studies in Hawaii Volcanoes National Park. Univ. Hawaii, Hawaii Bot. Sci. Pap. 2: 341–390.

2751. ———. 1968. A preliminary survey of the phytogeography of Kipahulu Valley: in Warner, R. E. (ed.), Scientific report of the Kipahulu Valley Expedition. Nature Conservancy, San Francisco, pp. 55–86.

2752. ———. 1972. Invasion, early succession and recovery of vegetation on the 1959 Kilauea volcanic surfaces, Hawaii Volcanoes National Park, Hawaii. Ph.D. dissertation, Univ. Hawaii, Honolulu, 326 pp.

2753. ———. 1976. Fifteen years of vegetation invasion and recovery after a volcanic eruption in Hawaii. Proc. 1st Conf. Nat. Sci., Hawaii Volcanoes Natl. Park, pp. 207–211.

2754. ——— and D. E. Gardner. 1978. Stand analysis of an invading firetree (*Myrica faya* Aiton) population, Hawai'i. Proc. 2nd Conf. Nat. Sci., Hawaii Volcanoes Natl. Park, pp. 274–288.

2755. ——— and ———. 1979 [1980]. Stand analysis of an invading firetree (*Myrica faya* Aiton) population, Hawaii. Pacific Sci. 33: 239–255.

2756. ——— and D. Mueller-Dombois. 1972. Invasion and recovery of vegetation after a volcanic eruption in Hawaii. U.S. IBP Island Ecosystems IRP Techn. Rep. 10: 1–172.

2757. ——— and ———. 1974. Invasion and recovery of vegetation after a volcanic eruption in Hawaii. Natl. Park Serv. Sci. Monogr. Ser. 5: 1–129.

2758. Smith, A. C. 1970. The Pacific as a key to flowering plant history. Univ. Hawaii Harold L. Lyon Arbor. Lecture 1: 1–26.

2759. ———. 1973. Studies of Pacific island plants. XXVI. *Metrosideros collina* (Myrtaceae) and its relatives in the southern Pacific. Amer. J. Bot. 60: 479–490.

2760. ———. 1974. Studies of Pacific island plants. XXVII. The genus *Gardenia* (Rubiaceae) in the Fijian region. Amer. J. Bot. 61: 109–128.

2761. ——— and E. S. Ayensu. 1964. The identity of the genus *Calyptosepalum* S. Moore. Brittonia 16: 220–227.

2762. Smith, C. W. 1976. Haleakalā RBI. Proc. 1st Conf. Nat. Sci., Hawaii Volcanoes Natl. Park, pp. 217–222.

2763. ———. 1978. Haleakala National Park crater district resources basic inventory: introduction and general overview. Proc. 2nd Conf. Nat. Sci., Hawaii Volcanoes Natl. Park, pp. 289–291.

2764. ——— (ed.). 1980. Resources base inventory of Kīpahulu Valley below 2000 feet. Coop. Natl. Park Resources Stud. Unit, Univ. Hawaii, Honolulu, 175 pp.

2765. ———. 1980. Proposed native ecosystem restoration program for Halapē, Keauhou, and Apua Point, Hawaii Volcanoes National Park. Coop. Natl. Park Resources Stud. Unit, Hawaii, Techn. Rep. 28: 1–35.

2766. ———. 1980. Haleakalā crater: its resources and management problems. Proc. 3rd Conf. Nat. Sci., Hawaii Volcanoes Natl. Park, pp. 311–317.

2767. ———. 1982. Towards a resource management plan for Kipahulu Valley. Proc. 4th Conf. Nat. Sci., Hawaii Volcanoes Natl. Park, pp. 152–155.
2768. ———. 1985. Impact of alien plants on Hawai'i's native biota: in Stone, C. P., and J. M. Scott (eds.), Hawai'i's terrestrial ecosystems: preservation and management. Coop. Natl. Park Resources Stud. Unit, Univ. Hawaii, Honolulu, pp. 180–250.
2769. Smith, H. H. 1960. Wood quality studies to guide Hawaiian forest industries. U.S.D.A. Forest Serv. Misc. Pap. 48: 1–19.
2770. Smith, J. E. 1809. *Dodonaea*: in Rees, A., The Cyclopaedia; or, universal dictionary of arts, sciences, and literature. Vol. 12. Longman, Hurst, Rees, Orme, & Brown, London, 2 pp.
2771. ———. 1811. *Hedyotis*: in Rees, A., The Cyclopaedia; or, universal dictionary of arts, sciences, and literature. Vol. 17. Longman, Hurst, Rees, Orme, & Brown, London, 6 pp.
2772. ———. 1814. *Pyrus*: in Rees, A., The Cyclopaedia; or, universal dictionary of arts, sciences, and literature. Vol. 29. Longman, Hurst, Rees, Orme, & Brown, London, 22 pp.
2773. ———. 1816. *Smilax*: in Rees, A., The Cyclopaedia; or, universal dictionary of arts, sciences, and literature. Vol. 33. Longman, Hurst, Rees, Orme, & Brown, London, 4 pp.
2774. ———. 1817. *Vaccinium*: in Rees, A., The Cyclopaedia; or, universal dictionary of arts, sciences, and literature. Vol. 36. Longman, Hurst, Rees, Orme, & Brown, London, 8 pp.
2775. ———. 1818. *Astelia*: in Rees, A., The Cyclopaedia; or, universal dictionary of arts, sciences, and literature. Vol. 39. Longman, Hurst, Rees, Orme, & Brown, London, 2 pp.
2776. Smith, J. G. 1904. The black wattle. (*Acacia decurrens*, Willd.). Hawaiian Forester Agric. 1: 151.
2777. ———. 1906. The black wattle (*Acacia decurrens*) in Hawaii. Hawaii Agric. Exp. Sta. Bull. 11: 1–16.
2778. ——— and Q. Q. Bradford. 1908. The ceara rubber tree in Hawaii. Hawaii Agric. Exp. Sta. Bull. 16: 1–30.
2779. Smith, L. B., D. C. Wasshausen, J. Golding, and C. E. Karegeannes. 1986. Begoniaceae, part I: illustrated key; part II: annotated species list. Smithsonian Contr. Bot. 60: 1–584.
2780. Smith, L. L. 1977. Development of emergent vegetation in a tropical marsh (Kawainui, O'ahu). Newslett. Hawaiian Bot. Soc. 16: 37–56.
2781. ———. 1978. Development of emergent vegetation in a tropical marsh. Master's thesis, Univ. Hawaii, Honolulu, 107 pp.
2782. ———. 1978. Development of emergent vegetation in a tropical marsh (Kawainui, O'ahu). Newslett. Hawaiian Bot. Soc. 17: 2–27.
2783. Smith, S. C. 1943. Silversword—rare jewel of an Hawaiian crater. Nature Mag. 36(1): 31–32.
2784. Sohmer, S. H. 1971. A revision of the genus *Charpentiera* (Amaranthaceae). Ph.D. dissertation, Univ. Hawaii, Honolulu, 241 pp.

2785. ———. 1971. The lectotype of the genus *Charpentiera* (Amaranthaceae). Taxon 20: 345–348.
2786. ———. 1972. Revision of the genus *Charpentiera* (Amaranthaceae). Brittonia 24: 283–312.
2787. ———. 1973. Microsporogenesis in *Charpentiera* (Amaranthaceae). Chromosome Information Serv. 14: 14–16.
2788. ———. 1973. A preliminary report of the biology of the genus *Charpentiera* (Amaranthaceae). Pacific Sci. 27: 399–405.
2789. ———. 1976. Kaluaa Gulch revisited. Newslett. Hawaiian Bot. Soc. 15: 23–24.
2790. ———. 1976. The genera *Charpentiera* and *Chamissoa* (Amaranthaceae): distant cousins at best. Newslett. Hawaiian Bot. Soc. 15: 49–52.
2791. ———. 1977. *Psychotria* L. (Rubiaceae) in the Hawaiian Islands. Lyonia 1: 103–186.
2792. ———. 1978. Morphological variation and its taxonomic and evolutionary significance in the Hawaiian *Psychotria* (Rubiaceae). Brittonia 30: 256–264.
2793. ——— and R. T. Hirano. 1972. *Charpentiera* (Amaranthaceae) as an ornamental. Newslett. Hawaiian Bot. Soc. 11: 46–47.
2794. Solbrig, O. T., L. C. Anderson, D. W. Kyhos, P. H. Raven, and L. Rüdenberg. 1964. Chromosome numbers in Compositae V. *Astereae* II. Amer. J. Bot. 51: 513–519.
2795. ———, D. W. Kyhos, M. Powell, and P. H. Raven. 1972. Chromosome numbers in Compositae VIII: *Heliantheae*. Amer. J. Bot. 59: 869–878.
2796. Solereder, H. 1892–1895. Loganiaceae. Nat. Pflanzenfam. IV. 2: 19–48 (1892), 49–50 (1895).
2797. Sorenson, J. C. 1977. *Andropogon virginicus* (broomsedge). Newslett. Hawaiian Bot. Soc. 16: 7–22.
2798. ———. 1979. Fire, lightning, and *'ohi'a* dieback. Newslett. Hawaiian Bot. Soc. 18: 9–23.
2799. ———. 1980. Phenology of *Andropogon virginicus* in Hawaii. Master's thesis, Univ. Hawaii, Honolulu, 196 pp.
2800. Southwood, T. R. E. 1960. The abundance of the Hawaiian trees and the number of their associated insect species. Proc. Hawaiian Entomol. Soc. 17: 299–303.
2801. Sparre, B. 1964. Wikstroemiae novae Hawaiienses—a supplement. Svensk Bot. Tidskr. 58: 497.
2802. Spatz, G. 1973. Some findings on vegetative and sexual reproduction of *koa*. U.S. IBP Island Ecosystems IRP Techn. Rep. 17: 1–45.
2803. ——— and D. Mueller-Dombois. 1972. The influence of feral goats on *koa* (*Acacia koa* Gray) reproduction in Hawaii Volcanoes National Park. U.S. IBP Island Ecosystems IRP Techn. Rep. 3: 1–16.
2804. ——— and ———. 1972. Succession patterns after pig digging in grassland communities on Mauna Loa, Hawaii. U.S. IBP Island Ecosystems IRP Techn. Rep. 15: 1–44.

2805. ——— and ———. 1973. The influence of feral goats on *koa* tree reproduction in Hawaii Volcanoes National Park. Ecology 54: 870–876.
2806. ——— and ———. 1975. Succession patterns after pig digging in grassland communities on Mauna Loa, Hawaii. Phytocoenologia 3: 346–373.
2807. ———, ———, and R. G. Cooray. 1972. Studies on the autecology of two native Hawaiian tree species along IBP transects. U.S. IBP Island Ecosystems IRP Techn. Rep. 2: 71–85.
2808. Speare, A. T. 1915. Weeds. Hawaiian Pl. Rec. 12: 218–223.
2809. ———. 1915. Weeds. Second paper. Hawaiian Pl. Rec. 12: 312–318.
2810. ———. 1915. Weeds. Third paper. Hawaiian Pl. Rec. 12: 400–405.
2811. ———. 1916. Weeds. Fourth paper. Hawaiian Pl. Rec. 13: 11–16.
2812. ———. 1916. Weeds. Fifth paper. Hawaiian Pl. Rec. 13: 81–86.
2813. ———. 1916. Weeds. Sixth paper. Hawaiian Pl. Rec. 13: 140–145.
2814. Spence, G. E., and S. L. Montgomery. 1976. Ecology of the dryland forest at Kānepuʻu, island of Lānaʻi. Newslett. Hawaiian Bot. Soc. 15: 62–80.
2815. Spieth, H. T. 1966. Hawaiian honeycreeper, *Vestiaria coccinea* (Forster), feeding on lobeliad flowers, *Clermontia arborescens* (Mann) Hillebr. Amer. Naturalist 100: 470–473.
2816. Sprague, T. A. 1914. *Hibiscus waimeae*. Bot. Mag. 140: *pl. 8547*.
2817. ———. 1914. *Hibiscus arnottianus*. Kew Bull. 1914: 45–47.
2818. ——— and V. S. Summerhayes. 1927. *Santalum*, *Eucarya*, and *Mida*. Bull. Misc. Inform. 5: 193–202.
2819. Sprengel, K. P. J. 1824–1828. Caroli Linnaei, equitis stellae polaris, archiatri regii, prof. med. et rei herb. in Univers. Upsal. Systema vegetabilium. Vols. 1–5. 16th ed. Librariae Dieterichianae, Göttingen.
2820. Stapf, O. 1893. The genus *Trematocarpus*. Ann. Bot. (London) 7: 396–399.
2821. Starbird, E. A. 1977. The island that's still Hawaii. Kauai. Natl. Geogr. Mag. 152: 584–613.
2822. Staudt, G. 1962. Taxonomic studies in the genus *Fragaria*. Typification of *Fragaria* species known at the time of Linnaeus. Canad. J. Bot. 40: 869–886.
2823. Stauffer, H. U. 1959. Revisio Anthobolearum eine morphologische Studie mit Einschluss der Geographie, Phylogenie und Taxonomie. Santalales-Studien IV. Mitt. Bot. Mus. Univ. Zürich 213: 1–261.
2823a. Stearn, W. T. 1947. Endlicher's "Genera Plantarum," "Iconographia Generum Plantarum" and "Atakta Botanika." J. Arnold Arbor. 28: 424–429.
2824. Stearns, H. T. 1940. Geology and ground-water resources of the islands of Lanai and Kahoolawe, Hawaii. Hawaii, Div. Hydrogr. Bull. 6: 1–177.
2825. ——— and K. N. Vaksvik. 1935. Geology and ground-water resources of the island of Oahu, Hawaii. Hawaii, Div. Hydrogr. Bull. 1: 1–479.

2826. Stebbins, G. L. 1941. Additional evidence for a holarctic dispersal of flowering plants in the Mesozoic era. Proc. 6th Pacific Sci. Congr., California 3: 649–660.
2827. Steenis, C. G. G. J. van (ed.). 1963. Pacific plant areas, Vol. 1. Monogr. Natl. Inst. Sci. Technol. (Manila) 8, Vol. 1: 1–297.
2828. ——— and M. M. J. van Balgooy (eds.). 1966. Pacific plant areas, Vol. 2. Blumea Suppl. 5: 1–312.
2829. Stein, J. D. 1980. Recovery of *koa* trees defoliated by the *koa* moth, *Scotorythra paludicola* (Lepidoptera: Geometridae) (Abstr.). Proc. 3rd Conf. Nat. Sci., Hawaii Volcanoes Natl. Park, p. 319.
2830. ———. 1983. Insects associated with *Acacia koa* seed in Hawaii. Environmental Entomol. 12: 299–302.
2831. ——— and P. G. Scowcroft. 1984 [1985]. Growth and refoliation of *koa* trees infested by the *koa* moth, *Scotorythra paludicola* (Lepidoptera: Geometridae). Pacific Sci. 38: 333–339.
2832. Steiner, W. W. M. 1972. Technical developments and genetic aspects of variation in *Metrosideros*. U.S. IBP Island Ecosystems IRP Techn. Rep. 2: 181–191.
2833. ———. 1973. Preliminary report on electrophoretic variation in *Acacia koa*. U.S. IBP Island Ecosystems IRP Techn. Rep. 21: 6.58.
2834. Stemmerik, J. F. 1964. Florae Malesianae precursores XXXVIII. Notes on *Pisonia* L. in the Old World (Nyctaginaceae). Blumea 12: 275–284.
2835. Stemmermann, R. L. 1976. Distribution and vegetative anatomy of Hawaiian sandalwood. Proc. 1st Conf. Nat. Sci., Hawaii Volcanoes Natl. Park, pp. 223–226.
2836. ———. 1977. Studies of the vegetative anatomy of the Hawaiian representatives of *Santalum* (Santalaceae), and observations of the genus *Santalum* in Hawaii. Master's thesis, Univ. Hawaii, Honolulu, 179 pp.
2837. ———. 1978. Haleakala National Park crater district resources basic inventory: the vascular flora of Haleakala. Proc. 2nd Conf. Nat. Sci., Hawaii Volcanoes Natl. Park, pp. 297–303.
2838. ———. 1980 [1981]. Observations on the genus *Santalum* (Santalaceae) in Hawai'i. Pacific Sci. 34: 41–54.
2839. ———. 1980 [1981]. Vegetative anatomy of the Hawaiian species of *Santalum* (Santalaceae). Pacific Sci. 34: 55–75.
2840. ———. 1981. A guide to Pacific wetland plants. U.S. Army Corps of Engineers, Honolulu, 118 pp.
2841. ———. 1982. Research on ecotypes of *Metrosideros*. Proc. 4th Conf. Nat. Sci., Hawaii Volcanoes Natl. Park, p. 156.
2842. ———. 1983 [1984]. Ecological studies of Hawaiian *Metrosideros* in a successional context. Pacific Sci. 37: 361–373.
2843. ———. 1984. Congeneric succession of *Metrosideros* in Hawaii (Abstr.). Proc. 5th Conf. Nat. Sci., Hawaii Volcanoes Natl. Park, p. 90.
2844. ———. 1986. Request for observations on *Rubus*. Newslett. Hawaiian Bot. Soc. 25: 72–73.

2845. ———. 1986. Ecological studies of 'ōhi'a varieties (*Metrosideros polymorpha*, Myrtaceae), the dominants in successional communities of Hawaiian rain forests. Ph.D. dissertation, Univ. Hawaii, Honolulu, 293 pp.
2846. ———, P. K. Higashino, W. Char, and L. Yoshida. 1986. Botanical survey of the Kahuku Training Area, O'ahu, Hawai'i. Newslett. Hawaiian Bot. Soc. 25: 90–117.
2847. ———, ———, and C. W. Smith. 1981. Haleakala National Park crater district resources basic inventory: conifers and flowering plants. Coop. Natl. Park Resources Stud. Unit, Hawaii, Techn. Rep. 38: 1–57.
2848. ———, C. W. Smith, and W. J. Hoe. 1979. Haleakala National Park Crater District resources basic inventory: 1976–77. Coop. Natl. Park Resources Stud. Unit, Hawaii, Techn. Rep. 24: 1–92.
2849. Stephens, S. G. 1958. Salt water tolerance of seeds of *Gossypium* species as a possible factor in seed dispersal. Amer. Naturalist 92: 83–92.
2850. ———. 1963. Polynesian cottons. Ann. Missouri Bot. Gard. 50: 1–22.
2851. ———. 1963. Cotton in the Hawaiian Islands. Newslett. Hawaiian Bot. Soc. 2: 49–52.
2852. ———. 1964. Native Hawaiian cotton (*Gossypium tomentosum* Nutt.). Pacific Sci. 18: 385–398.
2853. Steudel, E. G. 1840. Nomenclator botanicus seu: synonymia plantarum universalis, enumerans ordine alphabetico nomina atque synonyma, tum generica tum specifica, et a Linnaeo et a recentioribus de re botanica scriptoribus plantis phanerogamis imposita. Vol. 1. 2nd ed. J. G. Cottae, Stuttgart, 852 pp.
2854. ———. 1841. Nomenclator botanicus seu: synonymia plantarum universalis, enumerans ordine alphabetico nomina atque synonyma, tum generica tum specifica, et a Linnaeo et a recentioribus de re botanica scriptoribus plantis phanerogamis imposita. Vol. 2. 2nd ed. J. G. Cottae, Stuttgart, 810 pp.
2855. ———. 1853–1854. Synopsis plantarum glumacearum . . . Part I. J. B. Metzler, Stuttgart, 474 pp.
2856. ———. 1855. Synopsis plantarum glumacearum . . . Part II. J. B. Metzler, Stuttgart, 348 pp.
2857. Stevens, F. L. 1925. Hawaiian fungi. Bernice P. Bishop Mus. Bull. 19: 1–189.
2858. Stewart, M. 1972. *Noni*, the lore of Hawaiian medicinal plants. Bull. Pacific Trop. Bot. Gard. 2: 37–39.
2859. ———. 1972. *Ieie*, native fiber plant. Bull. Pacific Trop. Bot. Gard. 2: 77.
2860. ———. 1973. The versatile *wiliwili*. Bull. Pacific Trop. Bot. Gard. 3: 13.
2861. ———. 1973. Bogs in Hawaii. Bull. Pacific Trop. Bot. Gard. 3: 55–58.
2862. ———. 1973. New species found on Kauai. Bull. Pacific Trop. Bot. Gard. 3: 71.
2863. ———. 1975. *Hi'aloa*. Bull. Pacific Trop. Bot. Gard. 5: 32.

2864. Stewart, W. S. 1973. Tropical plants adaptable to mainland landscapes. Bull. Pacific Trop. Bot. Gard. 3: 75–77.
2865. Stokes, J. F. G. 1906. Hawaiian nets and netting. Mem. Bernice P. Bishop Mus. 2(1): 105–162.
2866. ———. 1921. Fish-poisoning in the Hawaiian Islands with notes on the custom in southern Polynesia. Occas. Pap. Bernice P. Bishop Mus. 7(10): 219–236.
2867. ———. 1932. Spaniards and the sweet potato in Hawaii and Hawaiian-American contacts. Amer. Anthropol. 34: 594–600.
2868. Stone, B. C. 1957. Rediscovery of a rare lobelioid, *Brighamia insignis* forma *citrina*, in Kauai, Hawaiian Islands. Bull. Torrey Bot. Club 84: 175–177.
2869. ———. 1961. Studies in the Hawaiian Rutaceae, III: on the New Caledonian species of "*Pelea*," and a misunderstood species of "*Platydesma*." Adansonia 1: 94–99.
2870. ———. 1961. A note on chromosome number in *Pandanus*. J. Jap. Bot. 36: 279–284.
2871. ———. 1962. A monograph of the genus *Platydesma* (Rutaceae). J. Arnold Arbor. 43: 410–427.
2872. ———. 1962. Taxonomic and nomenclatural notes on *Platydesma* (Hawaii) and a new name for a *Melicope* (Solomon Islands). Madroño 16: 161–166.
2873. ———. 1962. Studies in Hawaiian Rutaceae, II. Identity of *Pelea sandwicensis*. Pacific Sci. 16: 366–373.
2874. ———. 1963. The genus *Portulaca* in the Hawaiian Islands: *in* Chandra, L. (ed.), Advancing frontiers of plant sciences. Vol. 4. Inst. Advancem. Sci. Cult., New Delhi, pp. 141–149.
2875. ———. 1963. Archipelagic refuge. Endemic floras abound in Hawaiian chain. Nat. Hist. Mag. 1963(11): 32–39.
2876. ———. 1963. Studies in the Hawaiian Rutaceae, IV. New and critical species of *Pelea* A. Gray. Pacific Sci. 17: 407–420.
2877. ———. 1966. Studies in the Hawaiian Rutaceae, VII. A conspectus of species and varieties and some further new taxa in the genus *Pelea* A. Gray. Occas. Pap. Bernice P. Bishop Mus. 23(10): 147–162.
2878. ———. 1967. Carpel number as a taxonomic criterion in *Pandanus*. Amer. J. Bot. 54: 939–944.
2879. ———. 1967. A review of the endemic genera of Hawaiian plants. Bot. Rev. (Lancaster) 33: 216–259.
2880. ———. 1967. Materials for a monograph of *Freycinetia* (Pandanaceae) I. Gard. Bull. Straits Settlem. 22: 129–152.
2881. ———. 1967. Notes on the Hawaiian flora. Pacific Sci. 21: 550–557.
2882. ———. 1968. Materials for a monograph of *Freycinetia* Gaud. IV. Subdivision of the genus, with fifteen new sections. Blumea 16: 361–372.
2883. ———. 1968. Identity of the *Pelea* species (Rutaceae) described by H.

Léveillé. Studies in Hawaiian Rutaceae, VIII. Notes Roy. Bot. Gard. Edinburgh 27: 259–264.
2884. ———. 1968. Theophrastaceae, a family wrongly attributed to the Hawaiian flora. Pacific Sci. 22: 425.
2885. ———. 1968. Morphological studies in Pandanaceae. I. Staminodia and pistillodia of *Pandanus* and their hypothetical significance. Phytomorphology 18: 498–509.
2886. ———. 1969. The genus *Pelea* A. Gray (Rutaceae: *Evodiinae*). A taxonomic monograph (studies in the Hawaiian Rutaceae, 10). Phanerogamarum monographiae tomus III. J. Cramer, Lehre, West Germany, 180 pp.
2887. ———. 1969. Materials for a monograph of *Freycinetia* Gaud. X. Chronological list of all binomials. Taxon 18: 672–680.
2888. ———. 1970. Gaudichaud's species of *Pandanus* in the atlas of the botany of the voyage of 'La Bonite'. Malayan Sci. 5: 14–19.
2889. ———. 1972. A reconsideration of the evolutionary status of the family Pandanaceae and its significance in monocotyledon phylogeny. Quart. Rev. Biol. 47: 34–45.
2890. ———. 1976. On the biogeography of *Pandanus* (Pandanaceae). Compt. Rend. Sommaire Séances Soc. Biogéogr. 458: 69–90.
2891. ———. 1981. East Polynesian species of *Freycinetia* Gaudichaud (Pandanaceae). Micronesica 17: 47–58.
2892. ———. 1981. *Pandanus tectorius* in the Hawaiian Islands. Notes Waimea Arbor. & Bot. Gard. 8(2): 4–10.
2893. ———. 1982. Botanical authors and the International Code of Botanical Nomenclature. Notes Waimea Arbor. & Bot. Gard. 9(2): 4–8.
2894. ———. 1985. New and noteworthy Paleotropical species of Rutaceae. Proc. Acad. Nat. Sci. Philadelphia 137: 213–228.
2895. ——— and I. Lane. 1958. A new *Hedyotis* from Kauai, Hawaiian Islands. Pacific Sci. 12: 139–145.
2896. Stone, C. P. 1985. Alien animals in Hawai'i's native ecosytems: toward controlling the adverse effects of introduced vertebrates: *in* Stone, C. P., and J. M. Scott (eds.), Hawai'i's terrestrial ecosystems: preservation and management. Coop. Natl. Park Resources Stud. Unit, Univ. Hawaii, Honolulu, pp. 251–297.
2897. ———, P. C. Banko, P. K. Higashino, and F. G. Howarth. 1984. Interrelationships of alien and native plants and animals in Kipahulu Valley, Haleakala National Park: a preliminary report. Proc. 5th Conf. Nat. Sci., Hawaii Volcanoes Natl. Park, pp. 91–105.
2898. ——— and J. M. Scott (eds.). 1985. Hawai'i's terrestrial ecosystems: preservation and management. Coop. Natl. Park Resources Stud. Unit, Univ. Hawaii, Honolulu, 584 pp.
2899. ——— and D. D. Taylor. 1984. Status of feral pig management and research in Hawai'i Volcanoes National Park. Proc. 5th Conf. Nat. Sci., Hawaii Volcanoes Natl. Park, pp. 106–117.
2900. Stoner, M. F. 1976. Proposed theory on *ohia* forest decline in Hawaii: a

precipitant phenomenon related to soil conditions and island maturation (Abstr.). Proc. Amer. Phytopathol. Soc. 3: 215.

2901. ———, G. E. Baker, and D. K. Stoner. 1973. Progress report on the occurrence and ecological roles of soil fungi associated with *Acacia koa* on the Mauna Loa Transect. U.S. IBP Island Ecosystems IRP Techn. Rep. 21: 6.17–6.20.

2902. Storey, W. B. 1937. A comparison of chromosome numbers in *Passiflora*. Master's thesis, Univ. Hawaii, Honolulu, 31 pp.

2903. ———. 1941. The botany and sex relationships of the papaya: *in* Papaya production in the Hawaiian Islands. Hawaii Agric. Exp. Sta. Bull. 87: 5–22.

2904. ———. 1950. Chromosome numbers of some species of *Passiflora* occurring in Hawaii. Pacific Sci. 4: 37–42.

2905. ———. 1953. Genetics of the papaya. J. Heredity 44: 70–78.

2906. ———. 1958. Additional observations on the genetics of flower color in *Spathoglottis*. Bull. Pacific Orchid Soc. Hawaii 16: 7–13.

2907. ———. 1966. Chromosome numbers in *Cyrtandra*: *in* St. John, H., Monograph of *Cyrtandra* (Gesneriaceae) on Oahu, Hawaiian Islands. Bernice P. Bishop Mus. Bull. 229: 31–33.

2908. Streets, T. H. 1877. Contributions to the natural history of the Hawaiian and Fanning Islands and lower California. Bull. U.S. Natl. Mus. 7: 1–172.

2909. Stschegleew. 1859. [Title unknown]. Bull. Soc. Nat. Mosc. 32: 10; not seen.

2910. Stubbs, W. C. 1901. Report on the agricultural resources and capabilities of Hawaii. U.S.D.A. Bull. 95: 1–100.

2911. Suessenguth, K. 1938. Amarantaceen-Studien. I. Amarantaceae aus Amerika, Asien, Australien. Repert. Spec. Nov. Regni Veg. 44: 36–48.

2912. ——— and A. Ludwig. 1950. Myrtaceae. Mitt. Bot. Staatssamml. München 1: 18.

2913. ——— and H. Merxmüller. 1952. Species novae vel criticae. Mitt. Bot. Staatssamml. München 4: 99–114.

2914. Sun, M., and F. R. Ganders. 1986. Female frequencies in gynodioecious populations correlated with selfing rates in hermaphrodites. Amer. J. Bot. 73: 1645–1648.

2915. Sutton, B. C., and C. S. Hodges, Jr. 1983. Hawaiian forest fungi. III. A new species, *Gloeocoryneum hawaiiense*, on *Acacia koa*. Mycologia 75: 280–284.

2916. Svenson, H. K. 1929. Mongraphic studies in the genus *Eleocharis*. Rhodora 31: 121–135, 152–163, 167–191, 199–219, 224–242. [Rep. in Contr. Gray Herb. 86 without change in pagination.]

2917. ———. 1939. Monographic studies in the genus *Eleocharis*—V. Rhodora 41: 1–19, 43–77, 90–110.

2918. Svihla, A. 1957. Observations on French Frigate Shoals, February 1956. Atoll Res. Bull. 51: 1–2.

2919. Swamy, B. G. L. 1949. The comparative morphology of the Santalaceae: node, secondary xylem, and pollen. Amer. J. Bot. 36: 661–673.
2920. Swanholm, C. E., H. St. John, and P. J. Scheuer. 1959. A survey for alkaloids in Hawaiian plants. I. Pacific Sci. 13: 295–305.
2921. ———, ———, and ———. 1960. A survey for alkaloids in Hawaiian plants. II. Pacific Sci. 14: 68–74.
2922. Swezey, O. H. 1919. Cause of scarcity of seeds of the *koa* tree. Hawaiian Pl. Rec. 21: 102–106.
2923. ———. 1919. Insects occurring on plants of the Lobelioideae in the Hawaiian Islands: *in* Rock, J. F., A monographic study of the Hawaiian species of the tribe *Lobelioideae*, family Campanulaceae. Mem. Bernice P. Bishop Mus. 7: 98–99.
2924. ———. 1925. The insect fauna of trees and plants as an index of their endemicity and relative antiquity in the Hawaiian Islands. Proc. Hawaiian Entomol. Soc. 6: 195–209.
2925. ———. 1931. Some observations on the insect faunas of native forest trees in the Olinda forest on Maui. Proc. Hawaiian Entomol. Soc. 7: 493–504.
2926. ———. 1936. Fruit-eating and seed-eating insects in Hawaii. Proc. Hawaiian Entomol. Soc. 9: 196–202.
2927. ———. 1954. Forest entomology in Hawaii. An annotated check-list of the insect faunas of the various components of the Hawaiian forests. Special Publ. Bernice P. Bishop Mus. 44: 1–266.
2928. Szymkiewicz, D. 1938. Czwarty przyczynek statystyczny do geografii florystycznej [Quatrième contribution statistique à la géographie floristique]. Acta Soc. Bot. Poloniae 15: 15–22.
2929. Tabata, R. S. 1979. An introduction to Hawaiian coastal plants. Univ. Hawaii, UNIHI-SEAGRANT-AB-80–01: 1–6.
2930. ———. 1980. The native coastal plants of Oʻahu, Hawaiʻi. Proc. 3rd Conf. Nat. Sci., Hawaii Volcanoes Natl. Park, pp. 321–346.
2931. ———. 1980. The native coastal plants of Oʻahu, Hawaiʻi. Newslett. Hawaiian Bot. Soc. 19: 2–44.
2932. ——— and J. Moriyama. 1982. Manoa Cliffs Trail, Honolulu, Hawaiʻi. Plant identification guide and brief descriptions for selected plants. Hawaii Dept. Land Nat. Resources, Div. Forestry Wildlife, pp. 1–9.
2933. Tabrah, F. L., and B. M. Eveleth. 1966. Evaluation of the effectiveness of ancient Hawaiian medicine. Hawaii Med. J. 25: 223–230.
2934. Tagawa, T. K. 1976. Endangered species in Hawaii. Effect on other resource management. Newslett. Hawaiian Bot. Soc. 15: 7–14.
2935. Takahashi, A. 1973. Holiday hikes. Awaawapuhi Trail. Bull. Pacific Trop. Bot. Gard. 3: 67–70.
2936. Takahashi, M. 1952. Tropical forage legume and browse plants research in Hawaii. Proc. 6th Int. Grassland Congr. 2: 1411–1417.
2937. ——— and J. C. Ripperton. 1949. *Koa haole* (*Leucaena glauca*). Its establishment, culture and utilization as a forage crop. Hawaii Agric. Exp. Sta. Bull. 100: 1–56.

2938. Takeuchi, W. N. 1984. *Vicia menziesii* recovery plan. Dept. Interior, U.S. Fish & Wildlife Serv., Portland, 54 pp.
2939. Tanabe, M. J. 1980. Studies on *maile* cultivation (Abstr.). Proc. 3rd Conf. Nat. Sci., Hawaii Volcanoes Natl. Park, p. 347.
2940. Taubert, P. 1891–1894. Leguminosae. Nat. Pflanzenfam. III. 3: 70–388.
2941. Taylor, D. D. 1980. Controlling exotic plants in Hawaii Volcanoes National Park. Proc. 3rd Conf. Nat. Sci., Hawaii Volcanoes Natl. Park, pp. 349–355.
2942. ———. 1982. Exotic plant reduction in Hawaii Volcanoes National Park: an update. Proc. 4th Conf. Nat. Sci., Hawaii Volcanoes Natl. Park, pp. 173–184.
2943. Taylor, W. 1900. List of palms in the Hawaiian Islands. Hawaiian Almanac and Annual for 1901: 49.
2944. Teho, F. G. 1981. Plants of Hawaii: how to grow them. Petroglyph Press, Hilo, Hawaii, 27 unnum. pp.
2945. Tempsky, L. von. 1904. *Pamakani*, a dangerous plant pest. Proc. Hawaiian Live Stock Breeders' Assoc., 2nd Ann. Meeting, Honolulu, pp. 62–63.
2946. Ter Welle, B. J. H., A. A. Loureiro, P. L. B. Lisboa, and J. Koek-Noorman. 1983. Systematic wood anatomy of the tribe *Guettardeae* (Rubiaceae). J. Linn. Soc., Bot. 87: 13–28.
2947. Teraoka, W., K. Nagata, and C. Corn. 1981. Predation of *Pipturus albidus* fruit by rodents. 'Elepaio 41: 134.
2947a. Tessene, M. F. 1969. Systematic and ecological studies on *Plantago cordata*. Michigan Bot. 8: 72–103.
2948. Thaman, R. R. 1974. *Lantana camara*: its introduction, dispersal and impact on islands of the tropical Pacific Ocean. Micronesica 10: 17–39.
2949. Theobald, W. L. 1977. Garden collections—heliconia. Bull. Pacific Trop. Bot. Gard. 7: 85–86.
2950. ———. 1980. Ethnobotany of the coconut. Bull. Pacific Trop. Bot. Gard. 10: 8–11.
2951. Thomas, D. 1986. Views from another world. Honolulu 21(5): 104–111, 164, 166, 168, 170.
2952. Thompson, A. R. 1913. Chemistry of *kukui* oil. Hawaii Agric. Exp. Sta. Press Bull. 39: 5–8.
2953. Thompson, G. C. K. 1895. The wild flowers of Hawaii. Overland Monthly 25: 157–164.
2954. Thomson, J. D., and S. C. H. Barrett. 1981. Selection for outcrossing, sexual selection, and the evolution of dioecy in plants. Amer. Naturalist 118: 443–449.
2955. Thorne, R. F. 1963. Biotic distribution patterns in the tropical Pacific: *in* Gressitt, J. L. (ed.), Pacific Basin biogeography. Bishop Mus. Press, Honolulu, pp. 311–350.
2956. Thrum, T. G. 1879. Varieties of taro, (*Arum esculentum*). Hawaiian Almanac and Annual for 1880: 28–29.

2957. ———. 1885. Fruits and their seasons in the Hawaiian Islands. Hawaiian Almanac and Annual for 1886: 49–50.
2958. ———. 1886. Taro—*Colocasia antiquorum*. Hawaiian Almanac and Annual for 1887: 63–65.
2959. ———. 1891. Fruits, indigenous and introduced, of the Hawaiian Islands. Hawaiian Almanac and Annual for 1892: 75–81.
2960. ———. 1891. Indigenous Hawaiian woods. A carefully prepared description of the woods of the Hawaiian Islands. Hawaiian Almanac and Annual for 1892: 88–99.
2961. ———. 1904. The sandalwood trade of early Hawaii as told by the pioneer traders, voyagers and others. Hawaiian Almanac and Annual for 1905: 43–74.
2962. ———. 1905. Early sandalwood trade: Hawaiian version. Hawaiian Almanac and Annual for 1906: 105–108.
2963. ———. 1910. Early attempt at silk culture on Kauai. Hawaiian Almanac and Annual for 1911: 67–71.
2964. ———. 1914. Flowering trees of Honolulu. Hawaiian Almanac and Annual for 1915: 38–43.
2965. ———. 1922. Leaf uses of the Hawaiians. Hawaiian Almanac and Annual for 1923: 71–73.
2966. Tieghem, P. van. 1896. *Korthalsella*, genre nouveau pour la famille des Loranthacées. Bull. Soc. Bot. France 43: 83–87.
2967. ———. 1896. Sur le groupement des espèces en genres dans les Ginalloées, Bifariées, Phoradendrées et Viscées, quatre tribus de la famille des Loranthacées. Bull. Soc. Bot. France 43: 161–194.
2968. Timmons, L. D. 1909. Cotton in the Hawaiian Islands. Hawaiian Almanac and Annual for 1910: 160–164.
2969. Titcomb, M. 1948. Kava in Hawaii. J. Polynes. Soc. 57: 105–171.
2970. Tizard, T. H., H. N. Moseley, J. Y. Buchanan, and J. Murray. 1885. Narrative of the cruise of H.M.S. *Challenger* with a general account of the scientific results of the expedition: *in* Report on the scientific results of the voyage of H.M.S. *Challenger* during the years 1873–76. Vol. 1, Parts 1–2. Longmans & Co., London, 1110 pp.
2971. Tomich, P. Q., N. Wilson, and C. H. Lamoureux. 1968. Ecological factors on Manana Island, Hawaii. Pacific Sci. 22: 352–368.
2972. Tomlinson, P. B. 1965. A study of stomatal structure in Pandanaceae. Pacific Sci. 19: 38–54.
2973. Torrance, S. J., J. J. Hoffmann, and J. R. Cole. 1979. Wikstromol, antitumor lignan from *Wikstroemia foetida* var. *oahuensis* Gray and *Wikstroemia uva-ursi* Gray (Thymelaeaceae). J. Pharm. Sci. 68: 664–665.
2974. Touw, M. 1984. Preliminary observations on *Korthalsella* (Viscaceae) with special reference to vascular patterns. Blumea 29: 525–545.
2975. Trinius, C. B. 1830. Graminum genera quaedam speciesque complures definitionibus novis illustrare pergit. Mém. Acad. Imp. Sci. St.-Pétersbourg, Sér. 6, Sci. Math. 1(4): 353–416.

2976. ———. 1835. Panicearum genera retractavit speciebusque compluribus illustravit. Mém. Acad. Imp. Sci. St.-Pétersbourg, Sér. 6, Sci. Math., Seconde Pt. Sci. Nat. 3: 89–355.
2977. Tsuda, R. T. 1965. Marine algae from Laysan Island with additional notes on the vascular flora. Atoll Res. Bull. 110: 1–31.
2978. Turczaninow, N. 1863. Verbenaceae et Myoporaceae nonnullae hucusque indescriptae. Bull. Soc. Imp. Naturalistes Moscou 36: 193–227.
2979. Udvardy, M. D. F. 1961. The Harold J. Coolidge Expedition to Laysan Island, 1961. Elepaio 22: 43–47.
2980. Unabia, C. C. 1984. Grazing of the seagrass *Halophila hawaiiana* (Hydrocharitaceae) by the snail *Smaragdia bryanae* (Neritidae). Master's thesis, Univ. Hawaii, Honolulu, 103 pp.
2981. Uphof, J. C. T. 1941. Halophytes. Bot. Rev. (Lancaster) 7: 1–58.
2982. Urban, I. 1896. Ueber einige Ternstroemiaceen-Gattungen. Ber. Deutsch. Bot. Ges. 14: 38–51.
2983. Vandercook, J. W. 1939. King cane. The story of sugar in Hawaii. Hawaiian Sugar Planters' Assoc., Honolulu, 63 pp.
2984. Vassal, J. 1969. A propos des Acacias *heterophylla* et *koa*. Bull. Soc. Hist. Nat. Toulouse 105: 443–447.
2985. Vatke, W. 1874. Notulae in Campanulaceas herbarii regii berolinensis. Linnaea 38: 699–735.
2986. ———. 1876. Descriptiones specierum novarum. Linnaea 40: 221–224.
2987. Vesque, J. 1893. Guttiferae. Monogr. phan. 8: 1–669.
2988. ———. 1895. Revision du genre *Eurya* Thunb. Bull. Soc. Bot. France 42: 151–161.
2989. Vilmorin, J. M. P. L. de. 1905. Hortus Vilmorinianus. Catalogue des plantes ligneuses et herbacées existant en 1905 dans les collections de M. Ph. L. de Vilmorin et dans les cultures de MM. Vilmorin-Andrieux et Cie à Verrières-le-Buisson. Bull. Soc. Bot. France 51 (Appendix): 1–371.
2990. Visher, S. S. 1925. Tropical cyclones and the dispersal of life from island to island in the Pacific. Amer. Naturalist 59: 70–78.
2991. Vitousek, P. M., K. van Cleve, N. Balakrishnan, and D. Mueller-Dombois. 1983. Soil development and nitrogen turnover in montane rainforest soils on Hawai'i. Biotropica 15: 268–274.
2992. Vogel, T. 1836. De plantis in expeditione speculatoria Romanzoffiana observatis disserere pergitur. Leguminosae, adjectis quas Cl. Ehrenberg in Hispaniola collegit. Linnaea 10: 582–603.
2993. ———. 1843. Leguminosae: *in* Meyen, F. J. F., Observationes botanicas in itinere circum terram institutas . . . Nov. Actorum Acad. Caes. Leop.-Carol. Nat. Cur. 19, Suppl. 1: 1–46.
2994. Vogl, R. J. 1969. The role of fire in the evolution of the Hawaiian flora and vegetation. Proc. 9th Annual Tall Timbers Fire Ecol. Conf., pp. 5–60.
2995. ———. 1971. General ecology of northeast outer slopes of Haleakala Crater, East Maui, Hawaii. Contr. Nature Conservancy 6: 1–8.

2996. ——— and J. Henrickson. 1971. Vegetation of an alpine bog on East Maui, Hawaii. Pacific Sci. 25: 475–483.
2997. Vriese, W. H. de. 1849–1850. Analecta Goodenoviearum ad auctoritatem herbariorum Musei Caesarei Vindobonensis, Lessertii, Hookeri, Lindleji, Preissii, aliorum. Ned. Kruidk. Arch. 2: 1–32, 137–171. (Rep., 1850, 67 pp.).
2998. ———. 1854. Goodenovieae ad auctoritatem Musei Caesarei Vindobonensis, Parisiensis, illustr. Roberti Brownei, Guil. J. Hookeri, Joan. Lindleji, Franc. Lessertii, Lud. Preissii, Fred. Lud. Splitgerberi, aliorumque. Natuurk. Verh. Holl. Maatsch. Wetensch. Haarlem II, 10: 1–194.
2999. Waage, J. K., J. T. Smiley, and L. E. Gilbert. 1981. The *Passiflora* problem in Hawaii; prospects and problems of controlling the forest weed *P. mollissima* (Passifloraceae) with Heliconiine butterflies. Entomophaga 26: 275–284.
3000. Wagner, J. P. 1986. The rape of the fragrant trees. Honolulu 21(5): 96–97, 148, 150, 152, 154, 156, 158.
3001. Wagner, W. L. 1984. Status and research of the flowering plants of the Hawaiian Islands (Abstr.). Amer. J. Bot. 71(5/2): 197.
3002. ———. 1986. A new look at *Cyrtandra* (Gesneriaceae) in Hawai'i (Abstr.). Amer. J. Bot. 73: 792–793.
3003. ———. 1986. Manual of the flowering plants of Hawai'i: the first step (Abstr.). Amer. J. Bot. 73: 793.
3004. ——— and W. C. Gagné. 1986. Trouble in paradise. Public Gard. 1(1): 6–8.
3005. ———, D. R. Herbst, and S. H. Sohmer. 1986. Contributions to the flora of Hawai'i I. Acanthaceae–Asteraceae. Bishop Mus. Occas. Pap. 26: 102–122.
3006. ———, ———, and R. S. N. Yee. 1985. Status of the native flowering plants of the Hawaiian Islands: *in* Stone, C. P., and J. M. Scott (eds.), Hawai'i's terrestrial ecosystems: preservation and management. Coop. Natl. Park Resources Stud. Unit, Univ. Hawaii, Honolulu, pp. 23–74.
3007. Waimea Arboretum and Botanical Garden. 1983. Checklist of Hawaiian endemic, indigenous, food plants & Polynesian introductions in cultivation in Hawaii. Waimea Arbor. Found. Educational Ser. 2: 1–29.
3008. Walker, E. H. 1947. A subject index to Elmer D. Merrill's "A botanical bibliography of the islands of the Pacific." Contr. U.S. Natl. Herb. 30: 323–404.
3009. Walker, L. R., P. M. Vitousek, L. D. Whiteaker, and D. Mueller-Dombois. 1986. The effect of an introduced nitrogen-fixer (*Myrica faya*) on primary succession on volcanic cinder (Abstr.). Proc. 6th Conf. Nat. Sci., Hawaii Volcanoes Natl. Park, p. 98.
3010. Walker, R. 1912. A visit to the volcano twenty-four years ago. Paradise Pacific 25(8): 19–22.
3011. Walker, R. L. 1985. Status, research and management needs for alien

biota: a summary and commentary: *in* Stone, C. P., and J. M. Scott (eds.), Hawai'i's terrestrial ecosystems: preservation and management. Coop. Natl. Park Resources Stud. Unit, Univ. Hawaii, Honolulu, pp. 372–373.

3012. Wallace, A. R. 1880. Island life: or, the phenomena and causes of insular faunas and floras, including a revision and attempted solution of geological climates. Macmillan & Co., London, 526 pp.

3013. Wallace, R. 1973. Hawaii. Time-Life Books, New York, 184 pp.

3014. Walpers, W. G. 1842–1847. Repertorium botanices systematicae. Vols. 1–6. F. Hofmeister, Leipzig.

3015. ———. 1843. Cruciferas, Capparideas, Calycereas et Compositas, quas Meyenius in orbis circumnavigatione collegit, enumerat novasque describit: *in* Meyen, F. J. F., Observationes botanicas in itinere circum terram institutas . . . Nov. Actorum Acad. Caes. Leop.-Carol. Nat. Cur. 19, Suppl. 1: 247–296.

3016. ———. 1848–1871. Annales botanices systemicae. Vols. 1–7. F. Hofmeister, Leipzig.

3017. Walsh, G. E. 1963. An ecological study of the Heeia mangrove swamp. Ph.D. dissertation, Univ. Hawaii, Honolulu, 219 pp.

3018. ———. 1967. An ecological study of a Hawaiian mangrove swamp: *in* Lauff, G. H. (ed.), Estuaries. Publ. Amer. Assoc. Advancem. Sci. 83: 420–431.

3019. Walter, H. (Transl. D. Mueller-Dombois). 1971. Ecology of tropical and subtropical vegetation. Oliver & Boyd, Edinburgh, 539 pp.

3020. Walters, G. A. 1969. Direct seeding of brushbox, lemon-gum eucalyptus, and cluster pine in Hawaii. U.S.D.A. Forest Serv. Res. Note PSW-199: 1–3.

3021. ———. 1970. Selecting timber species to replace killed firetree in Hawaii. U.S.D.A. Forest Serv. Res. Note PSW-211: 1–4.

3022. ———. 1970. Direct seeding of lemon-gum eucalyptus, redwood, and brushbox in Hawaii. U.S.D.A. Forest Serv. Res. Note PSW-212: 1–3.

3023. ———. 1972. Survival of tropical ash planted in tordon-treated soils in Hawaii. U.S.D.A. Forest Serv. Res. Note PSW-263: 1–4.

3024. ———. 1973. Growth of saligna eucalyptus. J. Forest. (Washington) 71: 346–348.

3025. ———. 1973. Tordon 212 ineffective in killing firetree in Hawaii. U.S.D.A. Forest Serv. Res. Note PSW-284: 1–3.

3026. ———. 1974. *Toona australis* Harms. Australian toon: *in* Schopmeyer, C. S., Seeds of woody plants in the United States. U.S.D.A. Agric. Handbook 450: 813–814.

3027. ———. 1978. Bringing back the monarch of Hawaiian forests—*Acacia koa*. Proc. 2nd Conf. Nat. Sci., Hawaii Volcanoes Natl. Park, pp. 333–336.

3028. ——— and D. P. Bartholomew. 1984. *Acacia koa* leaves and phyllodes: gas exchange, morphological, anatomical, and biochemical characteristics. Bot. Gaz. (Crawfordsville) 145: 351–357.

3029. ——, F. T. Bonner, and E. Q. P. Petteys. 1974. *Pithecellobium* Mart. Blackbead: *in* Schopmeyer, C. S., Seeds of woody plants in the United States. U.S.D.A. Agric. Handbook 450: 639–640.

3030. —— and W. S. Null. 1970. Controlling firetree in Hawaii by injection of Tordon 22K. U.S.D.A. Forest Serv. Res. Note PSW-217: 1–3.

3031. —— and T. H. Schubert. 1969. Saligna eucalyptus growth in a five-year-old spacing study in Hawaii. J. Forest. (Washington) 67: 232–234.

3032. —— and C. D. Whitesell. 1971. Direct seeding trials of three major timber species in Hawaii. U.S.D.A. Forest Serv. Res. Note PSW-234: 1–2.

3033. Warburg, O. 1900. Pandanaceae. Pflanzenr. IV. 9 (Heft 3): 1–97.

3034. Warner, R. E. 1960. A forest dies on Mauna Kea. Pacific Disc. 13(9): 6–14.

3035. ——. 1960. A forest dies on Mauna Kea. How feral sheep are destroying an Hawaiian woodland. Elepaio 20: 82–86.

3036. ——. 1961. The problem of native forest destruction in Hawaii. Abstr. Symp. Pap., 10th Pacific Sci. Congr., Honolulu, pp. 251–252.

3037. —— (ed.). 1968. Scientific report of the Kipahulu Valley expedition. Nature Conservancy, San Francisco, 184 pp.

3038. ——. 1968. Some observations on the birds of Kipahulu Valley: *in* Warner, R. E. (ed.), Scientific report of the Kipahulu Valley Expedition. Nature Conservancy, San Francisco, pp. 133–145.

3039. Warner, R. M. 1972. A catalog of plants in the Plant Science Instructional Arboretum. Misc. Publ. Univ. Hawaii Coll. Trop. Agric. 98: 1–34. (Rev. ed., 1980, 39 pp.).

3040. Warshauer, F. R. 1976. The Kala-pana extension: its variety, vegetation and value. Proc. 1st Conf. Nat. Sci., Hawaii Volcanoes Natl. Park, pp. 237–240.

3041. ——. 1977. The Kalapana Extension of Hawaii Volcanoes National Park: its variety, vegetation, and value. Newslett. Hawaiian Bot. Soc. 16: 57 60.

3042. ——. 1984. Cash in your chips—or—where have all the forests gone? 'Elepaio 45: 48–51.

3043. —— and J. D. Jacobi. 1982. Distribution and status of *Vicia menziesii* Spreng. (Leguminosae): Hawai'i's first officially listed endangered plant species. Biol. Conserv. 23: 111–126.

3044. ——, ——, A. M. LaRosa, J. M. Scott, and C. W. Smith. 1983. The distribution, impact and potential management of the introduced vine *Passiflora mollissima* (Passifloraceae) in Hawai'i. Coop. Natl. Park Resources Stud. Unit, Hawaii, Techn. Rep. 48: 1–39.

3045. Waterman, P. G. 1975. New combinations in *Zanthoxylum* L. (1753). Taxon 24: 361–366.

3046. Watson, D. P. 1966. The wood rose. Univ. Hawaii Coop. Extens. Serv. Circ. 414: 1–8.

3047. ———. 1973. Coconut as an ornamental. Univ. Hawaii Coop. Extens. Serv. Circ. 478: 1–12.

3048. ——— and W. L. Theobald. 1974. Ornamental gingers in Hawaii. Amer. Hort. (Mount Vernon) 53(4): 20–29.

3049. ——— and W. W. J. Yee. 1973. Hawaiian ti. Univ. Hawaii Coop. Extens. Serv. Circ. 481: 1–14.

3050. Watt, G. 1907. The wild and cultivated cotton plants of the world. A revision of the genus *Gossypium* framed primarily with the object of aiding planters and investigators who may contemplate the systematic improvement of the cotton staple. Longmans, Green, and Co., London, 406 pp.

3051. Wawra, H. 1872–1875. Beiträge zur Flora der Hawai'schen Inseln. Flora Vols. 55–58.

3052. ———. 1872. Beiträge zur Flora der Hawai'schen Inseln. Flora 55: 513–517.

3053. ———. 1872. Beiträge zur Flora der Hawai'schen Inseln. Flora 55: 529–533.

3054. ———. 1872. Beiträge zur Flora der Hawai'schen Inseln. Flora 55: 554–560.

3055. ———. 1872. Beiträge zur Flora der Hawai'schen Inseln. Flora 55: 562–569.

3056. ———. 1872–1873. Skizzen von der Erdumseglung S. M. Fregette "Donau." Oesterr. Bot. Z. 22: 222–227, 259–265, 297–302, 332–335, 362–368, 397–405 (1872); 23: 23–29, 60–64, 94–99 (1873).

3057. ———. 1873. Beiträge zur Flora der Hawai'schen Inseln. Flora 56: 7–11.

3058. ———. 1873. Beiträge zur Flora der Hawai'schen Inseln. Flora 56: 30–32.

3059. ———. 1873. Beiträge zur Flora der Hawai'schen Inseln. Flora 56: 44–48.

3060. ———. 1873. Beiträge zur Flora der Hawai'schen Inseln. Flora 56: 58–63.

3061. ———. 1873. Beiträge zur Flora der Hawai'schen Inseln. Flora 56: 76–80.

3062. ———. 1873. Beiträge zur Flora der Hawai'schen Inseln. Flora 56: 107–111.

3063. ———. 1873. Beiträge zur Flora der Hawai'schen Inseln. Flora 56: 137–142.

3064. ———. 1873. Beiträge zur Flora der Hawai'schen Inseln. Flora 56: 157–160.

3065. ———. 1873. Beiträge zur Flora der Hawai'schen Inseln. Flora 56: 168–176.

3066. ———. 1874. Beiträge zur Flora der Hawai'schen Inseln. Flora 57: 257–265.

3067. ———. 1874. Beiträge zur Flora der Hawai'schen Inseln. Flora 57: 273–278.

3068. ———. 1874. Beiträge zur Flora der Hawai'schen Inseln. Flora 57: 294–300.
3069. ———. 1874. Beiträge zur Flora der Hawai'schen Inseln. Flora 57: 321–331.
3070. ———. 1874. Beiträge zur Flora der Hawai'schen Inseln. Flora 57: 362–368.
3071. ———. 1874. Beiträge zur Flora der Hawai'schen Inseln. Flora 57: 521–527.
3072. ———. 1874. Beiträge zur Flora der Hawai'schen Inseln. Flora 57: 540–543.
3073. ———. 1874. Beiträge zur Flora der Hawai'schen Inseln. Flora 57: 545–549.
3074. ———. 1874. Beiträge zur Flora der Hawai'schen Inseln. Flora 57: 562–569.
3075. ———. 1875. Beiträge zur Flora der Hawai'schen Inseln. Flora 58: 145–150.
3076. ———. 1875. Beiträge zur Flora der Hawai'schen Inseln. Flora 58: 171–176.
3077. ———. 1875. Beiträge zur Flora der Hawai'schen Inseln. Flora 58: 184–192.
3078. ———. 1875. Beiträge zur Flora der Hawai'schen Inseln. Flora 58: 225–232.
3079. ———. 1875. Beiträge zur Flora der Hawai'schen Inseln. Flora 58: 241–252.
3080. ———. 1875. Beiträge zur Flora der Hawai'schen Inseln. Corrigenda. Flora 58: 285–288.
3081. ———. 1883. Itinera principum S. Coburgi. Vol. I. Carl Gerold's Sohn, Vienna, 182 pp.
3082. Webster, G. L. 1951. The Polynesian species of *Myoporum*. Pacific Sci. 5: 52–77.
3083. ———. 1975. Conspectus of a new classification of the Euphorbiaceae. Taxon 24: 593–601.
3084. ———, E. A. Rupert, and D. L. Koutnik. 1982. Systematic significance of pollen nuclear number in Euphorbiaceae, tribe *Euphorbieae*. Amer. J. Bot. 69: 407–415.
3085. Weddell, H.-A. 1854. Revue de la famille des Urticées. Ann. Sci. Nat. Bot., sér. 4, 1: 173–212.
3086. ———. 1856–1857. Monographie da la famille des Urticées. Arch. Mus. Hist. Nat. 9: 1–592.
3087. ———. 1869. Urticaceae. Prodr. 16(1): 32–235[64].
3088. Weissich, P. R. 1962. Foster Botanical Garden. Present status of botanical gardens in Hawaii. Gard. J. New York Bot. Gard. 12: 220–222, 227.
3089. ———, M. W. Babcock, and R. C. McCullough. 1965. Foster Garden Notes. Foster Bot. Gard., Honolulu, 83 pp.

3090. Weller, D. M., and N. P. Larsen. 1929. The black spore of Hawaii (Abstr.). Proc. Hawaiian Acad. Sci. 4: 5–6.
3091. Wendland, H. 1862. Beiträge zur Palmenflora der Südseeinseln. Bonplandia 10: 190–200.
3092. Wenkam, R. 1967. Kauai and the park country of Hawaii. Sierra Club, San Francisco, 160 pp.
3093. Wentworth, C. K. 1925. The desert strip of West Molokai. Iowa Stud. Nat. Hist. 11(4): 41–56.
3094. ———. 1935. Mauna Kea, the white mountain of Hawaii. Mid-Pacific Mag. 31: 290–296.
3095. Wentworth, J., C. Lyman, H. Kamasaki, A. M. Hieronymous, R. S. Nekomoto, and D. Young. 1955. Plant damage by volcanic fumes (Abstr.). Proc. Hawaiian Acad. Sci. 30: 18–19.
3096. Wenzig, T. 1874. Pomariae Lindley. Linnaea 38: 1–206.
3097. Werny, F., and P. J. Scheuer. 1963. Hawaiian plant studies—IX. The alkaloids of *Platydesma campanulata* Mann. Tetrahedron 19: 1293–1305.
3098. Wester, L. L. 1981. Introduction and spread of mangroves in the Hawaiian Islands. Yearb. Assoc. Pacific Coast Geogr. 43: 125–137.
3099. ——— and J. O. Juvik. 1983. Roadside plant communities on Mauna Loa, Hawaii. J. Biogeogr. 10: 307–316.
3100. ——— and H. B. Wood. 1977. Koster's curse (*Clidemia hirta*), a weed pest in Hawaiian forests. Environmental Conservation 4: 35–41.
3101. Westervelt, W. D. 1902. Silver sword of Haleakala. Paradise Pacific 15(1): 15–17.
3102. Wettstein, R. von. 1895. Myoporaceae. Nat. Pflanzenfam. IV. 3b: 354–360.
3103. Weymouth, C. 1904. Note on the Hawaiian Islands. J. Roy. Hort. Soc. 28: 552–553.
3104. Wheeler, L. C. 1939. Notes on the genus *Aleurites*. Bot. Mus. Leafl. 7: 119–122.
3105. Whistler, W. A. 1980. Coastal flowers of the tropical Pacific. Pacific Tropical Botanical Garden, Lawai, Hawaii, 83 pp.
3106. White, S. E. 1949. Processes of erosion on steep slopes of Oahu, Hawaii. Amer. J. Sci. 247: 168–186.
3107. ———. 1950. Reply [to a review of "Processes of erosion on steep slopes of Oahu, Hawaii" by S. E. White]. Amer. J. Sci. 248: 511–514.
3108. White, W. 1985. Endangered and threatened wildlife and plants; public hearing and reopening of comment period on proposed endangered status for "*Achyranthes rotundata*." Fed. Reg. 50: 28959–28960.
3109. Whiteaker, L. D. 1978. The vegetation and environment of the crater district of Haleakala National Park. Master's thesis, Univ. Hawaii, Honolulu, 158 pp.
3110. ———. 1978. Haleakala—a tropical alpine and subalpine area. Newslett. Hawaiian Bot. Soc. 17: 38–53.
3111. ———. 1978. Haleakala National Park crater district resources basic

inventory: vegetation map of the crater district. Proc. 2nd Conf. Nat. Sci., Hawaii Volcanoes Natl. Park, pp. 337–344.

3112. ———. 1980. The vegetation and environment in the crater district of Haleakala National Park. Coop. Natl. Park Resources Stud. Unit, Hawaii, Techn. Rep. 35: 1–81.

3113. ———. 1983. The vegetation and environment of the crater district of Haleakala National Park. Pacific Sci. 37: 1–24.

3114. ——— and D. E. Gardner. 1985. The distribution of *Myrica faya* Ait. in the state of Hawai'i. Coop. Natl. Park Resources Stud. Unit, Hawaii, Techn. Rep. 55: 1–31.

3115. Whitesell, C. D. 1964. Silvical characteristics of *koa* (*Acacia koa* Gray). U.S.D.A. Forest Serv. Res. Pap. PSW-16: 1–12.

3116. ———. 1974. Tree plantings on Kahoolawe. Newslett. Hawaiian Bot. Soc. 13: 4–5.

3117. ———. 1974. *Acacia* Mill. Acacia: *in* Schopmeyer, C. S., Seeds of woody plants in the United States. U.S.D.A. Agric. Handbook 450: 184–186.

3118. ———. 1974. *Lucaena* [*Leucaena*] *leucocephala* (Lam.) deWit. Leadtree: *in* Schopmeyer, C. S., Seeds of woody plants in the United States. U.S.D.A. Agric. Handbook 450: 491–493.

3119. ———. 1975. Growth of young saligna eucalyptus in Hawaii: 6 years after thinning. U.S.D.A. Forest Serv. Res. Note PSW-299: 1–3.

3120. ———. 1976. Performance of seven introduced hardwood species on extremely stony mucks in Hawaii. U.S.D.A. Forest Serv. Res. Note PSW-309: 1–5.

3121. ——— and M. O. Isherwood, Jr. 1971. Adaptability of 14 tree species to two hydrol humic latosol soils in Hawaii. U.S.D.A. Forest Serv. Res. Note PSW-236: 1–5.

3122. ——— and G. A. Walters. 1976. Species adaptability trials for man-made forests in Hawaii. U.S.D.A. Forest Serv. Res. Pap. PSW-118: 1–30.

3123. ———, H. L. Wick, and N. Honda. 1971. Growth response of a thinned tropical ash stand in Hawaii . . . after 5 years. U.S.D.A. Forest Serv. Res. Note PSW-227: 1–3.

3124. Whitney, L. D. 1937. A new species of Hawaiian *Eragrostis*. Occas. Pap. Bernice P. Bishop Mus. 13(8): 75–76.

3125. ———. 1937. A new species of *Trisetum* and a new variety of *Panicum imbricatum* from the Hawaiian Islands. Occas. Pap. Bernice P. Bishop Mus. 13(16): 171–173.

3126. ———. 1937. A new lawn grass for Hawaii. Paradise Pacific 49(1): 24.

3127. ———. 1937. Some facts about taro—staff of life in Hawaii. Paradise Pacific 49(3): 15, 30.

3128. ———, F. A. I. Bowers, and M. Takahashi. 1939. Taro varieties in Hawaii. Hawaii Agric. Exp. Sta. Bull. 84: 1–86.

3129. ——— and E. Y. Hosaka. 1936. New species of Hawaiian *Panicum* and *Eragrostis*. Occas. Pap. Bernice P. Bishop Mus. 12(5): 1–6.

3130. ———, ———, and J. C. Ripperton. 1939. Grasses of the Hawaiian ranges. Hawaii Agric. Exp. Sta. Bull. 82: 1–148.
3131. Whitten, H. 1977. Ohia trees are dying. 'Elepaio 38: 28.
3132. Wichman, C., Jr. 1978. Limahuli Valley botanical survey. Bull. Pacific Trop. Bot. Gard. 8: 1–8.
3133. Wichman, J. R. 1980. Early hibiscus hybridization on Kauai. Bull. Pacific Trop. Bot. Gard. 10: 65–67.
3134. Wick, H. L. 1970. Lignin staining . . .a limited success in identifying *koa* growth rings. U.S.D.A. Forest Serv. Res. Note PSW-205: 1–3.
3135. ———, R. E. Nelson, and L. K. Landgraf. 1971. Australian toon planted in Hawaii: tree quality, growth, and stocking. U.S.D.A. Forest Serv. Res. Pap. PSW-69: 1–10.
3136. ——— and G. A. Walters. 1974. *Albizia* Durazz. Albizzia: in Schopmeyer, C. S., Seeds of woody plants in the United States. U.S.D.A. Agric. Handbook 450: 203–205.
3137. ——— and C. D. Whitesell. 1969. Stump diameter affects sprout development of tropical ash. U.S.D.A. Forest Serv. Res. Note PSW-196: 1–3.
3138. Wiens, D., and B. A. Barlow. 1971. The cytogeography and relationships of the viscaceous and eremolepidaceous mistletoes. Taxon 20: 313–332.
3139. Wilbur, R. L. 1963. A prior name for the Hawaiian *Gouldia terminalis* (Rubiaceae). Pacific Sci. 17: 421–423.
3140. ———. 1964. The correct name for the Hawaiian *Gossypium*. Pacific Sci. 18: 101–103.
3141. ———. 1965. Nomenclatural notes on Hawaiian Myrsinaceae. Pacific Sci. 19: 522.
3142. ———. 1966. *Bleekeria compta*: a new binomial for a Hawaiian apocynaceous tree. Pacific Sci. 20: 260–261.
3143. ———. 1969. The correct name of an Hawaiian rubiaceous tree: *Gouldia affinis* vs. *Gouldia terminalis*. Brittonia 21: 224–226.
3144. ———. 1977. The correct name for the Hawaiian *Gossypium*: *G. sandvicense*. Taxon 26: 140.
3145. ———. 1981. The lectotypification of *Gossypium tomentosum* Seem. and the name of the Hawaiian endemic *Gossypium*. Taxon 30: 478–481.
3146. Wilcox, E. V. 1910. The algaroba in Hawaii. Hawaii Agric. Exp. Sta. Press Bull. 26: 1–8.
3147. ———. 1911. *No ka hooulu ana i ka maia*. Kahua Hooulu Oihana Mahiai Amelika, Honolulu, 11 pp.
3148. ———. 1913. Palms for Hawaii. Friend 71: 37–38.
3149. ———. 1917. The palms of Hawaii. Mid-Pacific Mag. 14: 246–249.
3150. ——— and F. A. Clowes. 1911. *No ka hooulu ana i ke kalo*. Kahua Hooulu Oihana Mahiai Amelika, Honolulu, 15 pp.
3151. ——— and V. S. Holt. 1913. Ornamental hibiscus in Hawaii. Hawaii Agric. Exp. Sta. Bull. 29: 1–60.

3152. ——— and A. R. Thompson. 1913. The extraction and use of *kukui* oil. Hawaii Agric. Exp. Sta. Press Bull. 39: 1–5.
3153. Wilder, G. P. 1905. A short trip to the Midway Islands with Captain A. P. Niblack in the U.S.S. "*Iroquois.*" Hawaiian Forester Agric. 2: 390–396.
3154. ———. 1906 [1907]. Fruits of the Hawaiian Islands. Vol. I. Hawaiian Gazette Co., Ltd., Honolulu, 77 pp. (Rev. ed., 1911, 247 pp.).
3155. ———. 1917. Hibiscus development in Hawaii. Hawaiian Almanac and Annual for 1918: 86–89.
3156. ———. 1932. Early plant life of Hawaii together with a partial list of some later introductions. Mid-Pacific Mag. 44: 2–17.
3157. Wilder, L. K. 1932. Hibiscus of Hawaii. Bull. Garden Club Amer. 19: 28–33.
3158. Williams, F. N. 1896. A revision of the genus *Silene*, Linn. J. Linn. Soc., Bot. 32: 1–196.
3159. Williams, J. E. 1980. Native vs. exotic woody vegetation recovery following goat removal in the eastern coastal lowlands of Hawaii Volcanoes National Park. Proc. 3rd Conf. Nat. Sci., Hawaii Volcanoes Natl. Park, pp. 373–382.
3160. ———. 1984. The lowland dry forest and scrub of Hawaii Volcanoes National Park: vegetation recovery in an historically stressed ecosystem (Abstr.). Proc. 5th Conf. Nat. Sci., Hawaii Volcanoes Natl. Park, p. 118.
3161. ———. 1985. The lowland dry forest and scrub in Hawaii Volcanoes National Park: current status and developmental trends in a stressed ecosystem. Master's thesis, Univ. Hawaii, Honolulu, 200 pp.
3162. Willis, J. C. 1936. Some further studies in endemism (Abstr.). Proc. Linn. Soc. London 148: 86–94.
3163. ———. 1949. The birth and spread of plants. Boissiera 8: 1–561.
3164. Wilsie, C. P., and M. Takahashi. 1934. Napier grass (*Pennisetum purpureum*): a pasture and green fodder crop for Hawaii. Hawaii Agric. Exp. Sta. Bull. 72: 1–17.
3165. Wilson, C. A. 1984. Endangered and threatened wildlife and plants; determination of endangered status and critical habitat for *Kokia drynarioides* (*koki'o*). Fed. Reg. 49: 47939–47401.
3166. ———. 1985. Endangered and threatened wildlife and plants; proposed endangered status for *Santalum freycinetianum* Gaud. var. *lanaiense* Rock (Lanai sandalwood or *'iliahi*). Fed. Reg. 50: 9086–9089.
3167. ———. 1985. Endangered and threatened wildlife and plants; proposed endangered status for *Argyroxiphium sandwicense* var. *sandwicense* (*'ahinahina* or Mauna Kea silversword). Fed. Reg. 50: 9092–9095.
3168. ———. 1985. Endangered and threatened wildlife and plants; proposed endangered status for *Achyranthes rotundata*. Fed. Reg. 50: 15764–15767.
3169. Wilson, K. A. 1953. A taxonomic study of the genus *Eugenia* (Myrtaceae) in Hawaii. Master's thesis, Univ. Hawaii, Honolulu, 38 pp.

3170. ———. 1957. A taxonomic study of the genus *Eugenia* (Myrtaceae) in Hawaii. Pacific Sci. 11: 161–180.
3171. Wilson, S. 1890. On some of the birds of the Sandwich Islands. The Ibis ser. VI, 2: 170–197.
3172. Wilson, W. F. (ed.). 1919. David Douglas, botanist at Hawaii. Publ. privately, Honolulu, 83 pp.
3173. ——— (ed.). 1920. Hawaii Nei 128 years ago, by Archibald Menzies. Publ. privately, Honolulu, 199 pp.
3174. ——— (ed.). 1922. With Lord Byron at the Sandwich Islands in 1825, being extracts from the MS diary of James Macrae, Scottish botanist. Publ. privately, Honolulu, 75 pp.
3175. Wimmer, F. E. 1929. Studien zu einer Monographie der Lobelioïdeen (Lobelioïdeae IV). Repert. Spec. Nov. Regni Veg. 26: 1–20.
3176. ———. 1943. Campanulaceae–Lobelioideae. I. Pflanzenr. IV. 276b (Heft 106): 1–260.
3177. ———. 1948. Vorarbeiten zur Monographie der Campanulaceae–Lobelioideae: II. Trib. *Lobelieae*. Ann. Naturhist. Mus. Wien 56: 317–374.
3178. ———. 1953. Campanulaceae–Lobelioideae. II. Pflanzenr. IV. 276b (Heft 107): 261–814.
3179. ———. 1968. Campanulaceae–Lobelioideae supplementum. Pflanzenr. IV. 276c (Heft 108): 815–916.
3180. Winner, W. E., and H. A. Mooney. 1980. Responses of Hawaiian plants to volcanic sulfur dioxide: stomatal behavior and foliar injury. Science 210: 789–791.
3181. Wirawan, N. 1972. Floristic and structural development of native dry forest stands at Mokuleia, N. W. Oahu. Master's thesis, Univ. Hawaii, Honolulu, 123 pp.
3182. ———. 1974. Floristic and structural development of native dry forest stands at Mokuleia, N. W. Oahu. U.S. IBP Island Ecosystems IRP Techn. Rep. 34: 1–56.
3183. ———. 1978. Vegetation and soil-water regimes in a tropical rain forest valley on Oahu, Hawaiian Islands. Ph.D. dissertation, Univ. Hawaii, Honolulu, 420 pp.
3184. Wise, J. H. 1965. Medicine. I. Ancient Hawaiian remedies known today: *in* Handy, E. S. C., K. P. Emory, E. H. Bryan, Jr., P. H. Buck, J. H. Wise, and others, Ancient Hawaiian civilization. A series of lectures delivered at The Kamehameha Schools. The Kamehameha Schools, Honolulu, pp. 253–255. (Also publ. in Handy et al., rev. ed., 1965, Charles E. Tuttle Co., Rutland, Vt., pp. 257–259).
3185. Wishard, L. 1978. Coconut varieties. Bull. Pacific Trop. Bot. Gard. 8: 33–39.
3186. Wold, M. L., and R. M. Lanner. 1965. New-stool shoots from a 20-year-old swamp-mahogany eucalyptus stump. Ecology 46: 755–756.
3187. Wolff, H. 1913. Umbelliferae-Saniculoideae. Pflanzenr. IV. 228 (Heft 61): 1–305.

3188. ———. 1921. *Spermolepis hawaiiensis* spec. nov. Repert. Spec. Nov. Regni Veg. 17: 440–441.
3189. ———. 1927. Umbelliferae–Apioideae–Ammineae–Carinae, Ammineae novemjugatae et genuinae. Pflanzenr. IV. 228 (Heft 90): 1–398.
3190. Wong, W. H. C., Jr. 1972. Propagating and planting the creeping *naupaka*. Newslett. Hawaiian Bot. Soc. 11: 41.
3191. ———. 1974. *Grevillea robusta* A. Cunn. Silk-oak: *in* Schopmeyer, C. S., Seeds of woody plants in the United States. U.S.D.A. Agric. Handbook 450: 437–438.
3192. ———, N. Honda, and R. E. Nelson. 1967. Plantation timber on the island of Lanai—1966. U.S.D.A. Forest Serv. Resource Bull. PSW-7: 1–18.
3193. ———, R. E. Nelson, and H. L. Wick. 1968. Plantation timber on the island of Molokai—1967. U.S.D.A. Forest Serv. Resource Bull. PSW-9: 1–25.
3194. ———, H. L. Wick, and R. E. Nelson. 1969. Plantation timber on the island of Maui—1967. U.S.D.A. Forest Serv. Resource Bull. PSW-11: 1–42.
3195. Wood, H. B., R. A. Merriam, and T. H. Schubert. 1969. Vegetation recovering . . . little erosion on Hanalei watershed after fire. U.S.D.A. Forest Serv. Res. Note PSW-191: 1–5.
3196. Woodson, R. E., Jr. 1957. The botany of *Rauwolfia*: *in* Woodson, R. E., Jr., H. W. Youngken, E. Schlittler, and J. A. Schneider, *Rauwolfia*: botany, pharmocognosy, chemistry & pharmacology. Little, Brown and Co., Boston, pp. 3–31.
3197. Woodward, P. W. 1972. The natural history of Kure Atoll, Northwestern Hawaiian Islands. Atoll Res. Bull. 164: 1–318.
3198. Woolliams, K. R. 1972. Propagation of endangered tropical plants. Bull. Pacific Trop. Bot. Gard. 2: 17–20.
3199. ———. 1972. From the nursery. Bull. Pacific Trop. Bot. Gard. 2: 40.
3200. ———. 1972. A report on the endangered species. Bull. Pacific Trop. Bot. Gard. 2: 46–49.
3201. ———. 1972. From the nursery. Bull. Pacific Trop. Bot. Gard. 2: 78.
3202. ———. 1973. Summary of the native species in the Pacific Garden. Bull. Pacific Trop. Bot. Gard. 3: 14–15.
3203. ———. 1974. Endangered speciies now established in the grounds of Pacific Garden. Bull. Pacific Trop. Bot. Gard. 4: 33.
3204. ———. 1975. The propagation of Hawaiian endangered species. Newslett. Hawaiian Bot. Soc. 14: 59–68.
3205. ———. 1976. Propagation. Notes Waimea Arbor. 3(1): 5–6.
3206. ———. 1976. The propagation of Hawaiian endangered species: *in* Simmons, J. B., R. I. Beyer, P. E. Brandham, G. L. Lucas, and V. T. H. Parry, Conservation of threatened plants. Plenum Press, New York, pp. 73–83.
3207. ———. 1977. Report from Waimea Arboretum: *in* Palmer, D. D. (ed.),

Hawaiian plants—notes and news. Newslett. Hawaiian Bot. Soc. 16: 75–76.
3208. ———. 1978. Propagation of some endangered Hawaiian plants at Waimea Arboretum. Notes Waimea Arbor. & Bot. Gard. 5(1): 3–4.
3209. ———. 1980. Oahu yellow hibiscus found. Notes Waimea Arbor. & Bot. Gard. 7(1): 9, 12.
3210. ———. 1982. *Kokia cookei*—more good news. Notes Waimea Arbor. & Bot. Gard. 9(1): 3–4.
3211. ———. 1982. Cultivate Hawaii's plants! Notes Waimea Arbor. & Bot. Gard. 9(2): 10–13.
3212. ———, O. Degener, and I. Degener. 1980. *Kokia cookei* Deg. . . . then there were two! Notes Waimea Arbor. & Bot. Gard. 7(1): 2–7.
3213. ———, ———, and ———. 1980. Cooke's *Kokia* again. Notes Waimea Arbor. & Bot. Gard. 7(2): 8–9.
3214. Wright, R. A. 1984. Population ecology and successional behavior of exotic and native shrubs in the devastation area, Hawaii Volcanoes National Park, Hawaii (Abstr.). Proc. 5th Conf. Nat. Sci., Hawaii Volcanoes Natl. Park, p. 119.
3215. ———. 1985. Shrub population dynamics and succession on volcanic cinder in Hawaii. Master's thesis, Univ. Hawaii, Honolulu, 369 pp.
3216. Wurdack, J. J. 1967. The cultivated glorybushes, *Tibouchina* (Melastomataceae). Baileya 15: 1–6.
3217. Yanamura, H. K. 1971. Flora observed on Kahoolawe, June 1970 and April 1971. Newslett. Hawaiian Bot. Soc. 10: 31–32.
3218. Ydrac, F.-L. 1905. Recherches anatomiques sur les Lobéliacées. Ph.D. dissertation, Univ. Paris, École Supér. Pharm., Paris, 165 pp.
3219. Yee, W. W. J. 1958. The mango in Hawaii. Univ. Hawaii Coop. Extens. Serv. Circ. 388: 1–26.
3220. ———, E. K. Akamine, G. M. Aoki, R. A. Hamilton, F. H. Haramoto, R. B. Hine, O. V. Holtzmann, J. T. Ishida, J. T. Keeler, and H. Y. Nakasone. 1970. Papayas in Hawaii. Univ. Hawaii Coop. Extens. Serv. Circ. 436: 1–56.
3221. Yen, D. E. 1960. The sweet potato in the Pacific: the propagation of the plant in relation to its distribution. J. Polynes. Soc. 69: 368–375.
3222. ———. 1974. The sweet potato and Oceania. An essay in ethnobotany. Bernice P. Bishop Mus. Bull. 236: 1–389.
3223. ———, P. V. Kirch, P. Rosendahl, and T. Riley. 1972. Prehistoric agriculture in the upper valley of Makaha, Oahu: *in* Ladd, E. J., and D. E. Yen (eds.), Makaha Valley Historical Project. Interim report no. 3. Pacific Anthropol. Rec. 18: 59–68.
3224. Yocom, C. F. 1967. Ecology of feral goats in Haleakala National Park, Maui, Hawaii. Amer. Midl. Naturalist 77: 418–451.
3225. Yoshida, J. 1973. [Vegetation on Kupehau Trail]. Elepaio 33: 108–109.
3226. Yoshinaga, A. Y. 1978. Vegetation of the Hana rain forest, Haleakala National Park. Proc. 2nd Conf. Nat. Sci., Hawaii Volcanoes Natl. Park, pp. 346–348.

3227. ———. 1980. Upper Kīpahulu Valley weed survey. Coop. Natl. Park Resources Stud. Unit, Hawaii, Techn. Rep. 33: 1–17.
3228. ———. 1980. Exotic plants in Kīpahulu Valley: 1945–1980. Proc. 3rd Conf. Nat. Sci., Hawaii Volcanoes Natl. Park, pp. 387–392.
3229. Young, R. A., and P. Popenoe. 1916. Saving the *kokio* tree. J. Heredity 7: 24–27.
3230. Younge, O. R., and J. C. Moomaw. 1960. Revegetation of stripmined bauxite lands in Hawaii. Econ. Bot. 14: 316–330.
3231. ———, D. L. Plucknett, and P. P. Rotar. 1964. Culture and yield performance of *Desmodium intortum* and *D. canum* in Hawaii. Hawaii Agric. Exp. Sta. Univ. Hawaii Techn. Bull. 59: 1–28.
3232. Youngs, R. L. 1960. Physical, mechanical, and other properties of five Hawaiian woods. U.S.D.A. Forest Serv. Forest Prod. Lab. Rep. 2191: 1–19.
3233. ———. 1964. Hardness, density, and shrinkage characteristics of silk-oak from Hawaii. U.S.D.A. Forest Serv. Res. Note FPL-074: 1–14.
3234. Yun, I. K. 1985. Studies of the comparative anatomy of some parents and their hybrids in the Hawaiian silversword complex. Master's thesis, Univ. Hawaii, Honolulu, 172 pp.
3235. Yuncker, T. G. 1921. Revision of the North American and West Indian species of *Cuscuta*. Univ. Illinois Biol. Monogr. 6: 1–141.
3236. ———. 1932. The genus *Cuscuta*. Mem. Torrey Bot. Club 18: 109–331.
3237. ———. 1933. Revision of the Hawaiian species of *Peperomia*. Bernice P. Bishop Mus. Bull. 112: 1–131.
3238. ———. 1933. A revision of the Hawaiian species of *Peperomia* (Abstr.). Proc. Hawaiian Acad. Sci. 8: 18–19.
3239. ———. 1934. Some botanical aspects of the Hawaiian Islands. Torreya 34: 29–36.
3240. ———. 1937. Observations on the teratology of the genus *Peperomia*. Occas. Pap. Bernice P. Bishop Mus. 13(2): 5–9.
3241. ———. 1937. Three additional species of *Peperomia* in Hawaii. Occas. Pap. Bernice P. Bishop Mus. 13(14): 161–165.
3242. ———. 1949. New species of Hawaiian peperomias. Occas. Pap. Bernice P. Bishop Mus. 19(14): 257–260.
3243. ———. 1958. The Piperaceae—a family profile. Brittonia 10: 1–7.
3244. ——— and W. D. Gray. 1934. Anatomy of Hawaiian peperomias. Occas. Pap. Bernice P. Bishop Mus. 10(20): 1–19.
3245. Yzendoorn, R. 1911. The introduction of the algaroba. 18th Annual Rep. Hawaiian Hist. Soc. 1910: 29–34.
3246. Zahlbruckner, A. 1891. Ueber einige Lobeliaceen des Wiener Herbariums. Ann. K.K. Naturhist. Hofmus. 6: 430–445.
3247. ———. 1893. The genus *Trematocarpus*. Ann. Bot. (London) 7: 289–290.
3248. ———. 1893. [Ueber die Gattung *Trematocarpus*]. Verh. K.K. Zool.-Bot. Ges. Wien 43, Sitz.: 6–7.

3249. Zentmyer, G. A. 1980. *Phytophthora cinnamomi* and the diseases it causes. Amer. Phytopathol. Soc. Monogr. 10: 1–96.
3250. Zimmerman, E. C. 1970. Adaptive radiation in Hawaii with special reference to insects. Biotropica 2: 32–38.
3251. Zones, C. P. 1961. Discharge of ground water by phreatophytes in the Waianae District, Oahu, Hawaii. U.S. Geol. Surv. Profess. Pap. 424-D: 240–242.
3252. Zschokke, T. C. 1931. The church in the garden. Publ. privately, Honolulu, 15 pp.
3253. ———. 1932. The *kukui* oil of Hawaiian Islands. Peanut J. and Nut World (Suffolk, Va.) 12(1): 15, 23, 36.
3254. ———. 1933. Hawaii's climate ideal for perfume plants. J. Pan-Pacific Res. Inst. 8(2): 15–16.
3255. ———. 1933. Trees in Hawaii of Australian origin. Mid-Pacific Mag. 46: 3–5.
3256. ———. 1933. Poisonous plants now in Hawaii (Abstr.). Proc. Hawaiian Acad. Sci. 8: 19–20.
3257. ———. 1933. Poisonous plants now found in the Hawaiian Islands. Univ. Hawaii Agric. Extens. Serv. Agric. Notes 49: 1–4.

SUBJECT INDEX

All subjects can be cross-checked to the references cited for a certain plant or locality.

ABIOTIC ENVIRONMENT

111, 152, 153, 154, 189, 277, 347, 348, 350, 363, 364, 369, 403, 405, 406, 631, 637, 680, 707, 718, 736, 739, 740, 741, 760, 768, 800, 801, 833, 853, 893, 931, 1003, 1004, 1008, 1027, 1113, 1216, 1241, 1243, 1294, 1375, 1376, 1384, 1414, 1424, 1457, 1458, 1586, 1609, 1647, 1688, 1690, 1702, 1718, 1877, 1893, 1898, 1911, 1913, 1916, 1918, 1919, 1924, 1925, 1927, 1928, 1995, 2098, 2163, 2216, 2244, 2268, 2273, 2392, 2569, 2679, 2732, 2756, 2780, 2781, 2798, 2814, 2842, 2845, 2981, 2991, 2994, 3017, 3018, 3093, 3095, 3106, 3107, 3110, 3112, 3120, 3180, 3183, 3195, 3251

ANATOMY

113, 115, 116, 170, 246, 254, 283, 289, 290, 360, 387, 410, 411, 412, 413, 414, 415, 416, 417, 418, 419, 420, 421, 424, 430, 431, 432, 437, 438, 440, 445, 451, 541, 552, 610, 611, 781, 803, 806, 822, 934, 946, 947, 998, 1023, 1116, 1117, 1134, 1139, 1140, 1141, 1264, 1266, 1299, 1383, 1439, 1514, 1598, 1621, 1629, 1733, 1795, 2067, 2068, 2117, 2129, 2147, 2194, 2530, 2531, 2532, 2533, 2558, 2559, 2683, 2740, 2761, 2784, 2823, 2835, 2836, 2839, 2919, 2940, 2946, 2974, 3028, 3189, 3196, 3218, 3234, 3244

AUTECOLOGY

2, 32, 36, 48, 59, 73, 74, 75, 76, 83, 84, 85, 98, 102, 103, 113, 134, 135, 136, 138, 139, 141, 152, 153, 154, 158, 159, 168, 181, 204, 209, 220, 227, 246, 259, 261, 262, 273, 274, 275, 276, 277, 278, 289, 341, 342, 343, 347, 349, 350, 352, 354, 404, 405, 424, 445, 453, 454, 455, 456, 457, 459, 464, 474, 475, 476, 480, 508, 523, 524, 526, 527, 528, 554, 563, 564, 566, 567, 568, 572, 573, 574, 575, 577, 580, 581, 582, 589, 595, 596, 597, 598, 610, 614, 623, 662, 670, 736, 739, 740, 741, 759, 781, 782, 797, 799, 801, 804, 826, 828, 852, 913, 919, 932, 933, 944, 954, 955, 958, 973, 974, 979, 999, 1002, 1005, 1007, 1022, 1027, 1057, 1058, 1079, 1089, 1105, 1118, 1119, 1137, 1137a, 1137b, 1139, 1141, 1155, 1181, 1182, 1183, 1189, 1190, 1205, 1215, 1216, 1222, 1224, 1241, 1249, 1254, 1256, 1282a, 1292, 1294, 1295, 1324, 1325, 1328, 1329, 1330, 1332, 1350, 1353, 1358, 1366, 1389, 1390, 1398, 1408, 1412, 1413, 1414, 1415, 1416, 1417, 1419, 1420, 1427, 1430, 1431, 1436, 1439, 1470, 1474, 1502, 1513, 1528, 1530, 1532, 1536, 1540, 1541, 1542, 1543, 1547, 1548, 1549, 1550, 1551, 1552, 1553, 1576, 1578, 1579, 1581, 1582, 1585, 1606, 1613, 1629, 1639, 1668, 1669, 1672, 1691, 1699, 1703, 1705, 1706, 1713, 1714, 1718, 1719, 1724, 1725, 1733, 1735, 1760, 1766, 1767, 1772, 1781, 1791, 1804, 1805, 1821, 1823, 1858, 1859, 1875, 1883, 1888, 1890, 1900, 1903, 1905, 1906, 1907, 1908, 1910, 1911, 1912, 1913, 1915, 1916, 1917, 1920, 1924, 1925, 1935, 1938, 2000, 2003, 2005, 2057, 2059, 2073, 2074, 2075, 2078, 2083, 2085, 2086, 2089, 2097, 2108, 2113, 2127, 2128, 2129, 2130, 2131, 2136, 2137, 2142, 2143, 2144, 2154, 2161, 2163, 2165, 2176, 2179, 2182, 2184, 2190, 2194, 2202, 2203, 2216, 2218, 2219, 2229, 2232, 2243, 2247, 2248, 2249, 2255, 2256, 2261, 2262, 2263, 2264, 2265, 2266, 2267, 2268, 2289, 2308, 2322, 2323, 2333, 2334, 2336, 2341, 2562, 2571, 2573, 2575, 2576, 2581, 2582, 2584, 2594, 2684, 2688, 2701, 2702, 2754, 2755, 2778, 2780, 2781, 2784, 2797, 2798, 2799, 2800, 2802, 2807, 2815, 2829, 2830, 2831, 2832, 2841, 2842, 2843, 2845, 2852, 2886, 2898, 2900, 2901, 2903, 2915, 2922, 2923, 2924, 2925, 2927, 2937, 2938, 2957, 2971, 2980, 2991, 3009,

3019, 3023, 3024, 3026, 3028, 3029, 3031, 3038, 3114, 3115, 3117, 3118, 3123, 3131, 3136, 3137, 3171, 3180, 3186, 3191, 3214, 3215, 3220, 3232, 3233, 3234, 3250, 3251

BIBLIOGRAPHY

16, 166, 167, 222, 237, 292, 296, 299, 314, 315, 433, 439, 519, 521, 650, 736, 1216, 1287, 1441, 1521, 1553, 1788, 1793, 1807, 2331, 2332, 2335, 2346, 2347, 2470, 2500, 2501, 2555, 2678, 2948, 3008

CHECKLIST/FLORA

19, 20, 21, 25, 26, 30, 31, 40, 51, 52, 56, 58, 87, 89, 94, 97, 112, 114, 220, 228, 231, 264, 266, 292, 309, 311, 321, 330, 331, 338, 407, 478, 482, 534, 539, 540, 547, 554, 583, 604, 717, 765, 768, 771, 774, 808, 809, 823, 836, 839, 840, 892, 905, 917, 949, 1055, 1109, 1113, 1132, 1151, 1152, 1155, 1178, 1187, 1191, 1243, 1256, 1377, 1394, 1406, 1453, 1459, 1511, 1517, 1518, 1519, 1522, 1525, 1527, 1616, 1675, 1680, 1681, 1688, 1691, 1729, 1738, 1739, 1773, 1862, 1921, 1924, 1927, 1928, 1932, 1986, 1988, 2038, 2040, 2049, 2177, 2220, 2351, 2352, 2356, 2379, 2383, 2411, 2415, 2434, 2439, 2444, 2449, 2453, 2456, 2458, 2460, 2461, 2487, 2507, 2521, 2540, 2689, 2704, 2711, 2732, 2737, 2738, 2764, 2768, 2781, 2782, 2814, 2846, 2847, 2918, 2931, 2932, 2943, 2977, 3007, 3037, 3039, 3197, 3217

CHEMICAL ANALYSIS

138, 139, 141, 176, 258, 260, 508, 524, 526, 528, 529, 536, 560, 597, 798, 826, 827, 830, 831, 832, 961, 982, 983, 984, 986, 1022, 1028, 1032, 1066, 1067, 1095, 1105, 1123, 1176, 1177, 1205, 1270, 1284, 1285, 1316, 1369, 1407, 1516, 1555, 1607, 1629, 1743, 1762, 1766, 1767, 1769, 1770, 1808, 1809, 1810, 1811, 1812, 1813, 1814, 2101, 2111, 2119, 2120, 2173, 2178, 2180, 2256, 2327, 2527, 2542, 2543, 2544, 2545, 2546, 2547, 2548, 2549, 2550, 2551, 2552, 2682, 2784, 2832, 2886, 2920, 2921, 2933, 2952, 2973, 3028, 3097, 3164

CLADISTICS/NUMERICAL TAXONOMY

245, 471, 565, 987, 1002, 1447, 1600, 1743, 1803, 1804, 1805

CYTOLOGY/GENETICS

120, 272, 409, 416, 458, 460, 462, 463, 464, 465, 466, 467, 468, 469, 470, 470a, 473, 512, 523, 525, 529, 559, 576, 601, 716, 804, 822, 942, 982, 983, 984, 985, 1014, 1016, 1019, 1022, 1028, 1029, 1066, 1068, 1100, 1104, 1124, 1125, 1189, 1192, 1264, 1308, 1373, 1469, 1476, 1477, 1555, 1601, 1602, 1603, 1607, 1624, 1625, 1628, 1629, 1630, 1631, 1735, 1736, 1739, 1781, 1782, 2063, 2067, 2068, 2070, 2072, 2077, 2109, 2110, 2111, 2176, 2208, 2209, 2210, 2224, 2234, 2235, 2254, 2263, 2264, 2270, 2333, 2337, 2392, 2532, 2534, 2741, 2784, 2787, 2794, 2795, 2832, 2833, 2870, 2902, 2904, 2905, 2906, 2907, 2947a, 3138, 3222

DISPERSAL

17, 18, 81, 91, 137, 166, 167, 189, 282, 283, 310, 324, 333, 363, 370, 371, 420, 423, 425, 426, 427, 428, 429, 433, 434, 436, 439, 441, 442, 443, 444, 572, 574, 594, 606, 631, 637, 675, 689, 702, 707, 712, 753, 759, 770, 866, 872, 874, 875, 887, 900, 937, 942, 1017, 1019, 1022, 1064, 1065, 1087, 1127, 1133, 1147, 1187, 1255, 1269, 1314, 1317, 1318, 1328, 1360, 1384, 1388, 1424, 1528, 1553, 1664, 1668, 1701, 1780, 1789, 1941, 1960, 2133, 2246, 2258, 2350, 2367, 2377, 2392, 2393, 2399, 2492, 2513, 2602, 2603, 2703, 2708, 2711, 2722, 2729, 2735, 2754, 2755, 2827, 2828, 2849, 2850, 2948, 2955, 2990, 3044, 3098, 3100, 3221

DISTURBANCE EVENT

1, 2, 8, 17, 18, 22, 36, 37, 38, 44, 55, 77, 87, 94, 109, 124, 143, 145, 147, 148, 155, 159, 180, 182, 201, 219, 220, 229, 263, 293, 294, 298, 301, 303, 310, 317, 322, 324, 326, 327, 328, 335, 338, 348, 350, 364, 370, 407, 433, 435, 439, 446, 447, 448, 475, 510, 511, 517, 543, 544, 553, 555, 582, 607, 608, 609, 619, 621, 622, 631, 637, 643, 645, 646, 648, 649, 651, 652, 653, 655, 659, 661, 665, 670, 672, 680, 683, 684, 687, 691, 694, 697, 701, 702, 704, 707, 708, 733, 734, 735, 736, 744, 766, 768, 769, 772, 774, 808, 809, 811, 812, 813, 814, 815, 816, 820, 821, 824, 825, 836, 844, 850, 852, 853, 857, 862, 863, 871, 873, 878, 882, 899, 903, 906, 911, 914, 930, 936, 942, 950, 952, 953, 956, 967, 976, 1003, 1006, 1008,

SUBJECT INDEX 177

1010, 1011, 1031, 1057, 1070, 1077, 1081, 1082, 1083, 1084, 1085, 1114, 1133, 1143, 1147, 1153, 1154, 1156, 1157, 1158, 1159, 1160, 1161, 1162, 1163, 1164, 1165, 1166, 1166a, 1167, 1168, 1169, 1170, 1171, 1172, 1173, 1174, 1180, 1188, 1197, 1220, 1223, 1239, 1240, 1250, 1255, 1257, 1260, 1263, 1289, 1292, 1293, 1296, 1304, 1309, 1320, 1323, 1326, 1331, 1333, 1337, 1338, 1340, 1350, 1351, 1354, 1358, 1360, 1361, 1366, 1375, 1376, 1379, 1384, 1400, 1402, 1403, 1404, 1405, 1410, 1425, 1427, 1428, 1430, 1431, 1436, 1437, 1438, 1502, 1512, 1524, 1529, 1531, 1537, 1544, 1552, 1553, 1581, 1582, 1609, 1622, 1623, 1632, 1647, 1650, 1651, 1652, 1653, 1655, 1659, 1660, 1678, 1689, 1732, 1742, 1757, 1765, 1772, 1773, 1789, 1819, 1820, 1861, 1870, 1876, 1879, 1882, 1883, 1884, 1885, 1888, 1892, 1899, 1900, 1901, 1902, 1904, 1907, 1910, 1911, 1912, 1913, 1914, 1916, 1919, 1920, 1921, 1924, 1929, 1931, 1932, 1933, 1934, 1937, 1938, 1948, 1949, 1951, 1965, 1987, 1990, 1996, 1998, 1999, 2054, 2055, 2061, 2066, 2069, 2090, 2091, 2098, 2100, 2107, 2113, 2114, 2115, 2116, 2135, 2139, 2164, 2190, 2192, 2219, 2230, 2236, 2249, 2273, 2307, 2344, 2372, 2477, 2513, 2514, 2570a, 2571, 2572, 2573, 2574, 2577, 2580, 2581, 2582, 2583, 2584, 2733, 2742, 2753, 2754, 2755, 2757, 2762, 2763, 2764, 2765, 2766, 2767, 2768, 2781, 2782, 2789, 2798, 2803, 2804, 2805, 2806, 2821, 2824, 2829, 2830, 2831, 2837, 2847, 2875, 2886, 2896, 2897, 2898, 2899, 2910, 2922, 2925, 2926, 2930, 2931, 2938, 2941, 2942, 2945, 2947, 2948, 2971, 2994, 2999, 3001, 3004, 3006, 3009, 3011, 3027, 3034, 3035, 3036, 3038, 3040, 3041, 3042, 3043, 3095, 3098, 3099, 3100, 3108, 3112, 3114, 3115, 3153, 3159, 3160, 3165, 3166, 3167, 3168, 3181, 3182, 3195, 3211, 3224, 3227, 3228, 3229, 3230, 3239

ECONOMICS

4, 13, 24, 34, 53, 70, 72, 73, 76, 78, 82, 93, 102, 103, 107, 108, 110, 125, 149, 161, 174, 177, 183, 209, 219, 223, 224, 234, 235, 249, 259, 260, 261, 262, 267, 296, 316, 317, 318, 319, 320, 321, 326, 327, 338, 341, 342, 343, 344, 446, 447, 448, 449, 477, 480, 508, 531, 535, 536, 537, 548, 549, 550, 558, 560, 592, 602, 607, 612, 628, 717, 719, 728, 797, 833, 867, 890, 930, 932, 943, 944, 949, 976, 980, 988, 989, 1010, 1028, 1029, 1031, 1053, 1077, 1081, 1082, 1083, 1084, 1085, 1089, 1090, 1110, 1111, 1135, 1136, 1181, 1182, 1183, 1184, 1185, 1188, 1190, 1197, 1220, 1221, 1222, 1225, 1226, 1240, 1241, 1242, 1249, 1250, 1252, 1253, 1254, 1255, 1257, 1258, 1259, 1261, 1262, 1273, 1298, 1306, 1321, 1323, 1324, 1326, 1327, 1328, 1329, 1330, 1331, 1335, 1337, 1339, 1340, 1341, 1342, 1343, 1344, 1346, 1350, 1356, 1357, 1359, 1361, 1362, 1392, 1409, 1437, 1438, 1443, 1468, 1469, 1470, 1474, 1487, 1501, 1573, 1574, 1575, 1580, 1582, 1586, 1609, 1613, 1614, 1615, 1626, 1637, 1643, 1648, 1649, 1650, 1652, 1654, 1655, 1656, 1657, 1658, 1660, 1661, 1672, 1678, 1684, 1692, 1695, 1699, 1703, 1713, 1719, 1720, 1724, 1725, 1727, 1734, 1735, 1742, 1745, 1757, 1759, 1760, 1765, 1766, 1799, 1810, 1811, 1812, 1813, 1814, 1820, 1862, 1867, 1868, 1869, 1870, 1871, 1874, 1875, 1912, 1950, 1963, 1964, 1967, 1968, 1969, 1971, 1972, 1973, 1974, 1976, 1978, 1979, 1980, 1981, 1982, 1983, 1984, 1985, 1988, 1997, 2002, 2003, 2004, 2005, 2052, 2053, 2054, 2055, 2056, 2057, 2058, 2059, 2065, 2083, 2084, 2086, 2097, 2102, 2136, 2137, 2138, 2139, 2143, 2153, 2154, 2171, 2172, 2173, 2174, 2175, 2177, 2178, 2179, 2180, 2240, 2241, 2243, 2250, 2251, 2253, 2289, 2295, 2296, 2297, 2301, 2328, 2336, 2351, 2371, 2496, 2514, 2572, 2575, 2585, 2680, 2681, 2686, 2692, 2693, 2694, 2695, 2696, 2697, 2698, 2699, 2700, 2768, 2769, 2776, 2777, 2778, 2793, 2808, 2809, 2810, 2811, 2812, 2813, 2835, 2860, 2864, 2898, 2903, 2910, 2927, 2936, 2937, 2939, 2944, 2956, 2957, 2958, 2959, 2960, 2961, 2962, 2963, 2968, 2983, 3000, 3007, 3020, 3021, 3022, 3023, 3024, 3025, 3026, 3027, 3029, 3030, 3031, 3032, 3039, 3044, 3046, 3047, 3049, 3105, 3115, 3116, 3117, 3118, 3119, 3120, 3121, 3122, 3123, 3126, 3128, 3130, 3133, 3134, 3135, 3136, 3137, 3146, 3147, 3150, 3151, 3152, 3164, 3191, 3192, 3193, 3194, 3199, 3201, 3219, 3220, 3221, 3231, 3232, 3233

ENDANGERED SPECIES

8, 22, 25, 27, 29, 30, 37, 39, 40, 41, 55, 56, 58, 112, 147, 229, 281, 467, 590, 618, 687, 694, 697, 811, 813, 814, 815, 816, 873, 899, 906, 917, 942, 1006, 1143, 1153, 1154, 1156, 1157, 1158, 1159, 1160, 1161, 1162, 1163, 1164, 1165, 1166, 1166a, 1167, 1168, 1169, 1170, 1171, 1172, 1173, 1213, 1239, 1296, 1304, 1305, 1400, 1534, 1535, 1632, 1670, 1768, 1820, 1943, 1944, 1973, 1998, 2091, 2190, 2217, 2219, 2477, 2479, 2896, 2898, 2930, 2931, 2934, 2938,

3006, 3043, 3108, 3165, 3166, 3167, 3168, 3198, 3200, 3202, 3203, 3204, 3206, 3209, 3212, 3229

ETHNOBOTANY

11, 45, 54, 62, 63, 64, 81, 82, 93, 95, 99, 100, 101, 102, 103, 104, 105, 117, 145, 155, 156, 160, 177, 189, 190, 193, 203, 208, 223, 224, 230, 234, 239, 251, 268, 269, 296, 297, 300, 306, 320, 330, 333, 335, 337, 352, 446, 485, 488, 493, 494, 515, 520, 561, 584, 588, 591, 599, 613, 619, 625, 631, 632, 634, 637, 638, 639, 640, 641, 678, 700, 707, 719, 725, 728, 753, 775, 777, 810, 942, 945, 946, 947, 1012, 1053, 1056, 1071, 1078, 1090, 1091, 1092, 1093, 1094, 1115, 1123, 1135, 1146, 1219, 1268, 1283, 1300, 1311, 1314, 1319, 1325, 1328, 1330, 1332, 1334, 1336, 1337, 1340, 1354, 1363, 1368, 1371, 1372, 1378, 1382, 1391, 1393, 1401, 1402, 1403, 1404, 1405, 1406, 1432, 1463, 1464, 1465, 1466, 1467, 1473, 1501, 1511, 1514, 1533, 1554, 1583, 1612, 1626, 1633, 1666, 1673, 1674, 1678, 1679, 1682, 1685, 1689, 1694, 1697, 1698, 1699, 1703, 1704, 1705, 1706, 1707, 1708, 1711, 1713, 1715, 1716, 1717, 1721, 1722, 1723, 1755, 1758, 1764, 1790, 1794, 1797, 1799, 1800, 1808, 1809, 1822, 1854, 1860, 1867, 1869, 1993, 1994, 2006, 2007, 2008, 2009, 2016, 2017, 2028, 2036, 2041, 2042, 2043, 2044, 2050, 2065, 2066, 2076, 2105, 2183, 2198, 2199, 2201, 2239, 2241, 2258, 2272, 2282, 2284, 2297, 2308, 2310, 2311, 2312, 2329, 2330, 2349, 2367, 2371, 2378, 2386, 2415, 2419, 2430, 2458, 2478, 2487, 2521, 2586, 2587, 2588, 2589, 2592, 2764, 2851, 2852, 2858, 2859, 2863, 2865, 2866, 2867, 2910, 2932, 2933, 2950, 2956, 2958, 2960, 2965, 2969, 3000, 3048, 3049, 3090, 3105, 3115, 3127, 3128, 3152, 3156, 3172, 3173, 3174, 3184, 3222, 3223, 3239, 3253, 3254, 3256, 3257

EVOLUTIONARY BIOLOGY

49, 134, 135, 140, 178, 179, 189, 272, 282, 289, 411, 415, 416, 421, 423, 425, 426, 427, 428, 432, 433, 434, 435, 436, 438, 439, 443, 444, 445, 458, 459, 464, 468, 470, 474, 594, 601, 616, 674, 727, 747, 770, 802, 803, 860, 866, 869, 872, 900, 942, 958, 963, 964, 982, 983, 984, 986, 987, 1017, 1019, 1020, 1021, 1022, 1024, 1116, 1117, 1125, 1147, 1153, 1154, 1264, 1266, 1269, 1276, 1279, 1516, 1531, 1537, 1599, 1600, 1601, 1607, 1627, 1630, 1631, 1642, 1663, 1669, 1754, 1916, 1952, 1956, 2095, 2117, 2119, 2120, 2208, 2209, 2210, 2223, 2231, 2232, 2264, 2267, 2294, 2308, 2343, 2516, 2685, 2703, 2707, 2727, 2735, 2758, 2792, 2823, 2826, 2850, 2885, 2889, 2924, 2928, 2954, 3162, 3163, 3243

GENERAL WORK

127, 128, 166, 167, 218, 387, 402, 423, 433, 436, 439, 495, 631, 634, 637, 707, 728, 745, 780, 789, 792, 793, 872, 991, 995, 1039, 1126, 1187, 1238, 1384, 1499, 1500, 1739, 1797, 1798, 1919, 2036, 2042, 2043, 2044, 2284, 2297, 2411, 2691, 2738, 2819, 2827, 2828, 2879, 3014, 3016, 3051

HISTORY

63, 208, 225, 240, 264, 293, 296, 299, 308, 312, 313, 315, 321, 322, 325, 328, 358, 516, 518, 519, 561, 578, 579, 650, 664, 699, 732, 742, 743, 774, 849, 1013, 1076, 1108, 1109, 1113, 1127, 1207, 1233, 1236, 1261, 1283, 1384, 1385, 1401, 1443, 1460, 1519, 1525, 1546, 1645, 1646, 1696, 1710, 1739, 1783, 1784, 1785, 1794, 1797, 1800, 1873, 1955, 1997, 2054, 2201, 2279, 2306, 2317, 2360, 2362, 2428, 2434, 2444, 2449, 2453, 2456, 2458, 2460, 2461, 2470, 2521, 2540, 2541, 2737, 2875, 2970, 3056, 3092, 3094, 3172, 3173, 3174

HYBRIDIZATION

49, 142, 144, 145, 146, 221, 425, 428, 436, 460, 461, 463, 464, 466, 467, 469, 470a, 523, 525, 526, 528, 565, 576, 668, 702, 856, 861, 869, 877, 950, 962, 963, 964, 986, 988, 989, 1002, 1014, 1017, 1018, 1019, 1020, 1021, 1022, 1029, 1068, 1100, 1183, 1189, 1373, 1384, 1389, 1429, 1447, 1469, 1590, 1607, 1627, 1628, 1629, 1631, 1735, 1781, 1782, 2005, 2068, 2109, 2110, 2208, 2209, 2210, 2231, 2333, 2334, 2392, 2518, 2524, 2617, 2618, 2619, 2622, 2624, 2636, 2641, 2655, 2656, 2657, 2663, 2672, 2717, 2728, 2730, 2786, 2788, 2791, 2904, 2906, 3002, 3005, 3133, 3151, 3155, 3157, 3234

SUBJECT INDEX 179

Management/Conservation

22, 23, 28, 29, 33, 35, 39, 44, 46, 47, 48, 57, 61, 142, 143, 145, 148, 180, 220, 229, 250, 281, 294, 322, 327, 328, 342, 448, 461, 480, 531, 544, 553, 555, 558, 571, 590, 592, 609, 620, 652, 653, 655, 680, 687, 694, 697, 708, 734, 772, 807, 808, 811, 812, 813, 814, 815, 816, 825, 857, 873, 878, 899, 902, 903, 906, 912, 917, 949, 950, 951, 952, 953, 956, 957, 966, 967, 972, 976, 977, 1006, 1011, 1081, 1082, 1083, 1084, 1142, 1153, 1154, 1156, 1157, 1158, 1159, 1160, 1161, 1162, 1163, 1164, 1165, 1166, 1166a, 1167, 1168, 1169, 1170, 1171, 1172, 1173, 1174, 1178, 1179, 1180, 1220, 1223, 1255, 1263, 1289, 1292, 1293, 1296, 1304, 1305, 1309, 1326, 1331, 1333, 1337, 1338, 1340, 1341, 1347, 1354, 1355, 1361, 1363, 1366, 1379, 1380, 1381, 1399, 1409, 1410, 1427, 1428, 1430, 1512, 1529, 1534, 1538, 1539, 1544, 1545, 1553, 1609, 1622, 1623, 1632, 1647, 1650, 1651, 1653, 1654, 1660, 1670, 1684, 1729, 1742, 1747, 1773, 1774, 1861, 1870, 1874, 1876, 1884, 1887, 1892, 1894, 1899, 1901, 1904, 1906, 1913, 1914, 1915, 1916, 1919, 1923, 1931, 1932, 1937, 1942, 1944, 1949, 1953, 1956, 1959, 1963, 1964, 1965, 1966, 1967, 1969, 1970, 1971, 1972, 1973, 1974, 1975, 1976, 1977, 1978, 1979, 1980, 1981, 1982, 1983, 1984, 1985, 1986, 1999, 2002, 2069, 2083, 2084, 2086, 2088, 2089, 2100, 2106, 2134, 2138, 2139, 2140, 2141, 2190, 2191, 2192, 2230, 2236, 2237, 2283, 2295, 2301, 2305, 2572, 2573, 2574, 2578, 2580, 2699, 2700, 2733, 2742, 2754, 2755, 2764, 2765, 2766, 2767, 2768, 2780, 2781, 2803, 2805, 2814, 2821, 2824, 2829, 2837, 2844, 2847, 2848, 2896, 2898, 2899, 2929, 2930, 2931, 2934, 2938, 2941, 2942, 2999, 3011, 3021, 3027, 3030, 3034, 3035, 3036, 3043, 3044, 3088, 3100, 3108, 3112, 3159, 3161, 3165, 3166, 3167, 3168, 3190, 3198, 3200, 3202, 3203, 3204, 3205, 3206, 3207, 3208, 3209, 3210, 3211, 3212, 3213, 3224, 3227, 3228, 3229

Morphology

107, 108, 146, 204, 245, 246, 387, 411, 413, 414, 416, 419, 422, 426, 451, 452, 464, 469, 471, 481, 507, 512, 565, 575, 624, 781, 782, 802, 803, 806, 822, 861, 869, 934, 942, 985, 986, 1014, 1017, 1022, 1023, 1028, 1072, 1134, 1137b, 1264, 1266, 1267, 1439, 1506, 1555, 1598, 1599, 1600, 1621, 1629, 1667, 1669, 1714, 1717, 1770, 1781, 1803, 1804, 1805, 2063, 2068, 2095, 2117, 2147, 2148, 2182, 2184, 2194, 2308, 2333, 2353, 2378, 2392, 2401, 2525, 2532, 2533, 2683, 2738, 2739, 2740, 2784, 2786, 2788, 2792, 2820, 2823, 2852, 2878, 2885, 2889, 2919, 2940, 2972, 2984, 3028, 3102, 3189, 3234, 3237, 3238, 3240

Original Citation

67, 68, 71, 90, 92, 106, 118, 126, 127, 129, 131, 132, 144, 151, 184, 186, 187, 189, 191, 192, 205, 212, 213, 214, 215, 218, 221a, 229, 231, 232, 241, 242, 244, 247, 253, 265, 279, 280, 285, 288, 355, 372, 373, 374, 375, 377, 378, 379, 381, 382, 383a, 384, 387, 393, 394, 396, 399, 402, 411, 445, 462, 464, 479, 481, 482, 483, 490, 491, 492, 495, 496, 497, 498, 500, 506, 512, 515, 522, 533, 534, 541, 542, 551, 557, 562, 565, 593, 615, 627, 634, 636, 647, 654, 656, 657, 658, 666, 667, 671, 673, 676, 677, 678, 681, 682, 684, 685, 690, 695, 701, 703, 704, 705, 706, 707, 710, 713, 714, 715, 728, 729, 730, 731, 737, 745, 746, 748, 749, 751, 752, 757, 758, 761, 779, 780, 784, 786, 787, 788, 789, 791, 792, 793, 834, 835, 837, 840, 842, 843, 844, 845, 846, 848, 851, 858, 859, 861, 864, 865, 866, 867, 868, 869, 870, 876, 877, 883, 885, 888, 890, 894, 895, 897, 910, 918, 962, 986, 990, 991, 992, 993, 994, 996, 1001, 1025, 1036, 1037, 1038, 1039, 1040, 1041, 1042, 1043, 1044, 1045, 1046, 1047, 1048, 1049, 1050, 1051, 1059, 1072, 1072a, 1073, 1074, 1075, 1088, 1107, 1120, 1122, 1126, 1138, 1155, 1187, 1198, 1199, 1200, 1204, 1206, 1207, 1208, 1210, 1211, 1215, 1228, 1229, 1235, 1237, 1238, 1247, 1248, 1251, 1266, 1278, 1301, 1307, 1315, 1389, 1396, 1422, 1423, 1426, 1433, 1448, 1449, 1451, 1454, 1455, 1456, 1461, 1462, 1472, 1489, 1490, 1492, 1493, 1494, 1497, 1498, 1499, 1500, 1505, 1506, 1509, 1567, 1587, 1588, 1589, 1590, 1591, 1592, 1593, 1594, 1595, 1595a, 1596, 1604, 1608, 1610, 1611, 1619, 1620, 1628, 1629, 1737, 1739, 1746, 1750, 1752, 1775, 1777, 1796, 1797, 1798, 1801, 1802, 1806, 1806a, 1815, 1816, 1817, 1818, 1825, 1829, 1864, 1864a, 1865, 1945, 1946, 1991, 2045, 2047, 2048, 2071, 2080, 2081, 2082, 2093, 2094, 2095, 2096, 2104, 2112, 2121, 2124, 2125, 2126, 2133, 2151, 2155, 2156, 2157, 2158, 2159, 2167, 2187, 2192a, 2212, 2214, 2215, 2223, 2238, 2258, 2274, 2276, 2278, 2280, 2284, 2285, 2287, 2288, 2289, 2291, 2292, 2294, 2298, 2299, 2300, 2302, 2303, 2304,

2305, 2308, 2309, 2312, 2314, 2315, 2316, 2318, 2321, 2352, 2354, 2355, 2357, 2358, 2359, 2360, 2361, 2363, 2364, 2365, 2366, 2367, 2369, 2370, 2373, 2374, 2375, 2378, 2380, 2382, 2384, 2386, 2392, 2394, 2395, 2396, 2397, 2400, 2401, 2402, 2405, 2406, 2408, 2409, 2410, 2411, 2412, 2413, 2414, 2416, 2417, 2418, 2421, 2422, 2423, 2424, 2426, 2427, 2429, 2430, 2431, 2432, 2435, 2436, 2437, 2438, 2440, 2441, 2442, 2443, 2445, 2446, 2448, 2450, 2452, 2455, 2459, 2463, 2464, 2465, 2466, 2467, 2468, 2471, 2474, 2480, 2481, 2482, 2483, 2486, 2488, 2489, 2490, 2491, 2493, 2494, 2495, 2496, 2497, 2498, 2499, 2502, 2503, 2504, 2505, 2506, 2509, 2511, 2512, 2515, 2516, 2517, 2518, 2520, 2528, 2536, 2554, 2556, 2566, 2590, 2591, 2592, 2593, 2606, 2607, 2608, 2609, 2610, 2611, 2612, 2613, 2614, 2615, 2616, 2617, 2618, 2619, 2620, 2621, 2622, 2623, 2625, 2626, 2628, 2631, 2632, 2633, 2634, 2635, 2636, 2637, 2638, 2640, 2641, 2642, 2643, 2644, 2645, 2647, 2648, 2649, 2650, 2651, 2652, 2653, 2654, 2655, 2656, 2657, 2658, 2659, 2660, 2661, 2662, 2663, 2664, 2665, 2666, 2667, 2668, 2669, 2670, 2672, 2674, 2675, 2676, 2677, 2704, 2706, 2712, 2717, 2720, 2723, 2724, 2728, 2730, 2736, 2738, 2745, 2746, 2747, 2770, 2771, 2772, 2773, 2774, 2775, 2786, 2791, 2801, 2819, 2822, 2823, 2823a, 2836, 2838, 2853, 2854, 2856, 2872, 2873, 2876, 2877, 2881, 2886, 2892, 2894, 2895, 2909, 2911, 2912, 2913, 2966, 2967, 2975, 2982, 2986, 2992, 2993, 2997, 3005, 3033, 3045, 3051, 3052, 3053, 3054, 3055, 3057, 3058, 3059, 3060, 3061, 3062, 3063, 3064, 3065, 3066, 3067, 3068, 3069, 3070, 3071, 3072, 3073, 3074, 3076, 3077, 3078, 3079, 3082, 3085, 3086, 3087, 3091, 3124, 3125, 3129, 3141, 3142, 3170, 3175, 3176, 3177, 3178, 3179, 3187, 3188, 3236, 3237, 3241, 3242, 3246

PALEOBOTANY

679, 805, 898, 1367, 1662, 2330, 2595, 2600

PALYNOLOGY

189, 210, 413, 416, 464, 512, 616, 796, 806, 822, 1034, 1072, 1134, 1193, 1278, 1279, 1555, 1617, 1754, 2079, 2208, 2209, 2210, 2323, 2599, 2721, 2788, 3084, 3090

PATHOLOGY

28, 36, 48, 76, 96, 173, 200, 201, 342, 536, 573, 683, 739, 913, 915, 965, 968, 969, 970, 972, 973, 975, 976, 977, 978, 979, 981, 1130, 1216, 1217, 1218, 1224, 1271, 1280, 1281, 1282, 1282a, 1294, 1324, 1412, 1413, 1415, 1416, 1417, 1418, 1419, 1420, 1436, 1437, 1438, 1502, 1503, 1580, 1585, 1605, 1647, 1748, 1888, 1890, 1900, 1905, 1907, 1910, 1911, 1912, 1915, 1917, 2108, 2118, 2178, 2204, 2205, 2206, 2207, 2255, 2526, 2572, 2584, 2605, 2830, 2857, 2915, 3046, 3131, 3150, 3219, 3220, 3249

PHYTOGEOGRAPHY

88, 91, 137, 163, 164, 165, 166, 167, 184, 189, 211, 282, 283, 287, 291, 310, 333, 336, 357, 361, 362, 363, 364, 365, 366, 367, 368, 369, 371, 387, 402, 425, 426, 427, 435, 441, 487, 520, 530, 542, 547, 555, 569, 570, 581, 586, 587, 594, 616, 698, 750, 753, 756, 760, 768, 770, 773, 794, 795, 838, 847, 861, 862, 869, 872, 874, 875, 876, 888, 891, 900, 914, 922, 942, 964, 1003, 1004, 1015, 1017, 1021, 1030, 1060, 1082, 1083, 1084, 1085, 1107, 1132, 1186, 1187, 1198, 1213, 1245, 1265, 1267, 1317, 1318, 1345, 1346, 1347, 1349, 1384, 1386, 1387, 1388, 1389, 1406, 1421, 1424, 1425, 1447, 1456, 1506, 1508, 1528, 1531, 1537, 1553, 1590, 1595, 1632, 1652, 1664, 1665, 1668, 1680, 1687, 1688, 1689, 1690, 1691, 1700, 1703, 1706, 1707, 1726, 1740, 1741, 1753, 1780, 1792, 1825, 1863, 1877, 1885, 1892, 1897, 1941, 1995, 2229, 2246, 2258, 2277, 2284, 2286, 2289, 2290, 2297, 2304, 2308, 2312, 2313, 2316, 2341, 2367, 2369, 2371, 2377, 2389, 2392, 2393, 2397, 2399, 2437, 2479, 2511, 2519, 2523, 2586, 2588, 2592, 2602, 2603, 2685, 2687, 2703, 2708, 2710, 2711, 2713, 2715, 2719, 2722, 2725, 2726, 2727, 2729, 2735, 2743, 2744, 2758, 2760, 2784, 2823, 2826, 2827, 2828, 2840, 2850, 2852, 2867, 2875, 2879, 2884, 2890, 2908, 2928, 2948, 2955, 3012, 3043, 3056, 3138, 3163, 3237

POPULAR LITERATURE

1, 2, 3, 6, 7, 8, 9, 10, 11, 12, 13, 14, 15, 19, 24, 42, 43, 44, 45, 49, 50, 54, 69, 78, 79, 80, 86, 102, 103, 104, 105, 110, 117, 123, 124, 125, 150, 156, 157, 160, 161, 172, 175, 183, 190,

208, 219, 223, 224, 225, 226, 230, 239, 240, 251, 268, 269, 270, 293, 295, 296, 298, 300, 301, 302, 303, 304, 305, 306, 307, 308, 315, 322, 323, 328, 332, 333, 334, 335, 337, 353, 358, 359, 408, 423, 433, 434, 435, 439, 442, 443, 444, 449, 485, 488, 493, 494, 510, 517, 538, 543, 546, 548, 549, 550, 563, 564, 584, 588, 591, 612, 617, 629, 631, 637, 638, 639, 640, 641, 642, 643, 645, 646, 651, 659, 660, 694, 697, 698, 702, 707, 708, 719, 722, 723, 724, 725, 734, 773, 777, 807, 810, 812, 824, 825, 829, 844, 857, 900, 906, 927, 928, 929, 930, 953, 959, 989, 1011, 1031, 1054, 1056, 1069, 1070, 1071, 1076, 1078, 1079, 1080, 1085, 1092, 1096, 1097, 1098, 1099, 1115, 1133, 1135, 1136, 1147, 1149, 1150, 1154, 1185, 1194, 1195, 1196, 1214, 1219, 1223, 1256, 1258, 1260, 1262, 1268, 1272, 1283, 1300, 1304, 1305, 1306, 1309, 1318, 1319, 1320, 1321, 1322, 1324, 1325, 1327, 1328, 1329, 1330, 1331, 1332, 1334, 1336, 1340, 1344, 1347, 1351, 1360, 1364, 1365, 1370, 1371, 1372, 1374, 1391, 1392, 1393, 1397, 1400, 1459, 1466, 1467, 1473, 1485, 1486, 1487, 1488, 1511, 1513, 1520, 1524, 1526, 1531, 1533, 1537, 1546, 1573, 1576, 1583, 1614, 1615, 1616, 1633, 1634, 1635, 1636, 1639, 1640, 1641, 1642, 1643, 1644, 1646, 1659, 1664, 1665, 1671, 1672, 1673, 1674, 1677, 1679, 1686, 1692, 1693, 1695, 1697, 1698, 1699, 1704, 1711, 1744, 1756, 1758, 1763, 1764, 1783, 1784, 1785, 1786, 1787, 1817, 1824, 1860, 1866, 1867, 1871, 1872, 1873, 1948, 1951, 1955, 1959, 1963, 1964, 1967, 1973, 1978, 1989, 1994, 2006, 2007, 2008, 2010, 2011, 2012, 2013, 2014, 2015, 2016, 2017, 2018, 2019, 2020, 2021, 2022, 2023, 2024, 2025, 2026, 2028, 2029, 2030, 2031, 2032, 2033, 2034, 2036, 2037, 2038, 2039, 2040, 2041, 2042, 2043, 2044, 2050, 2060, 2069, 2087, 2103, 2169, 2170, 2171, 2196, 2198, 2199, 2200, 2201, 2221, 2225, 2226, 2240, 2272, 2277, 2290, 2296, 2319, 2324, 2338, 2339, 2340, 2350, 2371, 2383, 2478, 2492, 2535, 2564, 2585, 2586, 2587, 2588, 2589, 2592, 2691, 2783, 2821, 2825, 2858, 2859, 2860, 2863, 2865, 2875, 2929, 2932, 2935, 2949, 2950, 2951, 2953, 2959, 2960, 2961, 2962, 2963, 2964, 2970, 2983, 3000, 3004, 3010, 3012, 3013, 3034, 3035, 3048, 3089, 3092, 3094, 3101, 3103, 3105, 3127, 3148, 3149, 3153, 3154, 3155, 3156, 3157, 3171, 3172, 3184, 3185, 3239, 3245, 3252, 3253, 3254, 3255

REPRODUCTIVE BIOLOGY

76, 137, 138, 139, 140, 141, 145, 178, 179, 189, 221, 273, 274, 275, 349, 360, 425, 427, 428, 436, 453, 454, 455, 456, 464, 472, 523, 524, 526, 527, 572, 573, 574, 576, 577, 580, 589, 595, 596, 597, 702, 735, 770, 826, 828, 942, 963, 974, 1014, 1019, 1023, 1024, 1183, 1189, 1193, 1204, 1277, 1373, 1431, 1440, 1442, 1515, 1516, 1528, 1530, 1552, 1553, 1629, 1631, 1664, 1921, 1931, 2003, 2067, 2068, 2141, 2191, 2208, 2209, 2210, 2219, 2254, 2322, 2323, 2333, 2334, 2342, 2392, 2524, 2574, 2594, 2630, 2705, 2740, 2788, 2802, 2803, 2805, 2807, 2815, 2823, 2852, 2903, 2914, 2954, 3002, 3115, 3171, 3221

REVIEW

8, 351, 635, 649, 666, 667, 669, 671, 672, 673, 691, 698, 700, 767, 912, 1112, 1144, 1145, 1148, 1617, 1779, 1889, 2293, 2372, 2379, 3107

SYNECOLOGY

38, 65, 66, 77, 88, 89, 94, 109, 111, 153, 158, 211, 220, 263, 266, 333, 338, 339, 346, 348, 363, 364, 365, 369, 403, 405, 406, 433, 439, 447, 475, 511, 566, 567, 568, 600, 604, 607, 608, 609, 718, 726, 735, 736, 760, 763, 766, 767, 768, 769, 771, 805, 808, 809, 823, 833, 838, 841, 850, 853, 857, 879, 880, 881, 884, 893, 901, 907, 912, 918, 919, 931, 936, 949, 954, 955, 991, 1000, 1003, 1004, 1006, 1008, 1027, 1033, 1061, 1065, 1081, 1082, 1083, 1084, 1085, 1109, 1113, 1126, 1132, 1142, 1178, 1186, 1187, 1197, 1203, 1233, 1236, 1241, 1243, 1244, 1246, 1253, 1288, 1290, 1291, 1293, 1294, 1296, 1318, 1343, 1344, 1362, 1375, 1376, 1384, 1424, 1425, 1457, 1458, 1460, 1501, 1511, 1519, 1523, 1531, 1537, 1623, 1653, 1664, 1676, 1678, 1684, 1688, 1689, 1690, 1701, 1702, 1712, 1728, 1729, 1730, 1731, 1732, 1739, 1740, 1741, 1744, 1757, 1772, 1773, 1859, 1872, 1878, 1879, 1880, 1881, 1882, 1883, 1884, 1885, 1886, 1889, 1890, 1891, 1892, 1893, 1894, 1895, 1896, 1898, 1899, 1901, 1902, 1903, 1904, 1909, 1910, 1914, 1918, 1919, 1920, 1921, 1922, 1923, 1924, 1925, 1926, 1927, 1928, 1929, 1930, 1932, 1933, 1934, 1936, 1938, 1995, 1998, 2011, 2024, 2061, 2090, 2092, 2098, 2114, 2115, 2116, 2185, 2186, 2193, 2216, 2220, 2244, 2248, 2249, 2250, 2252, 2272,

2273, 2284, 2286, 2342, 2368, 2372, 2376, 2458, 2521, 2526, 2569, 2570, 2570a, 2579, 2580, 2595, 2597, 2598, 2600, 2711, 2713, 2731, 2732, 2734, 2737, 2744, 2749, 2750, 2751, 2752, 2753, 2756, 2757, 2762, 2765, 2768, 2780, 2781, 2782, 2804, 2806, 2814, 2840, 2846, 2848, 2861, 2898, 2930, 2931, 2994, 2995, 2996, 3006, 3009, 3017, 3018, 3019, 3037, 3040, 3041, 3056, 3093, 3099, 3107, 3109, 3110, 3111, 3112, 3113, 3132, 3159, 3160, 3161, 3171, 3181, 3182, 3183, 3195, 3214, 3215, 3223, 3226, 3230, 3239

Taxonomy/Systematics

5, 7, 67, 68, 71, 83, 90, 92, 106, 108, 116, 118, 119, 121, 122, 125, 126, 127, 128, 129, 130, 131, 132, 133, 142, 144, 145, 151, 162, 169, 171, 177, 184, 185, 186, 187, 188, 189, 191, 192, 194, 195, 196, 197, 198, 199, 202, 205, 206, 207, 211, 212, 213, 214, 214a, 215, 216, 217, 218, 221a, 229, 231, 232, 233, 236, 238, 241, 242, 243, 244, 247, 248, 252, 253, 255, 256, 257, 265, 271, 279, 280, 284, 285, 286, 288, 324, 340, 345, 351, 355, 372, 373, 374, 375, 376, 377, 378, 379, 380, 381, 382, 383, 383a, 384, 385, 386, 387, 388, 389, 390, 391, 392, 393, 394, 395, 396, 397, 398, 399, 400, 401, 402, 411, 413, 445, 450, 462, 463, 464, 466, 471, 479, 481, 482, 483, 484, 486, 489, 490, 491, 492, 495, 496, 497, 498, 499, 500, 501, 502, 503, 504, 505, 506, 512, 513, 514, 515, 522, 532, 533, 534, 541, 542, 545, 551, 557, 562, 565, 578, 585, 587, 593, 603, 605, 611, 615, 616, 626, 627, 632, 633, 634, 635, 636, 639, 644, 647, 654, 656, 657, 658, 662, 663, 666, 667, 668, 669, 671, 672, 673, 674, 676, 677, 678, 680, 681, 682, 684, 685, 686, 688, 690, 691, 692, 693, 695, 696, 700, 701, 702, 703, 704, 705, 706, 710, 711, 713, 714, 715, 716, 720, 721, 727, 728, 729, 730, 731, 737, 738, 745, 746, 747, 748, 749, 751, 752, 753, 754, 755, 757, 758, 761, 762, 764, 765, 778, 779, 780, 783, 784, 785, 786, 787, 788, 789, 790, 791, 792, 793, 795, 817, 818, 819, 834, 835, 837, 840, 842, 843, 844, 845, 846, 848, 851, 854, 855, 856, 858, 859, 861, 864, 865, 866, 867, 868, 869, 870, 871, 876, 877, 883, 885, 886, 888, 889, 890, 894, 895, 896, 897, 904, 908, 909, 910, 916, 918, 920, 921, 922, 923, 924, 925, 926, 934, 935, 937, 938, 939, 940, 941, 942, 962, 971, 982, 984, 986, 990, 991, 992, 993, 994, 996, 997, 1001, 1014, 1015, 1016, 1021, 1022, 1025, 1026, 1032a, 1035, 1036, 1037, 1038, 1039, 1040, 1041, 1042, 1043, 1044, 1045, 1046, 1047, 1048, 1049, 1050, 1051, 1052, 1059, 1062, 1063, 1066, 1068, 1072, 1072a, 1073, 1074, 1075, 1086, 1087, 1088, 1100, 1101, 1102, 1103, 1106, 1107, 1116, 1117, 1120, 1121, 1122, 1126, 1128, 1129, 1131, 1134, 1138, 1144, 1148, 1155, 1175, 1187, 1198, 1199, 1200, 1201, 1202, 1204, 1206, 1207, 1208, 1209, 1210, 1211, 1212, 1213, 1215, 1227, 1228, 1229, 1230, 1231, 1232, 1234, 1235, 1237, 1238, 1241, 1247, 1248, 1251, 1264, 1266, 1270, 1274, 1275, 1276, 1278, 1279, 1286, 1297, 1299, 1301, 1302, 1303, 1307, 1309, 1310, 1312, 1313, 1314, 1315, 1383, 1384, 1389, 1395, 1396, 1411, 1421, 1422, 1423, 1426, 1433, 1434, 1435, 1439, 1440, 1441, 1444, 1445, 1446, 1447, 1448, 1449, 1450, 1451, 1452, 1454, 1455, 1456, 1461, 1462, 1471, 1472, 1475, 1478, 1482, 1483, 1484, 1489, 1490, 1491, 1492, 1493, 1494, 1495, 1496, 1497, 1498, 1499, 1500, 1504, 1505, 1506, 1507, 1508, 1509, 1510, 1555, 1556, 1557, 1558, 1559, 1560, 1561, 1562, 1563, 1564, 1565, 1566, 1567, 1568, 1569, 1570, 1571, 1572, 1577, 1584, 1587, 1588, 1589, 1590, 1591, 1592, 1593, 1594, 1595, 1595a, 1596, 1603, 1604, 1608, 1610, 1611, 1619, 1620, 1628, 1629, 1634, 1638, 1663, 1683, 1685, 1687, 1706, 1708, 1709, 1737, 1738, 1739, 1741, 1746, 1749, 1750, 1751, 1752, 1754, 1770, 1771, 1775, 1776, 1777, 1778, 1779, 1796, 1797, 1798, 1800, 1801, 1802, 1803, 1804, 1805, 1806, 1806a, 1815, 1816, 1817, 1818, 1825, 1827, 1829, 1838, 1839, 1840, 1844, 1845, 1846, 1847, 1849, 1850, 1855, 1864, 1864a, 1865, 1939, 1940, 1945, 1946, 1947, 1991, 1992, 2001, 2035, 2036, 2042, 2043, 2044, 2045, 2046, 2047, 2048, 2051, 2062, 2063, 2064, 2068, 2070, 2071, 2079, 2080, 2081, 2082, 2093, 2094, 2095, 2096, 2099, 2104, 2112, 2117, 2121, 2122, 2123, 2124, 2125, 2126, 2132, 2133, 2145, 2146, 2147, 2149, 2150, 2151, 2152, 2155, 2156, 2157, 2158, 2159, 2160, 2162, 2165, 2166, 2167, 2168, 2181, 2187, 2188, 2189, 2192a, 2195, 2197, 2201, 2211, 2212, 2213, 2214, 2215, 2222, 2223, 2227, 2228, 2229, 2233, 2238, 2245, 2257, 2258, 2259, 2260, 2269, 2271, 2274, 2275, 2276, 2278, 2280, 2281, 2284, 2285, 2287, 2288, 2289, 2291, 2292, 2293, 2294, 2298, 2299, 2300, 2302, 2303, 2304, 2305, 2308, 2309, 2312, 2314, 2315, 2316, 2318, 2320, 2321, 2325, 2326, 2331, 2348, 2352, 2353, 2354, 2355, 2357, 2358, 2359, 2360, 2361, 2363, 2364, 2365, 2366, 2367, 2369, 2370, 2373, 2374, 2375, 2378, 2380, 2381, 2382, 2384, 2385, 2386, 2387, 2388, 2390, 2391, 2392, 2394, 2395,

2396, 2397, 2398, 2399, 2400, 2401, 2402, 2403, 2404, 2405, 2406, 2407, 2408, 2409, 2410, 2411, 2412, 2413, 2414, 2416, 2417, 2418, 2420, 2421, 2422, 2423, 2424, 2425, 2426, 2427, 2429, 2430, 2431, 2432, 2433, 2434, 2435, 2436, 2437, 2438, 2439, 2440, 2441, 2442, 2443, 2445, 2446, 2447, 2448, 2449, 2450, 2451, 2452, 2454, 2455, 2457, 2459, 2462, 2463, 2464, 2465, 2466, 2467, 2468, 2469, 2471, 2472, 2473, 2474, 2475, 2476, 2480, 2481, 2482, 2483, 2484, 2485, 2486, 2488, 2489, 2490, 2491, 2493, 2494, 2495, 2496, 2497, 2498, 2499, 2500, 2501, 2502, 2503, 2504, 2505, 2506, 2508, 2509, 2510, 2511, 2512, 2515, 2516, 2517, 2518, 2519, 2520, 2522, 2528, 2529, 2529a, 2536, 2537, 2538, 2539, 2553, 2554, 2556, 2557, 2558, 2559, 2560, 2561, 2563, 2565, 2566, 2567, 2568, 2589, 2590, 2591, 2592, 2593, 2601, 2604, 2606, 2607, 2608, 2609, 2610, 2611, 2612, 2613, 2614, 2615, 2616, 2617, 2618, 2619, 2620, 2621, 2622, 2623, 2624, 2625, 2626, 2627, 2628, 2629, 2630, 2631, 2632, 2633, 2634, 2635, 2636, 2637, 2638, 2639, 2640, 2641, 2642, 2643, 2644, 2645, 2646, 2647, 2648, 2649, 2650, 2651, 2652, 2653, 2654, 2655, 2656, 2657, 2658, 2659, 2660, 2661, 2662, 2663, 2664, 2665, 2666, 2667, 2668, 2669, 2670, 2671, 2672, 2673, 2674, 2675, 2676, 2677, 2687, 2690, 2704, 2706, 2709, 2710, 2712, 2714, 2715, 2716, 2717, 2718, 2719, 2720, 2721, 2723, 2724, 2725, 2726, 2728, 2730, 2736, 2738, 2745, 2746, 2747, 2748, 2759, 2770, 2771, 2772, 2773, 2774, 2775, 2779, 2784, 2785, 2786, 2790, 2791, 2792, 2796, 2801, 2816, 2817, 2819, 2820, 2822, 2823, 2823a, 2827, 2828, 2834, 2836, 2838, 2853, 2854, 2855, 2856, 2862, 2869, 2871, 2872, 2873, 2874, 2876, 2877, 2878, 2879, 2880, 2881, 2882, 2883, 2884, 2886, 2887, 2888, 2889, 2890, 2891, 2892, 2893, 2894, 2895, 2909, 2911, 2912, 2913, 2916, 2917, 2940, 2966, 2967, 2974, 2975, 2976, 2978, 2982, 2985, 2986, 2987, 2988, 2992, 2993, 2997, 2998, 3003, 3005, 3014, 3015, 3016, 3033, 3045, 3050, 3051, 3052, 3053, 3054, 3055, 3057, 3058, 3059, 3060, 3061, 3062, 3063, 3064, 3065, 3066, 3067, 3068, 3069, 3070, 3071, 3072, 3073, 3074, 3075, 3076, 3077, 3078, 3079, 3080, 3081, 3082, 3083, 3085, 3086, 3087, 3091, 3096, 3102, 3104, 3124, 3125, 3129, 3130, 3139, 3140, 3141, 3142, 3143, 3144, 3145, 3158, 3169, 3170, 3175, 3176, 3177, 3178, 3179, 3187, 3188, 3189, 3196, 3213, 3216, 3231, 3235, 3236, 3237, 3238, 3241, 3242, 3246, 3247, 3248

INDEX TO PLANT NAMES

Abrus precatorius 2311, 2386
Abutilon eremipetalum 533
  grandifolium 897
  incanum 133, 779, 937, 2360
  kauaiense 1207
  menziesii 55, 133, 1161, 1168
  mollissimum 1207
  sandwicense 533
Acacia 488, 1767, 2304, 3117, 3250
  decurrens 1208, 2776, 2777
  farnesiana 1208, 2993
  heterophylla 2993
  kauaiensis 1187, 1218, 1528, 2132, 2465
  koa 2, 8, 66, 77, 96, 134, 135, 136, 159, 181, 182, 201, 214a, 318, 342, 344, 357, 473, 474, 558, 566, 567, 568, 622, 675, 954, 955, 965, 969, 981, 1039, 1081, 1082, 1083, 1084, 1085, 1208, 1217, 1218, 1240, 1306, 1322, 1328, 1341, 1358, 1390, 1438, 1502, 1514, 1528, 1542, 1549, 1585, 1734, 1820, 1823, 1880, 1885, 1892, 1898, 1911, 1921, 1931, 1935, 2132, 2204, 2207, 2216, 2465, 2571, 2572, 2578, 2579, 2581, 2583, 2584, 2688, 2693, 2696, 2699, 2700, 2701, 2702, 2802, 2803, 2805, 2807, 2829, 2830, 2831, 2833, 2901, 2915, 2922, 2940, 2984, 3027, 3028, 3036, 3081, 3115, 3121, 3134
  koaia 182, 675, 1187, 1217, 1218, 1528, 1585, 2465, 3200
  [see also Racosperma]
Acaena 2722
  exigua 232, 1039
Acanthospermum australe 236
  brasilum 1210
  hispidum 236
Achillea millefolium 2311
Achyranthes 2463
  atollensis 583
  fruticosa 1865
  nelsonii 2429
  rotundata 61, 1165, 3108, 3168
  splendens 61, 618, 1187, 1865
Adenolepis pulchella 393
  [see also Bidens]
Adenostemma glutinosum 779, 2360
  latifolium 3015
  viscosum 3015
  [see also Lavenia]
Adinandra cordata 1435
Aerva sericea 1865

Agave americana 1950
  rigida 560, 1724, 1725
  sisalana 235
Ageratina adenophora 2945
  [see also Eupatorium]
Ageratum conyzoides 1310, 2269, 2809
  houstonianum 870
  sandwicense 1595
Agrostis fallax 1187
  kauaiensis 1187
  rockii 1074
  sandwicensis 879, 1187
Aira australis 2855
  hawaiiensis 2704
  nubigena 879, 2704
  pallida 2704
  [see also Deschampsia]
Albizia falcataria 3136
  moluccana 1656
Albizzia [see Albizia]
Alectryon macrococcus 2212, 2214, 2215, 2511
  mahoe 33, 2511
  [see also Mahoe]
Aleurites 2641
  moluccana 8, 203, 591, 678, 1081, 1082, 1083, 1115, 1306, 1323, 1325, 1362, 1514, 1673, 1674, 2310, 2319, 2656, 2881, 2952, 3104, 3152, 3253
  remyi 2633, 2653
  triloba 1084, 1085, 1421
Alocasia macrorrhiza 2066
Alphitonia ponderosa 252, 1187, 2087, 2430
Alpinia mutica 3048
  purpurata 3048
Alsinidendron 1737, 2657
  lychnoides 2641, 2652
  obovatum 2657, 2669
  trinerve 1741
  verticillatum 2641, 2652
  viscosum 2641, 2652
Alsinodendron [see Alsinidendron]
Alternanthera 3005
  echinocephala 889, 2433
  menziesii 889, 1771, 2382, 2433
Alyxia 2567
  myrtillifolia 1590
  oliviformis 922, 993, 1187, 1563, 2013, 2015, 2152, 2417, 2424, 2538, 2939
  sulcata 1238
  [see also Gynopogon]

Amaranthaceae 3006
Amaranthus 3005
  brownii 534, 959
  spinosus 2808
Ambrosia artemisiifolia 2311, 3005
Ammophila arenaria 304
Amsinckia intermedia 897
Anagallis arvensis 1522, 2202, 2311
Andropogon barbinodis 446
  byronis 2856
  glomeratus 885, 2941
  pertusus 341
  virginicus 341, 944, 980, 1375, 1376, 1882, 1883, 2797, 2799, 2941
Anecochilus [see Anoectochilus]
Anectochilus [see Anoectochilus]
Anisophyllum 1423
  [see also Chamaesyce and Euphorbia]
Anoda ovata 1797
  [see also Sida]
Anoectochilus apiculatus 870
  jaubertii 996
  sandvicensis 323, 1440, 1442, 1611, 1638, 1683
  [see also Odontochilus and Vrydagzynia]
Anthemis cotula 3005
Antidesma 1599, 2650
  crenatum 2408
  ×kapuae 2284
  platyphyllum 858, 1598, 1739, 2125, 2633
  pulvinatum 1187, 2085, 2495, 2633, 2668, 2675, 2677
  wawraeanum 191
Aphanopappus micranthus 1126
  [see also Lipochaeta]
Aquifoliaceae 1265
Arabis o-waihiensis 496, 2407
Araceae 2018
Aralia trigyna 994
  [see also Cheirodendron]
Arecaceae 21, 149, 155, 335, 2943, 3148, 3149
Argemone alba 2192a
  glauca 678, 885, 2104, 2881, 2933
  mexicana 498, 2311
×Argyrautia degeneri 2641
Argyrophyton [see Argyroxiphium]
Argyroxiphium 49, 80, 362, 408, 410, 415, 416, 417, 426, 459, 460, 461, 463, 464, 466, 468, 469, 470, 470a, 643, 1046, 1388, 1389, 1804, 1805, 1942, 1948, 2270, 2406, 2684, 3101
  caliginis 848
  grayanum 3234
  kauense 694, 697, 711, 1305, 1545, 1803, 1819, 2191, 2318
  macrocephalum 59, 322, 471, 697, 711, 734, 1035, 1036, 2191, 2783
  sandwicense 8, 322, 328, 393, 395, 471, 472, 694, 697, 711, 1035, 1036, 1164, 1187, 1234, 1427, 1428, 1429, 1430, 1431, 1545, 1803, 1804, 1805, 2190, 2191, 2318, 2340, 3167
  virescens 950, 1187, 1803

Aristolochia elegans 638, 1637
Artemisia 1046, 2706
  australis 394, 424, 1210, 2724, 2933
  eschscholtziana 221a
  hillebrandii 2724
  kauaiensis 424, 2724
  mauiensis 424, 2724
  microcephala 1187
Arum esculentum 2956
Arundina bambusifolia 1374
  graminifolia 897, 1440, 1442
Arundo donax 638
Asarina erubescens 885
Asclepias physocarpa 3005
Aspidosperma tuberculatum 2151
Astelia 2717, 2718, 2719, 2722, 2726, 3250
  menziesiana 2775
  veratroides 992, 2704, 2738, 3079
  waialealae 3079
Aster subulatus 1315
Asteraceae 1561, 1587
Asystasia gangetica 871, 1373
Atriplex oahuensis 1797
  semibaccata 641
Axonopus affinis 513

Bacopa monnieri 1564
  [see also Herpestes and Herpestis]
Baltimora recta 3005
Bamburanta 175
Bambusa vulgaris 1339
Batis maritima 437, 640, 871
Biancaea sepiaria 633
Bidens 426, 442, 630, 632, 962, 964, 1017, 1018, 1019, 1020, 1021, 1022, 1046, 1124, 1125, 1607, 1743, 1780, 2231, 2606, 2610, 2629, 2650, 2914
  amplectens 2618
  angustifolia 2080
  asymmetrica 2609, 2610, 2642
  awaluana 2677
  campylotheca 2518, 2610
  coartata 2611
  conjuncta 2607, 2653
  cosmoides 963, 1210, 2672
  ctenophylla 1781, 2610
  cuneata 1158, 1173, 2881
  degeneri 2610
  ×dimidiata 2656, 2663
  distans 2613
  fecunda 2616
  ferax 2618
  fulvescens 2611, 2655
  gardneri 3005
  gracilis 2080
  graciloides 2607
  halawana 2618
  hawaiensis 2613, 2653
  helianthoides 2565
  hillebrandiana 2610

hirsuta 2080
macrocarpa 2610
magnidisca 2616, 2653
mauiensis 2414, 2608, 2611, 2636
menziesii 1781, 2254, 2610, 2636, 2656
micrantha 993, 2608, 2610, 2623, 2657
micranthoides 2610
mutica 2080
napaliensis 2653, 2676
nematocera 2623
obtusiloba 2612
peduncularis 393
pentamera 2635, 2636
personans 2615, 2618
perversa 2642, 2668
pilosa 441, 639, 1210, 1622, 2808
populifolia 2611
remyi 749, 2608
salicoides 2611
sandvicensis 1210, 2607, 2610, 2612, 2616, 2657, 2672, 2675
schizoglossa 2612
skottsbergii 2614, 2615, 2668
torta 1762, 2657
waianensis 2618, 2657
waimeana 2607, 2676
wiebkei 2611
[see also Adenolepis, Campylotheca, and Coreopsis]
Bifaria 2967
[see also Korthalsella]
Bixa orellana 639
Bleekeria compta 3142
[see also Ochrosia]
Blumea laciniata 2222
sessiliflora 3005
Bobea 994
brevipes 1042
elatior 909, 994, 1042, 1187, 1514, 2194, 2946
gaudichaudii 2512
hookeri 1187, 2946
mannii 1187, 2946
sandwicensis 1187, 2946
timonioides 1187
[see also Burneya, Chomelia, Obbea, and Rytidotus]
Bocconia frutescens 885, 923
Boehmeria 946
albida 1238
glabra 779, 2360
grandis 1126, 2738
ovata 779, 2360
stipularis 3086, 3087
[see also Urtica]
Boerhaavia [see Boerhavia]
Boerhavia 442
albiflora 910
diffusa 1122, 1519
herbstii 910
tetrandra 910
Bonamia 1086

menziesii 1048, 1155, 1991
Borreria laevis 870
Brachiaria mutica 1089, 2780, 2781
Brassaia actinophylla 98, 638, 904, 1033, 1275
[see also Schefflera]
Brassica 1011
Brighamia 130, 431, 1127, 1309, 1635, 1739, 2141, 2308, 2338, 2560, 3178, 3179
citrina 531, 1308, 1309, 2384, 2395
insignis 421, 844, 1741, 2226, 2384, 2395, 2868
remyi 2395
rockii 2395
Broussaisia 994, 1275, 3250
arguta 865, 994, 2422, 3038
pellucida 996
Broussonetia papyrifera 64, 1421
Bruguiera conjugata 871
gymnorrhiza 641, 885, 3098
sexangula 641, 3017, 3018
Bryophyllum calycinum 1672
pinnatum 641
[see also Kalanchoë]
Buddleia asiatica 3214, 3215
Buddleja [see Buddleia]
Burneya gaudichaudii 506
Byronia anomala 1126
helleri 1590
sandwicensis 779, 1992, 2360
[see also Ilex]

Caesalpinia 2304
kavaiensis 1739
[see also Mezoneuron]
Cajan [see Cajanus]
Cajanus indicus 1468, 1469, 1470, 1649
Callisia fragrans 871
Calonyction album 1210
[see also Ipomoea]
Calophyllum inophyllum 640, 1703, 2987
Calpidia brunoniana 1120
forsteriana 1120
[see also Pisonia]
Calyptocarpus vialis 3005
Campanulaceae 1515, 1668, 1669, 2195, 2594, 2923, 3218
Camphusia 2998
glabra 2997
kauaiensis 647
[see also Scaevola]
Campylotheca cosmoides 1187
dichotoma 1187
grandiflora 393
hawaiensis 1187
macrocarpa 1187
mauiensis 1187
menziesii 1187
micrantha 393, 1187, 2704
molokaiensis 1187
mutica 1126
pulchella 1187

remyi 1187
rutifolia 1589
sandvicensis 1187
[see also Bidens]
Canavali [see Canavalia]
Canavalia 1911, 2400
　cathartica 2536
　galeata 890, 1039, 1187, 1980, 2162, 2304, 2360, 2536
　gaudichaudii 779, 2360
　hawaiiensis 2536
　kauaiensis 1904, 2536
　kauensis 970, 1381, 1538, 2409
　lanaiensis 1981
　maritima 2536
　molokaiensis 2536
　pubescens 1238, 2536
　sericea 2304, 2536
Canthium odoratum 1341, 2740, 3181, 3182
　[see also Plectronia]
Capparis sandwichiana 388, 639, 656, 996, 1958, 2017, 2391
　spinosa 1297
Cardamine flexuosa 2367, 2566
　hirsuta 2367
　konaensis 2367
Carex 1471, 1472
　alligata 1472
　brunnea 1451, 1489
　hawaiiensis 2494
　kauaiensis 1472
　montis-eeka 1187, 1472, 1490, 2738
　nealae 1472
　ovalis 923
　pluvia 1472
　sandwicensis 1490, 1493, 2722
　svenonis 2738
　wahuensis 1801, 2051
Carica papaya 96, 183, 612, 822, 1183, 2005, 2078, 2176, 2179, 2903, 2905, 3220
Cassia chamaecrista 2811
　floribunda 897
　gaudichaudii 216, 1238, 1352, 2023, 2519, 2993
　glauca 2023
　lechenaultiana 1858, 1904
　mimosoides 1208
　[see also Senna]
Cassytha filiformis 246, 699, 858, 2203, 2526
Casuarina 262, 821, 960, 1322, 1951, 2564, 2918, 3255
　equisetifolia 94, 304, 639, 2097
　glauca 2097
　quadrivalvis 1320
Catharanthus roseus 871
Cecropia obtusifolia 1033
Cedrela australis 1329
　toona 3232
　[see also Toona]
Cenchrus agrimonioides 534, 583, 716
　echinatus 285, 641, 716, 2809

hillebrandianus 3090
laysanensis 2416
Centaurium sebaeoides 754
　[see also Erythraea and Schenkia]
Ceodes brunoniana 2721
　umbellifera 1383, 2721
　[see also Pisonia]
Cereus martinii 211
　peruvianus 211
Cestrum diurnum 639
　nocturnum 926
Chamaecrista lechenaultiana 638
　nictitans 1286
　[see also Cassia]
Chamaesyce 1147, 1447, 3084
　celastroides 1448
　clusiifolia 106
　cordata 106
　deppeana 1815
　hookeri 106
　rockii 1448
　skottsbergii 1448
　sparsiflora 1448
　[see also Anisophyllum and Euphorbia]
Charpentiera 2501, 2784, 2786, 2789, 2790
　densiflora 2788, 2793
　elliptica 1126, 2787, 2788, 2911, 3005
　obovata 993, 1187, 2403, 2425, 2550, 2785, 2788, 2793, 2911, 3077
　ovata 993, 2785, 2787, 2793, 2911
　tomentosa 2787, 2788, 2793
Cheirodendron 134, 958, 1101, 1119, 1134, 1367, 2670, 2688, 3250
　dominii 1456
　fauriei 1208, 2659
　gaudichaudii 1118, 1407, 1408, 2256, 2591
　helleri 2659, 2661
　kauaiense 803, 1456, 2659
　platyphyllum 1208, 1456, 1780, 2591
　trigynum 135, 136, 200, 803, 1126, 1542, 1590, 2386, 2659, 2676, 2738
　wahiawense 1456
　[see also Aralia, Hedera, and Panax]
Chenopodium ambrosioides 638
　murale 641
　oahuense 68, 885
　pekeloi 699, 3205
　sandwicheum 67, 191, 1864, 1948, 1967, 1974
Chloris divaricata 1242
　inflata 999
Chomelia sandwicensis 1042
　[see also Bobea]
Chrysophyllum 786
　polynesicum 1187
　[see also Nesoluma]
Chrysopogon aciculatus 1904
Cirsium arvense 1597
　hawaiensis 1595a
Citharexylum caudatum 1033, 1840, 1841, 1846, 1855
Cladium angustifolium 284, 1495, 2722

leptostachyum 1797, 2780, 2781
meyenii 1495
Cladocarpa 2437
hispida 2441
[see also Sicyos]
Claoxylon 2633, 2650
helleri 2628, 2656
insigne 752
remyi 2633
sandwicense 1187, 1946, 1947, 2122, 2127, 2128, 2261, 2266, 2628, 2633, 2675, 2738
tomentosum 1126, 2122
Cleome sandwicensis 1039
spinosa 3015
Clermontia 431, 474, 484, 993, 1047, 1118, 1119, 1187, 1408, 1516, 2308, 2398, 2406, 2560, 3176, 3179
arborescens 1187, 2815
carinifera 1596
clermontioides 1126, 1210, 2314
coerulea 1187, 2314, 2738
drepanomorpha 2284
earina 2493
epiphytica 2359
fauriei 1590, 2281
forbesii 2358
fulva 1596
gaudichaudii 1187, 2284
grandiflora 396, 796, 993, 2358
haleakalensis 2284
hanaensis 2358
hawaiiensis 2284, 2314
hirsutinervis 2358
kakeana 589, 1797, 2985
kohalae 2280, 2284
konaensis 2429
leptoclada 2284, 2314
lindseyana 2316
loyana 2314
macrocarpa 996, 1187, 2985
macrophylla 2081
mannii 2493
molokaiensis 2358
montis-loa 2280
multiflora 1187, 2284
munroi 2358
oblongifolia 396, 409, 993, 2284, 2493, 2985
pallida 1187
paradisia 2472
parviflora 1187, 2284, 2472, 2738
peleana 2284
persicifolia 396, 993
pyrularia 1187
reticulata 2358
rockiana 3175
samuelii 848
subpetiolata 2358
tuberculata 837, 2445
wailauensis 2358
waimeae 2280, 2314

Clerodendrum fragrans 1829, 1832, 1833, 1834, 1842
Clidemia hirta 57, 653, 1223, 3100
Cnicus hawaiensis 1589
Coccinia grandis 1613
Cocculus ferrandianus 720, 994
integer 1187
lonchophyllus 1187
trilobus 854
virgatus 1187
Cocos 3185
nucifera 641, 679, 759, 1298, 1809, 1951, 2007, 2950, 3047
Coffea chamissonis 1238
kaduana 392, 505
mariniana 392, 505
[see also Psychotria]
Coix lachryma-jobi 689
Colocasia antiquorum 1694, 1713, 1714, 1715, 1716, 1717, 1718, 1719, 1720, 1721, 1722, 1723, 1808, 1809, 2585, 2958
esculenta 20, 62, 63, 82, 174, 222, 354, 1053, 1091, 1093, 1392, 1398, 1869, 2006, 2335, 2525, 3127, 3128, 3150
Colubrina asiatica 1314, 2076
oppositifolia 1314, 1354, 1355, 1356, 1357, 1363, 1514, 2088, 2467
Commelina 1580
Conanthus sandwicensis 1126
[see also Nama]
Connarus kavaiensis 1739
[see also Zanthoxylum]
Conocarpus erectus 688, 3098
Convolvulaceae 1736
Convolvulus arvensis 477, 480, 1651
sandwicensis 752
Conyza 3005
[see also Erigeron]
Coprosma 2095, 2722
cymosa 1187
elliptica 870
ernodeoides 868, 1042, 2095
fauriei 868, 1590
foliosa 1042
grayana 2284
kauensis 1126
longifolia 870, 1042
menziesii 1042
molokaiensis 2095
montana 1187, 2095
ochracea 2095
parvifolia 1590
pubens 870, 1042
rhynchocarpa 1042
serrata 2095
skottsbergiana 2096
stephanocarpa 1187
ternata 2095
vontempskyi 2284
waimeae 3069
Cordia subcordata 486, 775, 1514

Cordyline terminalis 619, 1033, 1063, 1755, 3049
  [see also Pleomele]
Coreopsis 193, 751, 1046
  molokaiensis 2627
  tinctoria 2627
  [see also Bidens]
Coronopus didymus 1522
Corynocarpus laevigatus 1779
Costus speciosus 3048
Crassocephalum crepidioides 3005
Crepis capillaris 3005
  japonica 1210, 1597
  molokaiensis 1589, 2281
  [see also Youngia]
Crotalaria 2601
  longirostrata 1208
  pallida 923
  saltiana 2811
Cryptocarya mannii 1187
  oahuensis 858
Cuphea carthagenensis 119
Curcuma domestica 3048
  longa 638
Cuscuta sandwichiana 522, 783, 2203, 2319, 2526, 3235, 3236
Cyanea 419, 431, 484, 993, 1047, 1187, 1516, 2308, 2398, 2406, 2560, 3176, 3179
  aculeatiflora 2284
  acuminata 1187, 1251
  angustifolia 409, 589, 1187, 2292, 2314, 2472
  arborea 1187
  arborescens 1739
  asarifolia 2414
  asplenifolia 1187
  atra 1187, 2284
  baldwinii 851
  bishopii 2284
  blinii 1590
  bondiana 2316
  bryanii 2314
  carlsonii 709, 2314, 2316
  chockii 2314
  comata 1187
  communis 2280
  copelandii 2292
  coriacea 1187, 2314
  densiflora 2316
  eriantha 2704
  fauriei 1590, 2281
  feddei 1590
  fernaldii 2292
  ferox 1187, 2292, 2738
  fissa 1187
  floribunda 2316
  gayana 1210, 2284, 2314
  gibsonii 1187
  giffardii 2300
  glabra 2483
  grimesiana 396, 491, 993, 1251, 2081, 2292, 2314, 2445
  hamatiflora 2284
  hardyi 2292
  hirtella 1187, 2705
  holophylla 1187
  humilis 3059
  juddii 843
  kunthiana 1187
  larrisonii 2287
  leptostegia 1210, 2704
  linearifolia 2314
  lobata 1739
  longifolia 1126
  longipedunculata 2314, 2316
  longissima 2445
  macrostegia 796, 1187, 2280
  mannii 1187
  marksii 2314
  mceldowneyi 2314
  megacarpa 2316
  membranacea 2314
  montis-loa 2704
  multispicata 1590
  nelsonii 2429
  noli-me-tangere 2292
  obtusa 1187
  occultans 2483
  palakea 843
  parviflora 2704
  pilosa 2284
  platyphylla 1187
  procera 1187
  profuga 843
  pulchra 2314
  recta 1187
  regina 2316
  remyi 2292
  rivularis 2284
  rollandioides 2300
  salicina 1596
  scabra 1187
  shipmanii 2314
  solanacea 1187
  solenocalyx 1187
  spathulata 1126
  stictophylla 2284, 2314
  superba 1187
  truncata 2292, 2316, 2359
  undulata 837
  wailauensis 2280
  [see also Macrochilus]
Cyathodes banksii 399, 2081
  douglasii 1048
  imbricata 484, 1049, 1187, 1590, 2909
  macraeana 399
  tameiameiae 399, 498, 1048, 1187, 2081, 3060
  [see also Styphelia]
Cynodon dactylon 1904
Cyperaceae 2046, 2047
Cyperus 1492, 2537
  auriculatus 1797
  caricifolius 1238
  compressus 885

cylindrostachys 2738
decipiens 1187
fauriei 1490
halpan 885, 2881
hawaiensis 1739
hillebrandii 244
hypochlorus 1187, 2738
laevigatus 1311, 1519
mauiensis 1187
multiceps 1238
niger 923
owahuensis 1797
pennatiformis 534, 1670
phleoides 1493
polystachyos 482, 1187
prescottianus 1238
remyi 242
rockii 1490
sandwicensis 242, 1490
trachysanthos 1238
wawraeanus 2238
[see also Mariscus]
Cyrtandra 227, 351, 542, 694, 697, 698, 934,
    1015, 1048, 1187, 1299, 1564, 2224, 2291,
    2298, 2302, 2303, 2368, 2377, 2392, 2406,
    2520, 2907, 3002, 3250
  arguta 2720
  asaroides 1589
  begoniifolia 1187
  biserrata 1187
  caulescens 2298
  cladantha 2738
  clarkei 2720
  clypeata 2464
  congesta 2468
  conradtii 2302
  cordifolia 400, 993, 2298
  crassifolia 2298
  cyaneoides 2280
  degenerans 1126
  elliptica 2468
  elstonii 1210
  endlicheriana 400
  fauriei 1589
  filipes 1187
  garnotiana 400, 993
  gayana 1126, 2738
  georgiana 848
  giffardii 2303
  glauca 752
  gracilis 1187
  grandiflora 400, 993, 2322, 2323
  grayana 1187, 2302
  halawensis 2302
  hashimotoi 2303
  haupuensis 2464
  hii 848
  hillebrandii 2986
  hobdyi 2408
  honolulensis 3055
  kalihii 2302, 3055
  kamoloensis 1589
  kauaiensis 3055
  kealiae 3055
  knudsenii 2298
  kohalae 2302
  latebrosa 1187
  laxiflora 1739, 2303
  lessoniana 400, 993, 1187, 2720
  limosiflora 2303
  longifolia 2291
  lydgatei 1187
  lysiosepala 1187, 2302, 2720
  macraei 2302
  macrocalyx 1187
  malacophylla 2298
  mauiensis 2298
  megistocalyx 2394
  menziesii 400
  montis-loa 2303
  munroi 848
  nutans 2380
  oahuensis 1589
  oenobarba 1126, 1739, 3055
  oliveri 2302
  olona 848
  paludosa 400, 993, 1187, 2291, 2303, 2393,
    3054
  paritiifolia 1187
  peltata 3055
  pickeringii 1187, 2298
  platyphylla 2298, 2738
  polyantha 2303
  procera 1187
  propinqua 848
  ramosissima 2303
  ruckiana 400, 1797
  sandwicensis 2322
  tintinnabula 2298
  triflora 400, 993, 1187, 3055
  umbracculiflora 2302
  vanioti 1590
  waianuensis 2291
  waihoiensis 2468
  wainihaensis 1589
  waiolani 1187, 3055
  [see also Cyrtandropsis]
Cyrtandropsis kaululuensis 1209, 1210
  [see also Cyrtandra]

Datura metel 641
  stramonium 2310
Delissea 130, 993, 1047, 1187, 2308, 2386, 2398,
    2501, 2560, 3176, 3179
  acuminata 396, 491, 993, 3057
  ambigua 396
  angustifolia 396, 730
  arborea 1739
  asplenifolia 1739
  calycina 396
  clermontioides 996

fallax 1187
fauriei 484, 1596
filigera 3058
fissa 1739
hirtella 1739
honolulensis 3057
kealiae 3057
konaensis 2503
kunthiana 996
laciniata 1187
occultans 706
parviflora 1187
pinnatifida 396
racemosa 1739
recta 3058
regina 3057
rhytidosperma 1739
sinuata 1187
subcordata 396, 993, 2432, 3057
undulata 396, 421, 993, 3057
waihiae 3057
Deschampsia australis 2738
hawaiiensis 2366
nubigena 1187
pallens 1187
[see also Aira]
Desmodium 278
canum 524, 526, 527, 529, 797, 889, 1249, 2334, 2336, 2337, 3231
heterocarpon 889
intortum 523, 524, 525, 526, 527, 528, 529, 797, 895, 2333, 2334, 2336, 3231
sandwicense 523, 524, 525, 526, 527, 528, 529, 895, 1277, 1802, 2109, 2110, 2111, 2333, 2334, 2336, 2555, 2557
triflorum 1904
uncinatum 523, 524, 525, 526, 527, 1208, 1277, 2333, 2334, 2336, 2813
[see also Meibomia]
Dianella 2725
sandwicensis 897, 1238, 2539, 2558
Dianthus armeria 923
prolifer 923
Dichanthelium 289, 541
hillebrandianum 290
imbricatum 290
isachnoides 290
[see also Panicum]
Dichrocephala integrifolia 925
Dicliptera chinensis 871
Digitaria 1131
consanguinea 992
decumbens 1858
henryi 871, 3126
oblongo-ovata 2093
pruriens 1904
[see also Syntherisma]
Dioscorea 176
pentaphylla 2378
Diospyros ferrea 109, 151, 866, 1446, 1542
hillebrandii 858, 866, 2504

sandwicensis 858, 1446, 2504
[see also Maba]
Dipanax 2591
dipyrena 1126
gymnocarpa 1126
kavaiensis 1126
manni 2591
[see also Tetraplasandra]
Diplomorpha bicornuta 1126
buxifolia 1126
elongata 1126
hanalei 1126
oahuensis 1126
phillyreifolia 1126
sandwicensis 1126
uva-ursi 1126
villosa 1126
[see also Wikstroemia]
Dissochondrus biflorus 5, 1073, 1155
[see also Setaria]
Dissotis rotundifolia 923
Dodonaea 979, 2833
angustifolia 1584
eriocarpa 389, 657, 968, 1584, 2181, 2643, 2653, 2656, 2668, 2675, 2676, 2677, 2688, 2770, 3180
sandwicensis 1584, 2643, 2669
spathulata 2770
stenoptera 1187, 1584, 2214, 2643
viscosa 341, 1187, 2181, 2214, 2647, 2652, 2668
×Dodonaea fauriei 1558, 1590
Dolichos galeatus 994
lablab 640, 2811
[see also Canavalia]
Dracaena aurea 1739
hawaiiensis 885
terminalis 1612
[see also Pleomele]
Drosera anglica 712
Drymaria cordata 1825
Drypetes 3083
forbesii 2633, 2639
phyllanthoides 327, 957, 2633, 2761
[see also Neowawraea and Xylosma]
Dubautia 49, 415, 416, 417, 426, 459, 460, 461, 463, 464, 466, 468, 469, 470, 470a, 472, 643, 950, 994, 1046, 1389, 2264, 2268, 2270, 2373, 2624, 2650
ciliolata 2262, 3234
fauriei 1589
herbstobatae 462
imbricata 2506
knudsenii 414, 1187, 2263, 2622, 3200, 3234
laevigata 2617
laxa 393, 1187, 1238, 2284, 2617, 2619, 2622, 2634, 2653, 2656, 2672, 2673, 2677, 2704
magnifolia 2617
×media 2617, 2672
×mendacoides 2663
×mendax 2617

menziesii 414, 1429, 2263
microcephala 2617, 2704
nagatae 2450
paleata 404, 405, 2622
×paludicola 2663
pauciflorula 2506
plantaginea 393, 994, 1187, 2366, 2617, 2646, 2653, 2672, 2704
platyphylla 414
raillardioides 404, 405, 1187, 2656
reticulata 2263
scabra 2262, 3214, 3215, 3234
sherffiana 870, 1155
waialealae 2276, 2617, 2622
waianapanapaensis 462
[see also Railliardia]
×Dubautia fallax 2622
fucosa 2622
Dubrueilia peploides 994
[see also Pilea]

Edwardsia chrysophylla 390, 1039, 1395, 2528, 3036
[see also Sophora]
Eichhornia crassipes 2780, 2781
Elaeocarpus 569
bifidus 614, 1207, 1238, 1514
Eleocharis calva 819, 2386
obtusa 992, 2916
palustris 2917
Elephantopus mollis 870
[see also Pseudelephantopus]
Embelia hillebrandii 1806
pacifica 1187, 1806, 3200
Emilia 871
coccinea 2070
fosbergii 2070
javanica 890
sonchifolia 1444, 2070
[see also Hieracium]
Epidendrum ×obrienianum 1440, 1442
Epilobium adenocaulon 2228
billardierianum 2233
cinereum 2228
Episyzygium 2912
oahuense 2912, 3170
[see also Psidium]
Eragrostis 879, 1301
atropioides 1187
equitans 2975
fosbergii 3124
grandis 1187
hawaiiensis 1187
monticola 1187
niihauensis 3129
paupera 864, 873
phleoides 1187
tenella 1904
thyrsoidea 1187
variabilis 960, 1187, 1375, 1376, 1519

wahowensis 2975
whitneyi 864
Erechtites hieracifolia 202
valerianifolia 202
Erigeron bellioides 3005
canadense 2565
lepidotum 393
multiflorus 1238
pauciflorus 1238, 3015
remyi 749
tenerrimus 749
[see also Conyza and Tetramolopium]
Erythraea sebaeoides 1049
[see also Centaurium]
Erythrina 141, 2229
monosperma 994, 1320, 1330, 1484, 1963, 2304, 2381
sandwicensis 171, 276, 619, 640, 830, 831, 832, 1034, 1284, 1285, 1306, 1390, 1475, 1483, 1484, 1542, 1601, 1602, 1964, 1967, 2381, 2386, 3181, 3182
tahitensis 138, 139, 170, 171, 961, 1475, 1476, 1477, 1478, 1479, 1480, 1481, 1482, 1484, 1601, 1761, 2101, 2232, 2327, 2860
Escholtzia [see Eschscholzia]
Eschscholzia californica 897
Eucalyptus 326, 943, 1474, 1574, 1575, 1734, 1745, 2858, 3121, 3255
citriodora 3020, 3022, 3032
globulus 2696
microcorys 3120
pilularis 2055, 2696, 3120
robusta 77, 1225, 1306, 1547, 1548, 1550, 1950, 2154, 2692, 2695, 2696, 2698, 2769, 3186, 3232
saligna 2154, 2696, 3024, 3031, 3119
sideroxylon 2055
Eugenia 1685, 3169, 3170
cumini 1875, 1882
koolauensis 611
malaccensis 1514
molokaiana 3170
rariflora 1187
sandwicensis 193, 1039, 1187
[see also Syzygium]
Eupatorium 871
glandulosum 2102
[see also Ageratina]
Euphorbia 432, 438, 1130, 1140, 2129, 2130, 2261, 2267, 2630, 2650, 3250
arnottiana 779, 2360
atrococca 1126, 2626
celastroides 247, 2265, 2429, 2626, 2656, 2657, 2669, 2672, 2675, 2676, 2677, 2720
clusiifolia 247, 248, 1187, 1238
cordata 248, 1421, 1797
degeneri 1439, 2265, 2626, 2669
dentata 2653
deppeana 1312, 2636, 2641, 3104
festiva 2626

forbesii 1139, 1141, 2127, 2128, 2265, 2266, 2626, 2655
haeleeleana 1138, 2408
halemanui 2626
hillebrandii 1569, 1590, 2265, 2626
hirta 2033
hookeri 247, 248, 1187, 2853
kuwaleana 2653
lorifolia 1187, 1766, 2283, 2284
multiformis 247, 248, 1421, 1569, 1590, 2626, 2633, 2636, 2672, 2675, 2676
myrtifolia 1238
oahuensis 2720
olowaluana 2626, 2653, 2656
peplus 1522
pulcherrima 2310
remyi 2626
rivularis 1126
rockii 817, 834
skottsbergii 618, 813, 814, 1239, 2626
sparsiflora 1126
stokesii 840
[see also Chamaesyce]
Euphorbiaceae 1135, 1287
Eurya degeneri 1433
sandwicensis 858, 1039, 1433, 1434, 1435, 2982, 2988, 3065
[see also Ternstroemiopsis]
Evodia 751
[see also Pelea]
Excavatia sandwicensis 1746
[see also Rauvolfia]
Exocarpos 2823, 2919
brachystachys 1187
casuarinae 126
gaudichaudii 384, 806
luteolus 835
menziesii 806
sandwicensis 126
Exocarpus [see Exocarpos]

Fabaceae 83, 85, 155, 266, 1253, 1648, 2010, 2250, 2253, 2312, 3006
Fagara 787
bluettiana 791, 2674
dipetala 2366, 2674, 2720, 2738
glandulosa 2674
hawaiiensis 2674
kauaiensis 2674, 2738
mauiensis 2366, 2674, 2738
multifoliolata 2720
oahuensis 2674
semiarticulata 2545, 2674, 2720
waianensis 715
[see also Zanthoxylum]
Festuca hawaiiensis 2522, 2523
sandvicensis 2238
Ficus 1652
microcarpa 885, 1136
retusa 1654, 1660, 2170

Filago gallica 897
Fimbristylis cymosa 1187, 1491
hawaiiensis 1187
polymorpha 1187
pycnocephala 45, 1187
Fitchia speciosa 413, 2795
Flaveria trinervia 948, 2189
Fragaria chiloensis 897, 2822
sandwicensis 627, 2989
Frankenia grandiflora 923
Fraxinus uhdei 1225, 1306, 1548, 2154, 2769, 3023, 3120, 3123, 3137, 3232
Freycinetia 1753, 2889, 3250
arborea 63, 595, 596, 597, 598, 990, 992, 1118, 1147, 1749, 2688, 2859, 2880, 2882, 2887, 2891, 2972
arnotti 2887
Fuchsia paniculata 256
Furcraea gigantea 1724, 1725

Gahnia 1396
affinis 206, 2145
arenaria 206
aspera 205
beecheyi 204, 205, 206, 207, 1493, 1739
congesta 206, 241
gahniiformis 205, 207, 2385
gaudichaudii 204, 206, 207, 241, 2856
globosa 204, 1490, 1739
javanica 1490
kauaiensis 205, 206
lanaiensis 710
leptostachya 206, 241
mannii 204, 206, 1187
mucronata 206, 241
[see also Lampocarya]
Galeatella 692
[see also Lobelia]
Gardenia brighamii 327, 1162, 1172, 1739, 2517, 2760
mannii 2517, 2760
remyi 1739, 2517, 2760
weissichii 2443, 2451
Garnotia acutigluma 1032a
sandwicensis 1187, 2529a
Gastonia oahuensis 1039, 1187
[see also Tetraplasandra]
Geniostoma 129
gaudichaudii 557
[see also Labordia]
Geranium 445, 2278
arboreum 859, 1039
cuneatum 859, 868, 1039, 1187, 1235, 2340
humile 859, 1187
kauaiense 2495
multiflorum 859, 868, 873, 1039, 1187
ovatifolium 1039
retrorsum 450
tridens 1187
Gnaphalium 2652, 3005

consanguineum 994
hawaiiense 2656
purpureum 1210, 2810
sandwicensium 394, 994, 2651, 2656, 2668
[see also Leontopodium]
Gonocarpus chinensis 2099
Gossypium 2849, 2968
  barbadense 1760, 2850, 2851
  drynarioides 3050, 3229
  hirsutum 2850, 2851
  sandvicense 938, 939, 1276, 2112, 2259, 2260, 2850, 3140, 3144, 3145
  tomentosum 329, 938, 941, 942, 1100, 1276, 1311, 1387, 1780, 2685, 2850, 2851, 2852, 3050, 3140, 3144, 3145
Gouania 2397
  bishopii 1187
  gagnei 2410
  hillebrandii 1156, 1159, 1187
  integrifolia 1797
  meyenii 2853
  vitifolia 1039
Gouldia 666, 667, 856, 861, 870, 877, 1017, 1663
  affinis 3139, 3143
  antiqua 2738
  arborescens 1126
  axillaris 658, 3068
  cirrhopetiolata 1590
  cordata 658, 2738
  coriacea 1187
  elongata 1126
  gracilis 2738
  hirtella 1187, 1210
  kaala 2738
  lanceolata 1126
  macrocarpa 1187
  macrothyrsa 2738
  ovata 658, 2738
  purpurea 2738
  sambucina 1126
  sandwicensis 340, 1043, 3067, 3068
  terminalis 860, 870, 894, 1187, 3139, 3143
  [see also Kadua and Petesia]
Grevillea 3255
  banksii 100, 101
  robusta 101, 2052, 2696, 2769, 3191, 3233
Gunnera 2369, 2722
  dominii 1454
  kaalensis 1733
  kauaiensis 479
  petaloïdea 385, 994, 1039, 1454, 1687, 2553
Gymnelaea sandwicensis 1307
[see also Nestegis]
Gynopogon oliviformis 1210
[see also Alyxia]

Habenaria holochila 323, 1187, 1391, 1683, 2012
[see also Platanthera]
Halophila hawaiiana 737, 1137, 1137a, 1137b, 1769, 1770, 2348, 2980

ovalis 738, 1106, 1107, 1770
Haplostachys 271, 1187, 2625, 2650
  bryanii 2620
  grayana 1187, 2620
  haplostachya 37, 41, 1169
  linearifolia 2622
  munroi 843, 873
  rosmarinifolia 1187
  truncata 1187
Hedera gaudichaudii 1039, 3063
  platyphylla 1039
  [see also Cheirodendron]
Hedychium 2015
  coronarium 639, 3048
  flavescens 3048
  flavum 639
  gardnerianum 2942, 3048
Hedyotis 869, 877
  acuminata 894, 2853
  centranthoides 2853
  chamissoniana 2853
  conostyla 994, 2853
  cookiana 2853
  coriacea 2771, 2853
  corymbosa 897, 923
  hookeriana 2853
  littoralis 2895
  mannii 894
  menziesiana 2853
  st.-johnii 2895
  schlechtendahliana 894, 2853
  [see also Kadua and Wiegmannia]
Heimerlia 2721
  [see also Pisonia]
Heimerliodendron brunoniana 1383, 2736
  [see also Pisonia]
Heliconia 2949
Heliocarpus popayanensis 1572
Heliotropium anomalum 45, 923, 1048, 1313, 2319
  curassavicum 401, 486, 1519
  procumbens 924
Hemigraphis 3005
Heptapleurum dipyrenum 1739
  gymnocarpum 751
  kavaiense 1739
  waimeae 3064
  [see also Tetraplasandra]
Herpestes monnicria 502
  [see also Bacopa]
Herpestis fauriei 1590
  [see also Bacopa]
Hesperocnide 1275
  sandwicensis 785, 3087
Hesperomannia 411, 1127, 1754, 2826
  arborea 873
  arborescens 265, 411, 412, 847, 1051, 1187
  arbuscula 411, 1187, 2446
  lydgatei 411, 817, 834
  mauiensis 2490

Heteropogon contortus 446, 1259, 1320, 1375, 1376, 1403
Hibiscadelphus 8, 142, 145, 147, 873, 1204, 1275, 1632, 2230, 2862
　bombycinus 848
　crucibracteatus 1204
　distans 55, 229, 467, 1160, 1166
　giffardianus 142, 144, 146, 327, 467, 702, 807, 2215
　hualalaiensis 142, 144, 146, 467, 633, 702, 2072, 2215, 3157
　×puakuahiwi 144, 146, 467, 702
　wilderianus 2215, 3157
Hibiscus 177, 193, 703, 989, 2004, 2021, 2030, 2072, 2320, 2321, 3133, 3151, 3155, 3157
　arnottianus 1147, 1206, 2738, 2817
　brackenridgei 479, 571, 988, 1039, 1206, 1982, 2206, 2321, 2345, 3209
　clayi 2881
　esculentus 221
　fauriei 1557, 1589, 2281
　furcellatus 1782
　immaculatus 2321
　kahilii 837, 1154, 1304
　kokio 479, 703, 2817
　newhousei 2321, 2881
　oahuensis 703
　roeatae 2408
　rosa-sinensis 988
　saintjohnianus 2321
　tiliaceus 64, 98, 221, 624, 942, 1206, 1514, 1679, 1951, 1958, 2066, 3019
　ula 703
　waimeae 7, 1126, 1206, 2816, 2817
　youngianus 1238
　[see also Pariti]
Hieracium javanicum 1444
　[see also Emilia]
Hillebrandia 1275
　sandwicensis 6, 644, 998, 1231, 1624, 1700, 2094, 2738, 2779
Hippobroma longiflora 871
　[see also Isotoma]
Holcus lanatus 1622
Holopeira lonchophylla 1806a
Hydnocarpus anthelmintica 1335, 1346, 1362
　castanea 1335
Hydrocotyle asiatica 2311
　bowlesioides 3005
Hyparrhenia rufa 1904
Hypericum 897
　degeneri 2271
　gramineum 2271
　japonicum 2271
　mutilum 2271

Ilex 1408
　anomala 115, 116, 286, 958, 1119, 1238, 1620, 1621, 1938
　hawaiensis 1266, 1267

　sandwicensis 1619
　[see also Byronia and Polystigma]
Indigofera 640
　anil 2813
　disperma 932
　suffruticosa 4, 1904, 2241, 2328
Ipomoea batatas 536, 613, 1091, 1093, 1378, 2008, 2329, 2330, 2332, 2867, 3221, 3222
　brasiliensis 2393, 2399
　cairica 1210
　congesta 2853
　fauriei 1563, 1590
　forsteri 1187
　gracilis 2077
　indica 908, 1386, 1390
　insularis 2853
　koloaensis 1590
　littoralis 889
　pes-caprae 1386, 1519, 1563, 1867, 2393, 2399, 2918, 3019
　purpurea 1210
　stolonifera 885, 1563
　triloba 2077
　tuba 885
　[see also Calonyction and Pharbitis]
Isachne pallens 1187
Ischaemum lutescens 284, 1072a
Isodendrion 1039, 1275, 1708, 1778, 2375
　christensenii 2448
　fauriei 1556, 1595
　laurifolium 1037, 1039
　longifolium 1037, 1039
　pyrifolium 1037, 1039
　subsessilifolium 1126
Isotoma longiflora 2311
　[see also Hippobroma]

Jacaranda acutifolia 638
Jacquemontia ovalifolia 2257, 2258
　sandwicensis 1048, 1187, 1210, 2421
Jacquinia aurantiaca 2690, 2884
Jambosa malaccensis 1208
　[see also Syzygium]
Jatropha curcas 2311
　integerrima 890
Joinvillea 532, 2062, 2688
　ascendens 280, 996, 2063, 2064, 2447
　gaudichaudiana 280, 2063, 2064
Jussiaea suffruticosa 1208
　[see also Ludwigia]

Kadua 1043
　acuminata 392, 506
　affinis 392, 506, 3139, 3143
　arnottii 729
　cookiana 340, 392, 506, 1739
　cordata 392, 506, 1187, 3066
　fluviatilis 837
　foliosa 1187

formosa 1187
glaucifolia 1043
glomerata 1238, 3066
grandis 1043, 2738
herbacea 1590
kaalae 3067
knudsenii 1187
laxiflora 1739
littoralis 1187
longipedunculata 2738
menziesiana 392, 506
parvula 1043
petiolata 1043
remyi 1187
waimeae 3066
[see also Hedyotis]
Kalanchoë pinnata 1672
verticillata 871
[see also Bryophyllum]
Kalstroemia cistoides 779, 2360
[see also Tribulus]
Keysseria 2406
erici 355
helenae 355
maviensis 355
[see also Lagenifera]
Koeleria glomerata 1498
Kokia 873, 942, 1275, 1604, 2305
cookei 37, 39, 41, 46, 811, 1973, 3210, 3212, 3213
drynarioides 1157, 1604, 1973, 3165, 3229
lanceolata 699, 1604
rockii 8, 1604, 2305, 3229
Korthalsella 615, 868, 1155, 1577, 2026, 2967, 2974, 3138
complanata 788, 1420, 2203, 2572
cylindrica 788
platycaula 614
remyana 2966
[see also Bifaria and Viscum]

Labordia 557, 1411, 2632, 2650, 2796
baillonii 2357
cyrtandrae 2357
decurrens 2631, 2641
fagraeoidea 129, 383, 993, 1044, 1187, 2631, 2657
fauriei 1590, 2281
glabra 1187, 2631
grayana 1187, 2720
hedyosmifolia 129, 2357, 2631, 2656
helleri 2631, 2669
hirtella 1187, 1739, 2631, 2636
hymenopoda 2656
hypoleuca 2354
kaalae 843, 2631
lophocarpa 1187
lydgatei 843
mauiensis 2631
membranacea 1739, 2631

molokaiana 129, 2357, 2631, 2669
nelsonii 2429
nervosa 2488
olympiana 2631
pallida 3052
sessilis 1044
tinifolia 1044, 2495, 2631, 2641
triflora 1187
venosa 2631
waialealae 3052
waiolani 3052
wawrana 2631, 2641
[see also Geniostoma]
Lagenifera 2722
erici 846
helenae 846
maviensis 846, 1739
viridis 2406
[see also Keysseria]
Lagenophora [see Lagenifera]
Lamiaceae 17, 18
Lampocarya gaudichaudii 279
[see also Gahnia]
Lantana camara 8, 863, 1409, 1828, 1829, 1830, 1832, 1834, 1836, 1837, 1842, 1843, 1852, 1853, 1875, 2136, 2137, 2138, 2139, 2948
Lavenia glutinosa 994
[see also Adenostemma]
Leontopodium japonicum 1589
[see also Gnaphalium]
Lepechinia hastata 794, 795, 868, 3207, 3208
[see also Sphacele]
Lepidium 2326
arbuscula 1187
bidentatum 897
orbiculare 2481
o-waihiense 496, 896, 3015
remyi 751
serra 1739, 3200
Leptospermum scoparium 1961
Lepturus repens 876
Leucaena 258
glauca 72, 75, 1306, 2105, 2163, 2813, 2937, 3251
latisiliqua 259
leucocephala 259, 260, 261, 797, 885, 1028, 1029, 2142, 2942, 3118, 3181, 3182
Linociera ligustrina 897
Liparis hawaiensis 323, 829, 1155, 1391, 1440, 1442, 1638, 1683, 1739, 2012, 2245
Lipochaeta 426, 924, 982, 983, 984, 985, 986, 987, 1046, 2208, 2209, 2210, 2270, 2386, 2429, 2624, 2650
acris 2497, 2619, 2635, 2656, 2881
alata 2619, 2635, 2653, 2881
aprevalliana 749, 751
asymmetrica 1589
australis 3061
bryanii 2619
christophersenii 2497
connata 393, 1187

degeneri 2619
deltoidea 2408
dubia 2634
elliptica 2497
exigua 2635
fauriei 1589
flexuosa 749, 751
forbesii 2619
garberi 2497
hastata 1187
hastulata 393
heterophylla 2619, 2663
integrifolia 45, 583, 2619, 2656, 2676
intermedia 2619
kaenaensis 2497
kahoolawensis 2619
kamolensis 2656
lahainae 3061
lavarum 393, 415, 2497, 2619, 2634, 2653, 2669
lobata 393, 1187, 2497, 2619, 2675, 2676
mauiensis 2497
minuscula 2669
molokiniensis 2497
nesophila 2497
peduncularis 749, 751
perdita 2619
pinnatifida 2497
×procumbens 2619
profusa 2619, 2676
robusta 2497
rockii 2497, 2619, 2663
setosa 2497
subcordata 2619, 3005
succulenta 393, 2619, 2795
tenuis 2619
variolosa 1589
venosa 37, 41, 1169, 2619, 2656, 3005
vittata 2497
waimeaensis 2408
warshaueri 2497
[see also Aphanopappus, Lipotriche, Microchaeta, Schizophyllum, Verbesina, and Wollastonia]
Lipotriche australis 1588
[see also Lipochaeta]
Livistona martii 996
[see also Pritchardia]
Lobelia 431, 779, 1047, 1516, 1667, 2308, 2360, 2707, 2820, 3178, 3179
ambigua 491
angustifolia 491
calycina 491
cardinalis 441
costata 3177
gaudichaudii 396, 605, 918, 996, 2280, 2292, 2516
gloria-montis 796
grayana 3177
hypoleuca 1187, 2516, 3177
kauaensis 1126, 2292

macrostachys 686, 996, 1128, 1129, 1238, 3247, 3248
niihauensis 2352, 2359
oahuensis 2300, 2515
pinnatifida 491
superba 491
taccada 886
tortuosa 1126, 2359
villosa 2516
yuccoides 1187
[see also Galeatella and Neowimmeria]
Lobularia maritima 583
Lonicera japonica 871
Lotus hispidus 923
Ludwigia octovalvis 2227, 2234
palustris 2227
[see also Jussiaea]
Luteidiscus calcisabulorum 2412
capillaris 2412
rockii 2412
[see also Tetramolopium]
Lycium carolinianum 1200
sandwicense 1049, 2014
Lycopersicon esculentum 508
Lysimachia 1088, 1278, 1279
daphnoides 1187, 1562, 1590
filifolia 843
forbesii 2285, 2354
glutinosa 2276
hillebrandii 1048, 1187, 2126, 3071
kalalauensis 2738
kipahuluensis 2406
koolauensis 842
longisepala 817, 834
lydgatei 1187
remyi 1187
rotundifolia 1187
[see also Lysimachiopsis]
Lysimachiopsis daphnoides 1126
hillebrandii 1126
lydgatei 1126
ovata 1126
remyi 1126
[see also Lysimachia]
Lythrum maritimum 501

Maba hillebrandii 1175, 2117
sandwicensis 381, 1048, 1175, 1693, 1698, 2117, 2966
[see also Diospyros]
Machaerina angustifolia 1449
gahniiformis 1396
meyenii 1451
[see also Morelotia and Vincentia]
Macrochilus superbus 396
[see also Cyanea]
Mahoe 1187, 2212
[see also Alectryon]
Malachra alceifolia 870
Mallotus philippensis 316

INDEX TO PLANT NAMES

Malva hawaiensis 1557, 1589
Malvaceae 1205, 3006
Malvastrum coromandelianum 2808
Mangifera indica 100, 101, 183, 1182, 1727, 1882, 2003, 2105, 2172, 2178, 2696, 3219
Manihot glaziovii 2778
Manilkara 1508
  emarginata 1505, 1507
Mariscus kunthianus 992
  [see also Cyperus]
Maurandya erubescens 870
Medicago 278
  polymorpha 897
Meibomia limensis 2554
  [see also Desmodium]
Melaleuca leucadendra 2780, 2781
  leucadendron 74, 1548, 1657, 2055
  quinquenervia 3255
Melastoma decemfidum 2164
  malabathricum 1858, 1875, 2163, 2164
Melia azedarach 376, 638, 2311
Melicope 1039, 1737
  cinerea 3063
  [see also Pelea]
Melinis minutiflora 1254, 1904, 1949, 1950, 2578, 3181, 3182
Melochia umbellata 923, 2221
Messerschmida [see Messerschmidia]
Messerschmidia 821
  argentea 641, 871
Messerschmidtia [see Messerschmidia]
Metagonia calycina 2081
  [see also Vaccinium]
Metrosideros 36, 69, 134, 152, 153, 227, 488, 558, 572, 573, 574, 576, 577, 580, 594, 683, 739, 740, 800, 852, 915, 919, 1000, 1220, 1222, 1240, 1280, 1282, 1367, 1415, 1540, 1542, 1917, 1935, 2294, 2530, 2531, 2722, 2832, 2843, 2900, 3131
  collina 66, 95, 135, 136, 173, 200, 342, 344, 453, 454, 455, 456, 457, 473, 474, 535, 566, 567, 575, 700, 799, 913, 933, 954, 955, 1208, 1281, 1282a, 1306, 1323, 1413, 1414, 1416, 1417, 1419, 1502, 1582, 1823, 1858, 1875, 1885, 1888, 1890, 1892, 1898, 1900, 1924, 1925, 2000, 2073, 2074, 2075, 2108, 2144, 2161, 2182, 2184, 2204, 2255, 2294, 2325, 2696, 2759, 2798, 2807, 2815, 2991, 3180
  feddei 2281
  lutea 1039
  macropus 626, 1238, 2515
  polymorpha 8, 48, 109, 154, 347, 348, 349, 350, 404, 405, 434, 581, 582, 626, 741, 801, 994, 1005, 1007, 1081, 1082, 1083, 1084, 1085, 1216, 1224, 1294, 1295, 1341, 1362, 1647, 1653, 1699, 1734, 1903, 1905, 1906, 1907, 1908, 1909, 1911, 1912, 1913, 1915, 1916, 1920, 1938, 2071, 2107, 2216, 2268, 2319, 2455, 2527, 2579, 2688, 2720, 2738, 2769, 2833, 2841, 2842, 2845, 3009, 3215, 3232

pumila 1208
rugosa 626, 1039
tremuloides 626, 2284
  [see also Nania]
Mezoneuron 1274
  kavaiense 84, 816, 1167, 1187, 2304, 3200, 3207, 3208
  [see also Caesalpinia]
Microchaeta 2080
  [see also Lipochaeta]
Mimosa pudica 257, 638, 2993
Momordica charantia 1560
Montanoa hibiscifolia 1104
Moraceae 1652
Morelotia gahniiformis 279, 992
  [see also Machaerina]
Morinda citrifolia 203, 506, 619, 639, 775, 1123, 1371, 1705, 1822, 2858, 2933
  lanaiensis 2466
  sandwicensis 2466
  trimera 1187, 1705, 2466
  waikapuensis 2466
Morus 2963
Munroidendron 1275, 1984, 2147, 2148
  racemosum 958, 1973, 2666, 2672, 2881
Murraya paniculata 640
Musa 110, 1009, 1185, 1634, 1711, 2175, 2686, 3147
  acuminata 2687
  balbisiana 2687
  fehi 1726
  ×paradisiaca 2174
  sapientum 1181, 1704
Myoporum fauriei 1595
  sandwicense 1049, 1366, 1461, 1564, 2248, 2249, 2573, 2574, 2577, 2688, 3082, 3102, 3121
Myrica faya 544, 966, 972, 974, 1399, 2754, 2755, 2941, 2942, 3009, 3021, 3025, 3030, 3114
Myrsine 1247
  fauriei 1590
  gaudichaudii 379, 380, 3071
  helleri 2442
  hosakae 695, 696, 3141
  kauaiensis 1187
  lanaiensis 1187
  lanceolata 1126
  lessertiana 200, 379, 380, 1048, 1938
  meziana 3141
  molokaiensis 1562, 1567, 1590
  punctata 3141
  sandwicensis 379, 380, 1048, 1590, 3071
  vanioti 1590
  [see also Rapanea and Suttonia]
Myrtaceae 155

Najas major 255, 487
Nama sandwicensis 120, 253, 507, 559, 885, 1048, 1052, 1201, 1202, 1519

[see also Conanthus]
Nania ×fauriei 1590
  ×feddei 1590
  glabrifolia 1126
  lutea 1126
  macropus 1590
  polymorpha 1126, 1590
  pumila 1126
  tremuloides 1126
  [see also Metrosideros]
Nasturtium sarmentosum 2367, 2407
  [see also Rorippa]
Neowawraea 1599, 1600, 3083
  phyllanthoides 1116, 1117, 1345, 1348, 1349, 1352, 1354, 1363, 1598, 1632, 2123, 2284, 2343
  [see also Drypetes]
Neowimmeria 663, 692, 1779
  [see also Lobelia]
Nephroica ferrandiana 1039, 1806a
Neraudia 593, 785, 946, 994, 1275, 2501
  cookii 2429
  glabra 1797
  kahoolawensis 1187
  melastomifolia 191, 994, 1187, 2738, 3086, 3087
  ovata 994
  pyramidalis 2480
  sericea 996, 3073, 3086
Nertera depressa 2722
  granadensis 1210, 2738
Nesoluma 121, 1508, 1509
  polynesicum 131, 132, 1506, 1510
  [see also Chrysophyllum]
Nestegis sandwicensis 1095
  [see also Gymnelaea, Olea, and Osmanthus]
Nicandra physalodes 2311
Nicotiana glauca 639, 1320
Nopalea cochenillifera 638
  [see also Opuntia]
Nothocestrum breviflorum 1049, 1187
  inconcinnum 2505
  latifolium 1049, 2404
  longifolium 1049, 1187, 1563, 2881
  peltatum 2738
  subcordatum 1739
Nototrichium 1187, 2658
  fulvum 2556
  humile 1187, 2654
  sandwicense 1187, 2386, 2454, 2654, 2655, 2657, 2668
  viride 1187, 2474, 2655, 2672, 2676
  [see also Psilotrichum and Ptilotus]

Obbea timonioides 1228
  [see also Bobea]
Ochrosia 922, 2436, 2567
  compta 916, 3199
  sandwicensis 327, 382, 1048, 1316, 1746, 2150, 2549

tuberculata 2151
  [see also Bleekeria]
Ocimum basilicum 885
  gratissimum 885
Odontochilus sandvicensis 897, 2012
  [see also Anoectochilus]
Oenothera stricta 721
Olea sandwicensis 1048
  [see also Nestegis]
Operculina tuberosa 3046
Opuntia cochenillifera 211
  ficus-indica 211
  megacantha 2032
  vulgaris 211
  [see also Nopalea]
Oreobolus 2722
  furcatus 1494, 1739, 2145, 2738, 2744
Oreodoxa regia 1321
Osmanthus 890
  sandwicensis 1426, 2001
  [see also Nestegis]
Osteomeles 1274
  anthyllidifolia 391, 499, 1232, 1610, 1867, 2196, 3096
  [see also Pyrus]
Oxalis corymbosa 356
  martiana 975
Oxyspora paniculata 2089

Pachyrhizus erosus 545
Paederia foetida 871
Palafoxia callosa 3005
Panax 1101
  ovatum 1238
  platyphyllum 1238
  [see also Cheirodendron]
Pandanaceae 155
Pandanus 162, 488, 1750, 1753, 1867, 2389, 2878, 2885, 2889, 2890
  chamissonis 996, 2888
  douglasii 996, 2469, 2870, 2888
  menziesii 996, 2888
  odoratissimus 98, 360, 1576, 1752, 2387, 2469, 2888, 2972
  tectorius 109, 705, 706, 775, 1875, 2469, 2870, 2892, 2893, 3033
Panicum 289, 1264, 2406
  affine 1238
  alakaiense 2738
  barbinode 1660, 1950
  beecheyi 1199, 1238
  carteri 1170, 1171, 1248
  cinereum 1187
  colliei 779, 2360
  conjugens 2431, 2738
  cornae 2482
  cynodon 2160, 2238
  degeneri 2187
  dichotomiflorum 897
  gossypinum 1238

gracilius 2431
havaiense 2238
heupueo 2386
hillebrandianum 1264, 2160, 2738
imbricatum 1187, 1264, 2160, 2738, 3125
isachnoides 1264, 2160, 2704, 2738
kauaiense 1199
konaense 3129
koolauense 2515
montanum 992
monticola 1187
moomomiense 2427
nephelophilum 992, 1187
niihauense 2352
nubigenum 1155, 2482
oreoboloides 2738
pellitoides 2355
pellitum 2976
ramosius 1199, 2482
tenuifolium 1238
torridum 992
[see also Dichanthelium]
Parentucellia viscosa 923
Pariti tiliaceum 640
[see also Hibiscus]
Paspalam conjugatum 370, 1333, 1653, 1789, 1858
 dilatatum 1949
 orbiculare 1858
 urvillei 871
 vaginatum 864
Passiflora 76, 1682, 2180, 2902, 2904
 edulis 113, 640, 1023
 foetida 1390
 mollissima 1, 226, 1551, 1552, 1553, 2572, 2579, 2581, 2941, 2942, 2999, 3044
 quadrangularis 640
Pelea 661, 673, 787, 827, 1039, 1737, 2365, 2406, 2876, 2877, 2883, 2886, 3038
 acutivalvata 1590
 ahiaensis 2486
 anisata 826, 828, 1270, 2015, 2105, 2319, 2542
 auriculifolia 1038, 1039
 balloui 2284
 barbigera 1176, 1177, 1187
 cauliflora 2459
 christophersenii 1270, 2548
 cinerea 1187, 2284, 2299
 clusiifolia 1038, 1039, 1270, 2459, 2738, 3062
 cookeana 2284
 cruciata 1126
 elliptica 1187
 fauriei 1590
 feddei 1592
 foetida 1590
 gayana 2299
 glabra 2408
 grandipetala 1592
 hawaiensis 2876, 3062
 hiiakae 2876
 hillebrandii 1590

kaalae 3062
kavaiensis 1187, 1208
kilaueaensis 2486
knudsenii 1187
lakae 2876
leveillei 1592
lohiauana 2876
lydgatei 1187, 2459
macropus 1187
makahae 2876
mannii 1187
microcarpa 1126
molokaiensis 1187
multiflora 2278
nodosa 1592
oahuensis 1270, 1592, 2876
oblongifolia 1038, 1039
orbicularis 1187
pallida 1187
paludosa 2459
parvifolia 1187
peduncularis 1592, 2876
penduliflora 1592
pickeringii 2495
pseudoanisata 2284
punaensis 2486
recurvata 2299
resiniflora 2720
rotundifolia 1038, 1039
sandwicensis 1038, 1039, 1187, 2873
sapotifolia 1187, 2284
scandens 2486
semiternata 2459
sessilis 1590
sherffii 673
singuliflora 1592
stonei 673
subpeltata 1592
volcanica 1038, 1039, 1187, 2284
waialealae 2738, 3062
waianaiensis 1592
wailauensis 2459
wainihaensis 2486
wawraeana 2284
zahlbruckneri 2284
[see also Evodia and Melicope]
Pennisetum ciliare 446
 clandestinum 1622, 2942
 purpureum 3164
 setaceum 341, 2941, 2942
Peperomia 373, 377, 378, 1817, 2532, 2533, 2534, 2563, 3237, 3238, 3240, 3243, 3244
 acrostigma 378
 asperulata 375
 candollei 2352
 cookiana 373, 1568
 cranwelliae 3242
 dextrolaeva 2364
 fauriei 1568, 1590
 gaudichaudii 1816
 haupuensis 2364

helleri 374, 378, 1590
hesperomannii 3078
hiloana 378
hochreutineri 375
hypoleuca 1187, 1816, 3078
insularum 373, 1818
kahiliana 3241
kipahuluensis 2406
latifolia 1816, 3078
leptostachya 1187, 1238
ligustrina 1187
macraeana 372, 1187
mapulehuana 1568
mauiensis 1187, 3078
membranacea 1187, 1238
oahuensis 373, 2364
pachyphylla 374, 1187, 1816
parvula 1187
pellucida 3241
pleistostachya 1187
plinervata 2429
pololuana 3241
purpurascens 373
reflexa 373, 1187, 1816, 1818, 3078
refractifolia 1590
remyi 374
sandwicensis 375, 1816, 3077
treleasei 3242
waihoiana 2423
waipioana 3242
Perrottetia sandwicensis 713, 1039, 1618
Persea americana 161, 1184, 2173
Petesia coriacea 1238
  terminalis 1238
  [see also Gouldia]
Petrorhagia velutina 169
Peucedanum kauaiense 1187
  sandwicense 1187
Phaius 175
  tankarvilleae 1440, 1442
Pharbitis insularis 3070
  [see also Ipomoea]
Phaseolus 278
Phyllanthus 2650
  distichus 1238, 1598
  nivosus 2633
  sandwicensis 858, 1589, 1945, 1947, 2633
Phyllostegia 213, 215, 271, 1048, 2429, 2625, 2650
  ambigua 1187
  bracteata 2622
  brevidens 2620, 2622, 2635, 2668
  chamissonis 213
  clavata 212
  degeneri 2675
  dentata 212
  electra 843
  floribunda 214, 2620
  glabra 212, 858, 868, 2622
  grandiflora 212, 1187, 2653
  haliakalae 3053

  helleri 2620, 2622
  hillebrandii 1187
  hirsuta 212
  hispida 1187
  honolulensis 3053
  knudsenii 1187
  lantanoides 2622
  leptostachys 212
  linearifolia 746, 752
  longiflora 479
  macraei 215
  macrophylla 212, 2622, 2668
  mannii 2622
  mollis 213, 868, 2620, 2634, 2653, 2664, 2668
  parviflora 2620, 2622, 2669
  racemosa 212, 2620, 2657
  rockii 2622
  stachyoides 2620
  variabilis 231, 583
  vestita 212, 2657
  waimeae 3053
  wawrana 2622
  yamaguchii 1251, 2675
  [see also Prasium]
Phytolacca brachystachys 1864a, 2079
  octandra 2079
  sandwicensis 779, 2360
Pilea peploides 3086, 3087
  [see also Dubrueilia]
Piper 3243
  methysticum 599, 638, 1371, 2069, 2076, 2547, 2564, 2568, 2933, 2969
Pipturus 946, 1002, 1017, 1275, 2067, 2068, 2717, 3250
  albidus 1208, 1739, 2688, 2704, 2714, 2947, 3087
  brighamii 2712, 2738
  eriocarpus 2712
  forbesii 1455
  gaudichaudianus 3085
  hawaiiensis 1569, 1589, 2714, 2738
  helleri 2712
  kauaiensis 1126
  oahuensis 2712
  pachyphyllus 2712
  pterocarpus 2738
  rockii 2712
  ruber 1126
  skottsbergii 1455
  taitensis 3085, 3086
Pisonia 2293
  grandis 1383
  sandwicensis 1187, 2834
  umbellifera 441, 1566, 2834
  [see also Calpidia, Ceodes, Heimerlia, Heimerliodendron, and Rockia]
Pistia stratiotes 639
Pithecellobium dulce 640, 1445
  saman 2696, 3029, 3032, 3121
  [see also Samanea]
Pittosporum 440, 1072, 2197, 2435, 2638, 2650

acuminatum 440, 1739, 2635, 2636
acutisepalum 2636
amplectens 2636, 2653
argentifolium 2636, 2641
cauliflorum 1187, 1739, 2635, 2720, 3081
cladanthum 2635, 2636
confertiflorum 1039, 2635, 2636, 2653, 2657, 2668
dolosum 2636, 2668
fauriei 1556, 1589
flocculosum 2635
forbesii 2636
gayanum 440, 2284, 2653
glabrum 440, 1238, 2635, 2636
glomeratum 1187
halophiloides 2635
halophilum 2635, 2278
hawaiiense 1187
helleri 2635
hillebrandii 1556, 1589
hosmeri 440, 2276, 2284, 2636, 2653, 2656
insigne 1187, 2635, 2636
kahananum 2636
kauaiense 1187, 2635, 2641
kilaueae 2491
lanaiense 2491
napaliense 440, 2636
spathulatum 1739
sulcatum 2635, 2636, 2657
terminalioides 1039, 2635, 2636, 2720
Planchonella 1508, 1509
aurantium 2720
sandwicensis 245, 757, 1510, 1562, 1567, 2155
spathulata 757, 2155
[see also Pouteria]
Plantago 923, 1566, 2159, 2309
fauriei 1590
gaudichaudiana 1590
glabrifolia 2157
grayana 2157, 2738
hawaiiensis 2157
hillebrandii 2157
krajinai 2158
major 1371, 2156, 2858
melanochrous 2158
muscicola 2157
pachyphylla 254, 1049, 2157, 2309, 2738, 2947a, 3074
princeps 497, 1049, 1187, 2309, 2947a, 3074
queleniana 993
Platanthera holochila 1440, 1442, 1638, 2088, 2140
[see also Habenaria]
Platydesma 787, 1737, 2299, 2871, 2872
auriculifolia 1187, 2869
campanulata 1187, 1741, 2284, 2738, 3097
cornuta 1187
fauriei 1558, 1590
oahuensis 1558, 1590
remyi 2869
rostrata 1187

spathulata 2552, 2872
Plectranthus parviflorus 238
[see also Solenostemon]
Plectronia odorata 1210, 1332
[see also Canthium]
Pleiosmilax menziesii 2593
sandwichensis 2593
[see also Smilax]
Pleomele aurea 288
auwahiensis 2498
fernaldii 674, 2370
halapepe 2498
kaupulehuensis 2498
konaensis 2498
lanaiensis 674
rockii 2498
[see also Cordyline and Dracaena]
Pluchea ×fosbergii 565
indica 418, 565, 871
odorata 418, 565, 2242, 2354
Plumbago auriculata 871
zeylanica 639
Poa longe-radiata 1187
monticola 992
siphonoglossa 1075
variabilis 992
Poaceae 1198, 1252, 2029, 2048, 2250, 2251, 2331, 2341, 3130
Poinsettia 638
pulcherrima 2564
Polycoelium sandwicense 383a
Polygala paniculata 923
Polygonum capitatum 897
glabrum 503
punctatum 897
Polypogon monspeliensis 1522
Polypremum procumbens 885
Polystigma hookeri 1775
[see also Ilex]
Portulaca 2166, 2874
caumii 534, 762
cyanosperma 761, 2366
lutea 539, 2167, 2168
oleracea 2167
pilosa 1001, 1606
sclerocarpa 1039, 2167
villosa 490, 762, 1606, 2167
Potamogeton o-waihiensis 500, 1062
Pouteria 121, 885
sandwicensis 122
[see also Planchonella, Sapota, and Sideroxylon]
Prasium glabrum 993
grandiflorum 993
macrophyllum 993
parviflorum 993
[see also Phyllostegia]
Pritchardia 184, 185, 187, 188, 189, 324, 479, 562, 586, 679, 755, 1064, 1213, 1632, 1636, 1706, 1751, 1862, 1863, 1957, 2293, 2315, 2353, 2688, 2727

affinis 679, 1214, 1271
arecina 187
aylmer-robinsonii 2386
beccariana 2288
donata 2495
eriophora 187
eriostachya 187, 1214
gaudichaudii 1298, 1763, 3091
gracilis 2235
hillebrandi 184
insignis 187
lanigera 184, 227
macrocarpa 90
martii 1298, 1763, 3091
minor 186
napaliensis 2480
remota 184, 776, 959
rockiana 187
schattaueri 1215
thurstonii 1271
weissichiana 2315
[see also Livistona and Styloma]
Procris glabra 1238
[see also Urera]
Prosopis 619, 1364, 1365
chilensis 446, 3251
dulcis 3245
juliflora 8, 117, 345, 1081, 1082, 1083, 1084, 1085, 1208, 1320, 1321, 1322, 1323, 1327, 1362, 2240, 3146, 3245
pallida 345, 890, 1257, 1259, 1306
Pseudelephantopus spicatus 870
[see also Elephantopus]
Pseudomorus brunoniana 2738
pendulina 2823a
[see also Streblus]
Psidium cattleianum 867, 885, 967, 1189, 1192, 1193, 1692, 2767
cujavillus 1189, 1192
guajava 1189, 1192, 1208, 1692, 1858, 1875, 2858
littorale 867
[see also Episyzygium]
Psilotrichum humile 752
viride 752
[see also Nototrichium]
Psychotria 888, 2791, 2792
fauriei 885
grandiflora 1739, 2355
hawaiiensis 885
hexandra 1739, 3069
hirta 1126
hirtula 2738
hobdyi 2791
kaduana 885
longissima 2440
mariniana 885
mauiensis 2440
rosacea 2414
wawrae 2791
[see also Coffea and Straussia]

Pteralyxia 2152, 2567
elliptica 2480
kauaiensis 481
macrocarpa 481
[see also Vallesia]
Pterotropia 1187
dipyrena 1187
gymnocarpa 1187
kaalae 2720
kauaiensis 1187
[see also Tetraplasandra]
Ptilotus sandwicensis 1739
[see also Nototrichium]
Pueraria thunbergiana 70, 640
Pyrus anthyllidifolia 2772
[see also Osteomeles]

Racosperma kauaiense 2133
koa 2133
[see also Acacia]
Raillardia [see Railliardia]
Railliardia 362, 362, 643, 993, 1046, 1051, 1389, 2373, 2624, 2650
ciliolata 394, 1187, 1227, 2617, 2653, 2656
coriacea 2619
demissifolia 2406, 2617, 2619, 2653
×dolosa 2622
fauriei 1589
herbstobatae 2484
hillebrandi 1739
kohalae 2738
latifolia 2663
laxiflora 394
linearis 394, 994, 2617
lonchophylla 2617, 2635
menziesii 2617, 2619
molokaiensis 1187, 2617
montana 1739, 2617, 2619
platyphylla 2617, 2619, 2635
reticulata 2619
rockii 2619
scabra 394, 2617
sherffiana 2656
ternifolia 2617
thyrsiflora 2617
×vafra 2622
waianapanapaensis 2484
[see also Dubautia]
Ranunculus hawaiensis 1039
mauiensis 1039, 2738
Rapanea 682
helleri 695, 696
hendersonensis 695
hosakana 695
lessertiana 1562
pukooensis 1562
sandwicensis 1562
[see also Myrsine]
Rauvolfia 2223, 2475, 2567, 2648, 3196
degeneri 1032

mauiensis 1032, 2544, 2656, 2675
molokaiensis 2653, 2672
remotiflora 2881
sandwicensis 382, 1032, 2149, 2682, 3005, 3196
[see also Excavatia]
Rauwolffa [see Rauvolfia]
Rauwolfia [see Rauvolfia]
Remya 747, 3005
  kauaiensis 1187, 2669
  mauiensis 1187
Reynoldsia 958, 1039, 1101, 1102, 1103, 1275, 2147, 2661
  hosakana 2665
  mauiensis 2665
  sandwicensis 803, 1039, 1040, 1041, 1780, 2386, 2476
Rhamnus californicus 890
Rhizophora mangle 638, 871, 1578, 1579, 1695, 2529, 2990, 3017, 3018, 3098
Rhodomyrtus tomentosa 1858, 2163
Rhus sandwicensis 1039
  semialata 101, 784, 1324, 1709
Rhynchelytrum repens 1375, 1376
[see also Tricholaena]
Rhynchospora castanea 1797
  glauca 1497
  lavarum 992, 1451, 1497
  sclerioides 243, 1238, 1496
  spiciformis 1187, 2738
  thyrsoidea 1493
Richardia brasiliensis 870, 1603
Ricinus communis 2310
Rivina humilis 2079
Rivinia [see Rivina]
Rockia sandwicensis 1120, 1121, 1383, 2721
[see also Pisonia]
Rollandia 419, 431, 993, 2308, 2560, 3176, 3179
  ambigua 730
  angustifolia 1154, 1305, 2300
  bidentata 2359
  calycina 730
  crispa 396, 993
  delessertiana 996
  fauriei 1596
  grandifolia 1187
  humboldtiana 996, 2361, 3058
  kaalae 3059
  lanceolata 396, 752, 858, 993, 1187, 2515
  longiflora 1187, 3059
  parvifolia 837
  pedunculosa 3059
  pinnatifida 730
  purpurellifolia 2280
  racemosa 1187
  st.-johnii 2515
  scabra 3059
  truncata 2280
  waianaeensis 2359
Rorippa elstonii 1208
[see also Nasturtium]

Rubiaceae 616, 2739
Rubus 669, 723, 1597
  argutus 971, 1999
  damienii 1559, 1589
  discolor 2844
  ellipticus 923
  hawaiiensis 64, 781, 977, 1039, 2720, 3061
  hillebrandii 1559, 1589
  koehnei 1559, 1589
  macraei 64, 977, 1039, 2313
  moluccanus 923
  penetrans 971, 972, 977, 3214, 3215
  rosifolius 782, 871
  sandwicensis 1559, 1594
  ulmifolius 897
Ruellia brittoniana 871
  graecizans 870
  prostrata 3005
Rumex albescens 1187
  brownii 897
  giganteus 71, 503, 677, 897, 1625, 1776, 3077, 3201, 3214, 3215
  skottsbergii 677, 3201
Ruppia maritima 500, 2509
Rytidotus sandvicensis 1229
[see also Bobea]

Saccharum 107, 108
  officinarum 2983
  spontaneum 1735
Sagina hawaiensis 2121, 2124
Sagittaria sagittifolia 1026
Salicornia virginica 923
Samanea saman 1306
[see also Pithecellobium]
Sambucus mexicana 885
Sanicula 2502, 2604
  haleakalae 2502
  kauaiensis 1039, 2515, 2604
  lobata 2502
  purpurea 2515, 2604
  sandwicensis 2515, 2604, 3187
Santalum 13, 685, 764, 825, 1190, 1337, 1340, 1341, 1350, 1353, 1573, 1643, 2015, 2103, 2289, 2296, 2371, 2496, 2706, 2710, 2727, 2818, 2919, 2961, 2962, 3000
  album 74, 1357, 1359, 1362, 1795
  cuneatum 2289
  ellipticum 45, 209, 384, 885, 993, 1045, 2835, 2836, 2838, 2839
  freycinetianum 168, 209, 384, 993, 1045, 1163, 1187, 1359, 1362, 1795, 1796, 2284, 2709, 2836, 2838, 2839, 3166
  haleakalae 1187, 1622, 2835, 2836, 2838, 2839
  involutum 2496
  longifolium 1796
  majus 2496
  paniculatum 384, 685, 2716, 2835, 2836, 2838, 2839
  pilgeri 2289

pyrularium 897, 1045, 2738
salicifolium 1796
Sapindus lonomea 2408
  oahuensis 2211, 2213, 3181, 3182
  saponaria 623, 1542, 2214, 2216, 2431, 2704
  thurstonii 2215, 2431
Sapota sandwicensis 1048
  [see also Pouteria]
Sapotaceae 1565
Sarx alba 2437
  [see also Sicyos]
Scaevola 430, 452, 821, 1017, 1047, 1462, 2119, 2120, 2261, 2267, 2706, 2727, 2918, 2998
  ×blinii 1590
  cerasifolia 451, 2738
  chamissoniana 397, 451, 492, 552, 993, 1187, 1210, 1462, 1590
  ciliata 397, 731
  coriacea 55, 687, 815, 1142, 1166a, 2081, 3190
  cylindrocarpa 1187, 1949
  fauriei 1590
  frutescens 484, 647, 920, 1462, 1958, 2374
  gaudichaudiana 492, 668, 1014, 2231, 2683
  gaudichaudii 397, 1016, 1238, 1462
  glabra 397, 996, 1238
  kauaiensis 421, 2374
  kilaueae 676, 3200
  koenigii 397, 552, 920
  ligustrifolia 2081
  lobelia 552
  menziesiana 397, 492
  mollis 397, 451, 668, 1014, 1238, 2231, 2361, 2374
  montana 552, 993, 2997
  plumerioides 2081
  plumierii 920
  procera 484, 1187, 2231, 2738
  pubescens 2081
  sericea 492, 647, 654, 883, 920, 1303, 2997
  skottsbergii 2354
  swezeyana 2274, 2275
  taccada 45, 442, 451, 883, 886, 1303, 1519, 1867, 2388, 2418
  wahiawaensis 2471
  [see also Camphusia and Temminckia]
Scalesia menziesii 2649, 2652
Schefflera actinophylla 935
  [see also Brassaia]
Schenkia sebaeoides 1059, 2589
  [see also Centaurium]
Schiedea 674, 1737, 2124, 2524, 2644, 2650, 2652
  adamantis 1158, 1173, 2401
  apokremnos 2401
  diffusa 1039, 2640, 3065
  globosa 2641, 2675
  haleakalensis 2637
  hawaiiensis 1187
  helleri 2640
  hookeri 2641, 2655, 2669
  implexa 2640
  kaalae 2640, 3065
  kealiae 483, 858
  ligustrina 496, 778, 818, 2641, 2655, 2669
  lychnoides 1187
  lydgatei 1187, 2641, 2669
  mannii 2401
  membranacea 2408
  menziesii 1187, 2495, 2641
  nuttallii 1187, 1237, 2664, 2669
  oahuensis 3065
  pubescens 1187, 2640, 2669
  remyi 2641, 2668, 2669
  salicaria 1187, 2652
  sarmentosa 2645
  spergulina 1039, 2641, 2672
  stellarioides 1208, 2640, 2669
  verticillata 534, 959
Schinus dentata 92
  molle 638
  terebinthifolius 639, 871, 967, 2031, 3181, 3182
Schizophyllum micranthum 2080
  [see also Lipochaeta]
Scirpus 199, 1452
  californicus 194, 198, 2780, 2781, 2854
  erectus 196
  juncoides 1450
  lacustris 1450
  maritimus 1450
  nudissimus 2917
  paludosus 195, 197, 198
  rockii 1490
  validus 194, 197
Scytalis anomala 2993
Senecio capillaris 394, 994, 2390
  mikanioides 1789
  sandvicensis 394, 2462, 3005
Senna 1286
  gaudichaudii 1349
  [see also Cassia]
Sesban [see Sesbania]
Sesbania 512, 776
  hawaiiensis 704
  hobdyi 704
  molokaiensis 704
  sesban 1208
  tomentosa 28, 509, 704, 1143, 1154, 2652
Sesuvium portulacastrum 1519, 3019
Setaria biflora 1187
  geniculata 1858
  palmifolia 871
  sphacelata 897
  [see also Dissochondrus]
Sicyocarya 2437, 2438
  cucumerina 2441
  macrophylla 2441
  [see also Sicyos]
Sicyos 551, 1302, 2437, 2438
  atollensis 2402
  caumii 2402
  cucumerinus 1038, 1039

INDEX TO PLANT NAMES

fauriei 1590
hillebrandii 2355
hispidus 1187
laciniatus 1187
lamoureuxii 2402
lasiocephalus 2720
laysanensis 2402
macrophyllus 1038, 1039
maximowiczii 551
microcarpus 1739
nihoaensis 959, 2402
niihauensis 2386
pachycarpus 1238
remyanus 551
semitonsus 2402
[see also Cladocarpa, Sarx, Sicyocarya, and Skottsbergiliana]
Sida 1557, 2429
  cordifolia 624
  diellii 1039
  fallax 45, 133, 1207, 1867, 2319
  fauriei 1595
  ledyardii 2429
  meyeniana 133, 1207
  nelsonii 2429
  oahuensis 1589
  sandwicensis 1595
  sertum 1039
  spinosa 1207
Sidastrum paniculatum 940
Sideroxylon auahiense 2278, 2284
  ceresolii 2284
  rhynchospermum 2276
  sandwicense 1187
  spathulatum 1187, 2278, 2284
  [see also Pouteria]
Siegesbecka [see Sigesbeckia]
Siegesbeckia [see Sigesbeckia]
Sigesbeckia orientalis 2809
Silene alexandri 1187, 3158
  cryptopetala 1187, 3158
  degeneri 2645
  gallica 1208, 2645
  hawaiiensis 2652
  lanceolata 1039, 1187, 2645, 2652, 3158
  struthioloides 1039, 2645, 2652, 3158
Sisyrinchium acre 1739
Skottsbergiliana 2437
  lasiocephala 2413
  partita 2413
  [see also Sicyos]
Smilax melastomifolia 386, 1187, 2773
  sandwicensis 386, 1187
  [see also Pleiosmilax]
Solanaceae 17, 18
Solanum 2014, 2590
  aculeatissimum 2310
  carterianum 2284, 2881
  fauriei 1590
  haleakalaense 2396
  hillebrandii 2396

incompletum 758, 1049, 1187
kavaiense 1187, 3200
laysanense 231
nelsonii 534, 758, 1049, 2386, 2590
nigrum 534, 1563, 1822, 2310
nodiflorum 700, 1858
repandum 923
sandwicense 233, 1049, 1238
seaforthianum 639
sodomeum 2310
trifolium 2310
woahense 233
Solenostemon scutellarioides 1566
  [see also Plectranthus]
Solidago altissima 870
Sonchus asper 890
  ciliatus 3015
  laevis 2565
  oleraceus 80, 2810
Sophora 514, 2722
  chrysophylla 74, 80, 317, 515, 610, 681, 1081, 1082, 1083, 1084, 1085, 1292, 1306, 1323, 1345, 1362, 1366, 1369, 1538, 1542, 1911, 2113, 2161, 2248, 2249, 2304, 2573, 2574, 2575, 2576, 2577, 2580, 2582, 2815, 3207, 3208
  grisea 2660
  lanaiensis 681
  molokaiensis 681
  unifoliata 681, 2660
  [see also Edwardsia]
Spathodea campanulata 639, 690, 1322, 2767
  nilotica 690
Spathoglottis 1374
  plicata 546, 603, 633, 871, 1440, 1442, 1858, 2012, 2906
Spergularia marina 870, 871
Spermolepis hawaiiensis 3188
  sandwicensis 3189
Sphacele hastata 1048
  [see also Lepechinia]
Sporobolus capensis 1858
  virginicus 1187
Stachytarpheta australis 1829, 1834, 1842, 1847
  cayannensis 1830, 1834, 1842, 1858, 2165
  dichotoma 2812
  gracilis 1832
  incana 1853
  jamaicensis 1829, 1834, 1842, 1852, 1856, 2165
  mutabilis 1843, 2165
  ×trimeni 1843
  urticifolia 1830, 1834, 1835, 1842, 2165
Stellaria media 2645
Stenogyne 215, 1048, 1187, 2429, 2625, 2650
  affinis 843, 2635, 2675
  angustifolia 37, 41, 1169, 2622, 2656
  bifida 1187
  calaminthoides 1187, 2620, 2653, 2668
  calycosa 2622
  cinerea 1187

cordata 214
cranwelliae 2634
diffusa 2668, 3053
fauriei 1566, 1590
glabrata 2620
haliakalae 3054
kaalae 2646, 2653, 3054
kamehamehae 2622, 3053
kanehoana 2088, 2636
kealiae 2656, 3054
longiflora 748, 752
macrantha 212, 1187, 2622, 2636
microphylla 214, 1187, 2634, 2656
mollis 2366, 2646
nelsoni 214
oxygona 2668
parviflora 1739
pohakuloaensis 2452
purpurea 1187, 1739, 2620, 2622, 3054
rotundifolia 217, 1187, 1230, 2620
rugosa 212, 2366, 2622, 2646
scandens 2622
scrophularioides 212, 923, 2620, 2622, 3053
serpens 1187
sessilis 214, 2622, 2656, 2675
sherffii 636
sororia 2622
vagans 1187
viridis 1187
Straussia fauriei 1590
glomerata 2300
grandiflora 479
hawaiensis 1042
hillebrandii 2284, 2738
kaduana 1042, 1187, 3069
leptocarpa 1187
longissima 2284
mariniana 1042, 1210
oncocarpa 1187, 2284
psychotrioides 1126
pubiflora 1126
sessilis 714
[see also Psychotria]
Streblus pendulinus 585, 587
[see also Pseudomorus]
Strongylodon ruber 2992
Styloma 562
[see also Pritchardia]
Stylosanthes guianensis 797
Styphelia 570, 2833
douglasii 1211, 2363, 2748
tameiameiae 918, 1858, 1939, 2363, 2748
[see also Cyathodes]
Suttonia 1806
angustifolia 696
apodocarpa 1592
cuneata 1592
fauriei 1591
flavida 1592
hillebrandii 2284
kauaiensis 790, 1208

knudsenii 2284
lanaiensis 2284
lanceolata 2284
lessertiana 790
mauiensis 1592
meziana 1592
molokaiensis 1591, 2281
pukooensis 1592
punctata 1592
sandwicensis 1208, 2284
spathulata 2284
vanioti 1593
volcanica 2284
[see also Myrsine]
Swietenia mahagoni 633
Syntherisma helleri 1126
[see also Digitaria]
Syzygium cumini 2559
jambos 2559, 2767
malaccense 2559
oahuense 2913
sandwicensis 1940, 2071
[see also Eugenia and Jambosa]

Tacca hawaiiensis 700, 1608, 1791
leontopetaloides 700, 753, 1791, 2551
oceanica 2082
pinnatifida 1791
Taraktogenos kurzii 1335, 1346, 1362
Temminckia 2998
chamissoniana 2997
ciliata 2997
gaudichaudii 2997
mollis 2997
[see also Scaevola]
Tephrosia 2866
piscatoria 775, 2310
purpurea 2066, 2076, 2933
Terminalia catappa 640
myriocarpa 316, 1658, 3121
Ternstroemiopsis sandwicensis 2982
[see also Eurya]
Tetramolopium 601, 1046, 1627, 1628, 1629, 1630, 1631, 2624, 2650
arbusculum 2621
arenarium 1187, 1629, 2621
bennettii 2621
capillare 2390
chamissonis 1187
consanguineum 1187, 1629, 2621, 2656
conyzoides 1187, 2621
filiforme 1629, 2086, 2621
humile 409, 1187, 1629, 2621, 2794
lepidotum 1629, 2621
polyphyllum 2621
remyi 1187
rockii 1629, 2621
sylvae 1629
tenerrimum 393, 2045, 2146

[*see also* Erigeron, Luteidiscus, *and* Vittadinia]
Tetraplasandra 803, 958, 1039, 1101, 1102, 1103, 1119, 1275, 1408, 2147, 2148, 2671
  bisattenuata 2662
  elstonii 1208
  gymnocarpa 802, 2665, 2667
  hawaiensis 227, 802, 1039, 1040, 1041, 1821, 2662
  kaalae 2667
  kahanana 2667
  kavaiensis 802, 2665, 2667
  kohalae 2738
  lanaiensis 2278
  lihuensis 2667
  lydgatei 2662
  meiandra 2495, 2662, 2665, 2667, 2738, 3199
  micrantha 2665, 3199
  munroi 2665
  oahuensis 2667, 2676
  pupukeensis 2662, 2665, 2667, 2720
  racemosa 845
  turbans 2665
  waialealae 2278, 2662, 2672
  waianensis 2667
  waimeae 2662, 3064
  [*see also* Dipanax, Gastonia, Heptapleurum, Pterotropia, *and* Triplasandra]
Thespesia populnea 641, 921, 2696
Thevetia neriifolia 2310
  peruviana 638, 923, 2422
Tibouchina multiflora 897
  semidecandra 871, 2164
  urvilleana 3216
Toona australis 2696, 3026, 3135
  ciliata 1225, 2154, 2769, 3120
  [*see also* Cedrela]
Torilis nodosa 3005
Touchardia 946, 1275
  angusta 2499
  latifolia 785, 947, 996, 1372, 1707, 3087
Trematocarpus 1129, 3247, 3248
  macrostachys 2561, 2820, 3246
  [*see also* Trematolobelia]
Trematolobelia 420, 431, 2308, 2489, 3038, 3178, 3179
  macrostachys 421, 438, 686, 1667, 2280, 2406
  sandwicensis 686
  wimmeri 671
  [*see also* Trematocarpus]
Tribulus cistoides 489, 539, 1519
  [*see also* Kalstroemia]
Trichachne insularis 871
Tricholaena repens 999, 1904
  rosea 925
  [*see also* Rhynchelytrum]
Tridax procumbens 2188
Trifolium repens 273, 274, 275, 277
Triplasandra kaalae 1187
  lydgatei 1187
  meiandra 1187

waimeae 1126
[*see also* Tetraplasandra]
Trisetum glomeratum 879, 2854, 2855
  inaequale 3125
Tristania conferta 74, 2055, 2143
Typha latifolia 2780, 2781

Ulex europaeus 1747
Uncinia uncinata 1087, 1489, 2722, 2738
Urera 946, 3072
  glabra 3086
  kaalae 2091, 3072
  konaensis 2429
  sandwicensis 2284, 3085, 3087
  [*see also* Procris]
Urtica grandis 1238
  sandwicensis 3086
  [*see also* Boehmeria *and* Hesperocnide]

Vaccinium 1064, 2706, 2723
  calycinum 398, 868, 1422, 2723, 2738, 2774
  cereum 398, 498, 2715
  dentatum 441, 868, 2715, 2738, 2774
  fauriei 484, 1590
  hamatidens 1590
  meyenianum 1422
  pahalae 868
  peleanum 968, 973
  penduliflorum 398, 484, 993, 1048, 1187, 1208
  reticulatum 968, 973, 978, 1048, 1208, 2715, 2723, 2774, 3060, 3214, 3215
  [*see also* Metagonia]
Vallesia macrocarpa 1187
  [*see also* Pteralyxia]
Verbascum thapsus 60
Verbena bonariensis 1829, 1834, 1842, 1844
  brasiliensis 1845, 1848
  litoralis 1210, 1829, 1832, 1834, 1842, 1844, 1848, 1851, 1854
  nudiflora 2978
Verbenaceae 1826
Verbesina connata 993
  encelioides 583, 871
  hastulata 1238
  lavarum 993
  lobata 993
  succulenta 1238
  [*see also* Lipochaeta]
Vernonia cinerea 2810
Veronica arvensis 1564
  hawaiensis 1589
Vicia menziesii 41, 1555, 1768, 1820, 1943, 1944, 2217, 2219, 2938, 3043
Vigna o-wahuensis 2992
  sandwicensis 1039
Vincentia 992
  angustifolia 992
  [*see also* Machaerina]
Viola 192, 817, 1037, 1054, 1708, 2278, 2730

chamissoniana 500, 1025, 2704, 3065
helenae 834, 2457, 3200
helioscopia 1187
kauaiensis 848, 1039, 1208
lanaiensis 2457
×luciae 2728
maviensis 1739, 2728
oahuensis 834
robusta 751, 1187, 2728
sandwicensis 1595, 2281
tracheliifolia 500, 1025
Viscum articulatum 1187
  moniliforme 504, 3063
  pendulum 1126
  [see also Korthalsella]
Vitex hawaiiensis 1504, 1832, 1850, 2035
  mollis 1850, 2035
  ovata 885, 1827, 1867
  trifolia 1829, 1831, 1832, 1833, 1834, 1837, 1838, 1839, 1842, 1849, 1852, 1857
Vittadinia 1046
  [see also Tetramolopium]
Vrydagzynia sandwicensis 1391
  [see also Anoectochilus]

Waltheria americana 193, 2420, 2812, 2863
  fauriei 1557, 1589
  indica 923, 1208, 2221, 2420
  pyrolifolia 1039, 2420
Wedelia trilobata 32, 890, 2208, 2210, 3005
Wiegmannia glauca 1797
  [see also Hedyotis]
Wickstroemia [see Wikstroemia]
Wikstroemia 476, 727, 804, 1066, 1067, 1068, 1275, 2076, 2310, 2688, 2866
  basicordata 2746
  bicornuta 1187
  buxifolia 1050
  caudata 2745
  caumii 2720
  degeneri 2746
  elongata 1050, 1187
  eugenioides 2745
  fauriei 1590
  foetida 1050, 2973, 3076
  forbesii 2745
  furcata 2284, 2738

  haleakalensis 2720
  hanalei 3077
  isae 2745
  lanaiensis 2745, 2746
  leptantha 2720
  macrosiphon 2746
  monticola 2746
  oahuensis 2284
  palustris 1208
  perdita 701
  phillyreifolia 1050, 2720, 2746
  pulcherrima 684, 2720
  recurva 2720, 2746
  sandwicensis 1050, 1187, 1208, 1568, 1777, 2066
  sellingii 2738
  skottsbergiana 2801
  uva-ursi 1050, 1187, 1867, 2746, 2973
  vacciniifolia 2745
  villosa 1187
  [see also Diplomorpha]
Wilkesia 49, 362, 410, 416, 417, 426, 459, 460, 461, 463, 464, 466, 468, 469, 470, 470a, 643, 1046, 1389, 2270, 3101
  grayana 1187
  gymnoxiphium 415, 472, 1035, 1036, 2405, 3200, 3234
  hobdyi 2405
Wollastonia 924
  [see also Lipochaeta]

Xanthium strumarium 1210, 2565, 2812
Xanthoxylum [see Zanthoxylum]
Xylosma crenatum 2422
  hawaiiense 2639, 2747
  hillebrandii 3065
Xyris 890

Youngia japonica 118
  [see also Crepis]

Zanthoxylum glandulosum 1187
  hawaiiense 1187
  kauaense 1039
  oahuense 1187
  [see also Fagara]

INDEX TO PLACE NAMES

Alaka'i Swamp 403, 404, 405, 406, 2450
'Amauulu 3186
Awa'awapuhi Trail 1149, 2935

Barber's Point 618, 948

Central Union Church 3252

Diamond Head Crater 1158, 1173, 1539

'Ēkahanui 715
'Ewa Coastal Plain 210

Foster Botanic Garden 1076, 1370, 2035, 3089
French Frigate Shoals 87, 1155, 1517, 2918

Gardner Pinnacles 539
Glenwood 2288
Green Island 583

Hālawa Valley 1406, 2069
Haleakalā 217, 220, 244, 276, 713, 734, 1074, 1132, 1292, 1427, 1428, 1429, 1430, 1431, 1622, 1773, 2263, 2340, 2566, 2762, 2766, 2783, 2995, 2996, 3024, 3110, 3226
Haleakalā National Park 59, 60, 1289, 1293, 1410, 2339, 2763, 2837, 2847, 3109, 3111, 3112, 3113, 3224
Haleakalā National Park Crater District 2848
Halema'uma'u Trail 725
Hāmākua 69
Hanakāpī'ai 2424
Hanalei Valley 1272, 2287, 3195
Harold L. Lyon Arboretum 20, 21, 1013, 1033, 1191, 2793, 2951
Hā'upu 2384, 2481, 2868
Hau'ula Forest Reserve 295
Hawai'i Island 2, 3, 9, 11, 12, 14, 15, 16, 43, 48, 65, 66, 69, 80, 81, 109, 124, 134, 136, 142, 143, 144, 147, 148, 153, 154, 157, 158, 159, 169, 181, 182, 200, 201, 227, 270, 315, 316, 317, 318, 319, 321, 329, 338, 342, 343, 346, 347, 348, 350, 448, 449, 453, 454, 455, 456, 509, 544, 566, 567, 568, 573, 575, 577, 596, 604, 608, 609, 622, 623, 631, 637, 651, 666, 667, 670, 670, 676, 683, 684, 685, 689, 707, 709, 711, 723, 724, 725, 732, 733, 735, 736, 739, 741, 742, 760, 800, 801, 807, 808, 809, 838, 841, 881, 885, 892, 905, 913, 915, 919, 933, 954, 955, 957, 965, 966, 1000, 1005, 1027, 1055, 1090, 1109, 1114, 1157, 1169, 1179, 1180, 1214, 1215, 1220, 1233, 1236, 1268, 1281, 1288, 1290, 1291, 1294, 1295, 1296, 1334, 1366, 1367, 1379, 1380, 1381, 1390, 1399, 1415, 1417, 1419, 1420, 1436, 1437, 1453, 1502, 1511, 1512, 1530, 1535, 1541, 1542, 1543, 1545, 1549, 1551, 1555, 1581, 1582, 1585, 1606, 1728, 1729, 1730, 1731, 1756, 1768, 1781, 1783, 1785, 1819, 1821, 1823, 1870, 1877, 1878, 1879, 1880, 1890, 1891, 1894, 1895, 1896, 1898, 1899, 1900, 1901, 1904, 1914, 1918, 1920, 1921, 1922, 1924, 1925, 1926, 1927, 1928, 1929, 1930, 1931, 1932, 1937, 1942, 1943, 1944, 1991, 2000, 2024, 2027, 2050, 2056, 2061, 2065, 2073, 2074, 2075, 2107, 2108, 2114, 2115, 2116, 2139, 2144, 2154, 2161, 2187, 2190, 2216, 2217, 2218, 2219, 2236, 2243, 2248, 2249, 2255, 2262, 2273, 2288, 2307, 2318, 2329, 2344, 2409, 2413, 2429, 2437, 2452, 2458, 2477, 2494, 2499, 2503, 2573, 2574, 2575, 2576, 2577, 2579, 2580, 2581, 2582, 2614, 2700, 2712, 2732, 2744, 2749, 2750, 2752, 2753, 2754, 2755, 2756, 2757, 2765, 2802, 2803, 2804, 2805, 2806, 2807, 2833, 2843, 2845, 2899, 2901, 2938, 2941, 2942, 2991, 3009, 3022, 3025, 3027, 3030, 3032, 3034, 3035, 3040, 3041, 3042, 3043, 3094, 3095, 3099, 3120, 3121, 3123, 3135, 3137, 3159, 3160, 3161, 3165, 3167, 3180, 3186, 3214, 3215
Hawai'i National Park [*see* Hawai'i Volcanoes National Park]
Hawai'i Volcanoes National Park 9, 11, 12, 14, 15, 16, 124, 142, 143, 144, 148, 158, 159, 350, 544, 631, 637, 666, 667, 707, 723, 724, 736, 807, 808, 809, 892, 905, 913, 919, 955, 965, 966, 1005, 1179, 1180, 1268, 1334, 1379, 1380, 1381, 1390, 1511, 1512, 1530, 1535, 1606, 1870, 1877, 1878, 1890, 1894, 1896, 1898, 1899, 1901, 1904, 1922, 1927, 1928, 1931, 1932, 1937, 1942, 2061, 2114, 2161, 2216, 2236, 2344, 2749, 2750, 2752, 2755, 2765,

2802, 2803, 2899, 2941, 2942, 3160, 3161, 3214, 3215
Hāwī 2187
Heʻeia 1578, 1579
Heʻeia Swamp 3017, 3018
Hiʻiaka Crater 670
Hilina Pali 2115, 2116
Hilo 2050
Hilo Forest Reserve 319, 2579
Hōnaunau 1055
Hōnaunau Forest Reserve 449, 709, 3137
Honolulu 334, 335, 337, 563, 564, 927, 1258, 1321, 1327, 1486, 1665, 2131, 2171, 2964
Honolulu Botanic Gardens 26, 31, 114, 228
Honomalino 957
Hualālai 270, 1785
Hulu Island 2689

ʻIao Valley 2490
Iolani Palace grounds 2025, 2038, 2040

Kaʻena 2535
Kaʻena–Keālia Ridge 483
Kaʻena Point 104, 105, 660
Kahana 2394
Kahana Valley 3183
Kahe Point 1933, 1936
Kahoʻolawe Island 23, 86, 294, 839, 1257, 1259, 1320, 1365, 1527, 1990, 2345, 2397, 2482, 2824, 3116, 3217
Kahuku Training Area 2846
Kalalau Valley 645, 646
Kalapana 1914
Kalapana Extension 3040, 3041, 3159
Kalōpā State Park 43
Kaluaa Gulch 2789
Kamakou Preserve 617
Kānepuʻu 2814
Kapapa Island 1958, 1966
Kaʻū 1090, 2409
Kaʻū Forest Reserve 1291, 2318
Kauaʻi Island 113, 186, 226, 403, 404, 405, 406, 517, 531, 600, 645, 646, 649, 656, 695, 696, 712, 726, 761, 826, 828, 834, 835, 845, 956, 963, 1075, 1078, 1079, 1108, 1138, 1149, 1160, 1166, 1226, 1272, 1635, 1636, 1644, 1831, 1835, 1836, 1859, 1875, 2140, 2163, 2220, 2287, 2384, 2405, 2408, 2414, 2415, 2424, 2448, 2450, 2460, 2464, 2471, 2480, 2481, 2505, 2506, 2663, 2666, 2672, 2705, 2821, 2862, 2868, 2935, 2949, 3132, 3133, 3188, 3189, 3195, 3202, 3203, 3230
Kaʻula Islet 302, 482, 2507
Kaupō Gap 2637
Kawainui Marsh 2780, 2781, 2782
Kealakekua 596
Keauhou Ranch 456
Keauhou–Kīlauea 2700
Ke Kuaʻaina 1988

Kīhei 1772
Kīlauea 346, 676, 2754, 2757
Kīlauea Forest Reserve 338, 566, 567, 1730, 1731, 1880, 1921
Kīlauea Iki 2756
Kīpahulu Valley 407, 680, 718, 950, 967, 1465, 1525, 1544, 2406, 2751, 2764, 2767, 2897, 3037, 3038, 3227, 3228
Kīpapa Gulch 96, 1241, 1243, 1244, 1246
Kīpuka Kalawamauna 1169
Kīpuka Kulalio 622
Kīpukapuaulu 2, 3, 915
Kīpukakī 623
Kohala Mountains 227
Kōkeʻe 226, 517
Koloa 726
Kona 513, 1821
Kōnāhuanui 1680
Koʻolau Forest Reserve 3031, 3119
Koʻolau Mountains 918, 944, 1003, 1004, 2090, 2584, 3106
Kūkaʻiau Ranch 3025
Kūlani 2307
Kūlani Camp 3022
Kupehau Trail 3225
Kure Atoll 547, 583, 1517, 3197

Lānaʻi Island 648, 675, 710, 775, 851, 857, 1204, 1957, 1965, 2370, 2673, 2814, 2824, 3166, 3192
Lānaʻihale 851
Lanipō Trail 2844
Lapakahi 2329
Lāwaʻi Valley 2220
Laysan Island 231, 253, 303, 311, 332, 774, 823, 1011, 1133, 1155, 1519, 1523, 1670, 1951, 2416, 2540, 2977, 2979
Leahi Native Garden [see Na Laau Hawaii]
Lehua Island 482, 761, 2242
Lehua Maka Noe 600
Limahuli Valley 3132
Lisianski Island 301, 540

Mahanaloa Valley 1138
Mākaha Valley 1501, 3223
Makapuʻu 44, 45
Makiki Valley 2694, 2697
Makiki Valley Loop Trail 2324
Mākua Valley 1345, 1349, 1352, 2087
Mālaekahana 2443, 2451
Mānana Island 1962, 2971
Mānoa Cliffs Trail 2932
Mānoa Valley 904, 1688, 2342, 2372
Manowaialee Forest Reserv 318
Manukā Forest Reserve 1027
Maui Island 59, 60, 79, 97, 217, 220, 244, 276, 281, 407, 477, 680, 687, 713, 718, 734, 795, 815, 824, 849, 853, 950, 967, 1074, 1132, 1142, 1156, 1159, 1178, 1203, 1222, 1224, 1289, 1292, 1293, 1339, 1394, 1410, 1427, 1428,

INDEX TO PLACE NAMES

1429, 1430, 1431, 1438, 1465, 1525, 1544, 1605, 1622, 1647, 1651, 1772, 1773, 1784, 2263, 2339, 2340, 2380, 2390, 2406, 2423, 2490, 2518, 2566, 2637, 2675, 2689, 2751, 2762, 2763, 2764, 2766, 2767, 2783, 2815, 2829, 2831, 2837, 2847, 2848, 2897, 2925, 2945, 2995, 2996, 3024, 3031, 3037, 3038, 3109, 3110, 3111, 3112, 3113, 3119, 3185, 3190, 3194, 3224, 3226, 3227, 3228
Maunahui 657
Maunakapu–Palikea Trail 1520
Mauna Kea 80, 81, 157, 315, 1109, 1366, 1453, 2027, 2190, 2248, 2249, 2573, 2574, 2576, 2580, 2582, 2744, 3034, 3035, 3094, 3167
Mauna Kea Forest Reserve 317
Mauna Loa 65, 453, 568, 575, 577, 608, 609, 742, 838, 841, 881, 954, 1233, 1236, 1288, 1549, 1783, 1819, 1879, 1891, 1895, 1918, 1930, 2073, 2074, 2218, 2804, 2805, 2806, 2901, 3099, 3120, 3135
Midway Atoll 17, 18, 94, 304, 311, 331, 820, 821, 1077, 1522, 2049, 2356, 3153
Moanalua 1662
Moanalua Valley 2085
Mōkeʻehia Islet 1394
Mokoliʻi Islet 1170, 1171, 1248
Mokulēʻia 1113, 1355, 3181, 3182
Mokulēʻia Forest Reserve 3020
Mokulēʻia Trail 1526
Mokulua 1954
Mokumanu Islet 1962
Molokaʻi Island 242, 446, 447, 617, 657, 1406, 1609, 1987, 2096, 2134, 2135, 2193, 2412, 2427, 2474, 2498, 2541, 2651, 2728, 3093, 3193
Molokini Islet 478, 839
Moʻomomi 2412, 2427
Mount Kaʻala 3124, 1963, 1964, 1967, 1969, 1970, 1971, 1972, 1973, 1974, 1976, 1978, 1979, 1980, 1981, 1982, 1983, 1984, 1985, 1986

Nānāwale Forest Reserve 3032
Necker Island 776, 1151
Nihoa Island 554, 556, 776, 959, 1152, 1155
Niʻihau Island 840, 1351, 1364, 2352, 2386, 2386, 2461, 2487
Niu Valley 2274, 2275
Northwestern Hawaiian Islands 88, 291, 305, 308, 311, 314, 534, 555, 773, 776, 876, 1524, 2402
Nuʻalolo 2415
Nuʻuanu Valley 943

Oʻahu Island 4, 20, 21, 26, 27, 28, 31, 33, 39, 44, 45, 46, 54, 77, 92, 96, 98, 104, 105, 111, 114, 120, 172, 210, 228, 243, 295, 309, 334, 335, 337, 351, 372, 437, 483, 486, 487, 489, 491, 492, 495, 496, 498, 499, 500, 501, 502, 503, 504, 505, 506, 513, 538, 563, 564, 571, 589, 618, 619, 636, 660, 714, 715, 759, 763, 768, 769, 813, 814, 834, 842, 904, 918, 927, 932, 943, 944, 948, 949, 1003, 1004, 1008, 1009, 1013, 1025, 1031, 1033, 1059, 1076, 1080, 1104, 1113, 1158, 1170, 1171, 1173, 1191, 1200, 1239, 1241, 1243, 1244, 1246, 1248, 1258, 1299, 1321, 1327, 1335, 1342, 1343, 1345, 1348, 1349, 1352, 1355, 1356, 1358, 1375, 1376, 1397, 1459, 1486, 1501, 1505, 1507, 1518, 1520, 1526, 1539, 1578, 1579, 1593, 1655, 1662, 1665, 1671, 1675, 1680, 1681, 1686, 1688, 1765, 1776, 1777, 1796, 1800, 1828, 1845, 1846, 1848, 1855, 1883, 1933, 1936, 1940, 1954, 1955, 1958, 1962, 1963, 1964, 1966, 1967, 1968, 1969, 1970, 1971, 1972, 1973, 1974, 1976, 1978, 1979, 1980, 1981, 1982, 1983, 1984, 1985, 1986, 1988, 1989, 1994, 1995, 2025, 2035, 2038, 2040, 2045, 2059, 2069, 2083, 2085, 2086, 2087, 2089, 2090, 2092, 2127, 2128, 2131, 2171, 2201, 2234, 2244, 2265, 2266, 2269, 2274, 2275, 2322, 2323, 2324, 2342, 2353, 2354, 2361, 2368, 2371, 2372, 2383, 2392, 2394, 2410, 2443, 2446, 2451, 2462, 2467, 2488, 2512, 2520, 2535, 2538, 2565, 2569, 2584, 2587, 2588, 2616, 2618, 2655, 2660, 2694, 2697, 2780, 2781, 2782, 2789, 2793, 2825, 2844, 2846, 2907, 2912, 2921, 2930, 2931, 2932, 2947, 2951, 2964, 2971, 2975, 2978, 2986, 3017, 3018, 3020, 3039, 3089, 3106, 3108, 3124, 3168, 3181, 3182, 3183, 3205, 3207, 3208, 3209, 3210, 3223, 3225, 3251, 3252
ʻŌlaʻa 1551
Olinda 2815, 2925

Pacific Tropical Botanical Garden 1108, 1635, 1636, 2949, 3185, 3198, 3202, 3203
Pahole Gulch 1995, 2127, 2128, 2266, 2947
Paikō Lagoon 1518
Pālehua Trail 1397
Pauahi Crater 3180
Paʻupaʻu Ridge 2380
Peʻahināiʻa Trail 636
Pearl and Hermes Atoll 89, 960
Pōhakuloa Training Area 1114, 2452
Polihale 656
Popoiʻa Islet 309, 1954
Puaʻaluʻu 1178
Punaluʻu Mountains 834
Pūpūkea 1358
Puʻuhuluhulu 2262
Puʻu Ka Pele Forest Reserve 956, 1160
Puʻukoholā Heiau 604, 1728, 1729
Puʻukukui 97
Puʻulehua 2503
Puʻupane 1348
Puʻuwaʻawaʻa 1436, 1756

Queen's Beach 949

Royal Hawaiian Hotel 172

Sand Island 94
South Kona 1215

Tantalus 1104
Tern Island 1155, 1517, 2918

University of Hawaii/Manoa campus 538, 1459, 1994, 2383

Waʻahila Hawaiian Garden 1968
Wahiawā Bog 835
Wahiawā Botanic Garden 54
Wahiawā Mountains 834, 2506
Waiahewahewa Gulch 2651
Waiāhole Forest Reserve 1335
Waiākea Arboretum 343, 2243
Waiākea Forest Reserve 2494
Waiʻalae Nui Ridge 77
Waiʻanae 2912
Waiʻanae District 3251
Waiʻanae Mountains 1681, 2410, 2467
Waiau Lake 2024
Waiehu 1142
Waiheʻe 687
Waihoʻi 2423
Waihoʻi Valley 2518
Waikolu Valley 2728
Wailea Point 281
Wailua Game Reserve 3230
Waimānalo Research Station Arboretum 3039
Waimea 169, 3188, 3189
Waimea Arboretum 27, 28, 33, 39, 46, 571, 3205, 3207, 3208, 3210
Waimea Drainage Basin 2505
Wainiha–Mānoa drainage 2448

ABOUT THE AUTHORS

Susan W. Mill will receive her B.S. in biology at Stanford University in 1989. From 1983 to 1988 she was the technical editor and research assistant for the Bishop Museum research project to produce the *Manual of the Flowering Plants of Hawai'i*.

Donald P. Gowing received his Ph.D. in plant physiology at Cornell University, New York, in 1952, following his several years of military service. After a career of research in tropical crop production in Hawai'i, the Middle East, Africa, and the Far East, he retired in 1984. He is now a volunteer in the Herbarium Pacificum at the Bishop Museum, involved with the verification and curation of the Types Collection, and the preparation of its computerized database.

Derral R. Herbst received his Ph.D. from the University of Hawaii in Honolulu in 1971. He has worked as a botanist for the Pacific Tropical Botanical Garden, the University of Hawaii, the U.S. Army Corps of Engineers, and currently for the U.S. Fish and Wildlife Service, where he is principally concerned with the Endangered Species Program. He has over 20 years of field experience with the island floras of Micronesia and Polynesia. He is one of the authors of the *Manual of the Flowering Plants of Hawai'i*.

Warren L. Wagner received his Ph.D. from Washington University in St. Louis in 1981. From 1982 to 1988 he was associate botanist at the Bishop Museum in Honolulu. Currently he is an associate curator in the Botany Department, National Museum of Natural History, Smithsonian Institution. He is the senior author of the *Manual of the Flowering Plants of Hawai'i*. His research includes Pacific Basin floristics, plant geography, and systematics.